VOLUME ONE HUNDRED AND TEN

ADVANCES IN
APPLIED MICROBIOLOGY

VOLUME ONE HUNDRED AND TEN

ADVANCES IN
APPLIED MICROBIOLOGY

Edited by

GEOFFREY MICHAEL GADD
Dundee, Scotland, United Kingdom

SIMA SARIASLANI
Wilmington, Delaware, United States

ELSEVIER

ACADEMIC PRESS
An imprint of Elsevier

Academic Press is an imprint of Elsevier
50 Hampshire Street, 5th Floor, Cambridge, MA 02139, United States
525 B Street, Suite 1650, San Diego, CA 92101, United States
The Boulevard, Langford Lane, Kidlington, Oxford OX5 1GB, United Kingdom
125 London Wall, London, EC2Y 5AS, United Kingdom

First edition 2020

ISBN: 978-0-12-820703-1
ISSN: 0065-2164

For information on all Academic Press publications
visit our website at https://www.elsevier.com/books-and-journals

Publisher: Zoe Kruze
Acquisitions Editor: Ashlie M. Jackman
Editorial Project Manager: Peter Llewellyn
Production Project Manager: James Selvam
Cover Designer: Victoria Pearson

Typeset by SPi Global, India

Contents

Contributors

Estrella Duque
CSIC- Estación Experimental del Zaidín, Granada, Spain

Yuto Fukuyama
Laboratory of Marine Microbiology, Graduate School of Agriculture, Kyoto University, Kyoto, Japan

Thomas Heine
Environmental Microbiology, Faculty of Chemistry and Physics, TU Bergakademie Freiberg, Freiberg, Germany

Masao Inoue
Laboratory of Marine Microbiology, Graduate School of Agriculture, Kyoto University, Kyoto, Japan

Johan Larsbrink
Wallenberg Wood Science Center, Gothenburg and Stockholm; Division of Industrial Biotechnology, Department of Biology and Biological Engineering, Chalmers University of Technology, Gothenburg, Sweden

Anna C. Lienkamp
Microbial Biotechnology, Faculty of Biology and Biotechnology, Ruhr University Bochum, Bochum, Germany

Lauren Sara McKee
Wallenberg Wood Science Center, Gothenburg and Stockholm; Division of Glycoscience, Department of Chemistry, KTH Royal Institute of Technology, AlbaNova University Centre, Stockholm, Sweden

Lázaro Molina
CSIC- Estación Experimental del Zaidín, Granada, Spain

Kimiho Omae
Laboratory of Marine Microbiology, Graduate School of Agriculture, Kyoto University, Kyoto, Japan

Juan-Luis Ramos
CSIC- Estación Experimental del Zaidín, Granada, Spain

Yoshihiko Sako
Laboratory of Marine Microbiology, Graduate School of Agriculture, Kyoto University, Kyoto, Japan

Ana Segura
CSIC- Estación Experimental del Zaidín, Granada, Spain

Christopher R. Thornton
Biosciences, Hatherly Laboratories, University of Exeter, Exeter, United Kingdom

Dirk Tischler
Microbial Biotechnology, Faculty of Biology and Biotechnology, Ruhr University Bochum, Bochum, Germany

Takashi Yoshida
Laboratory of Marine Microbiology, Graduate School of Agriculture, Kyoto University, Kyoto, Japan

CHAPTER ONE

Detection of the 'Big Five' mold killers of humans: *Aspergillus*, *Fusarium*, *Lomentospora*, *Scedosporium* and Mucormycetes

Christopher R. Thornton*
Biosciences, Hatherly Laboratories, University of Exeter, Exeter, United Kingdom
*Corresponding author: e-mail address: C.R.Thornton@exeter.ac.uk

Contents

Abstract

Fungi are an important but frequently overlooked cause of morbidity and mortality in humans. Life-threatening fungal infections mainly occur in immunocompromised patients, and are typically caused by environmental opportunists that take advantage of a weakened immune system. The filamentous fungus *Aspergillus fumigatus* is the most important and well-documented mold pathogen of humans, causing a number

Advances in Applied Microbiology, Volume 110
ISSN 0065-2164
https://doi.org/10.1016/bs.aambs.2019.10.003

of complex respiratory diseases, including invasive pulmonary aspergillosis, an often fatal disease in patients with acute leukemia or in immunosuppressed bone marrow or solid organ transplant recipients. However, non-*Aspergillus* molds are increasingly reported as agents of disseminated diseases, with *Fusarium*, *Scedosporium*, *Lomentospora* and mucormycete species now firmly established as pathogens of immunosuppressed and immunocompetent individuals. Despite well-documented risk factors for invasive fungal diseases, and increased awareness of the risk factors for life-threatening infections, the number of deaths attributable to molds is likely to be severely underestimated driven, to a large extent, by the lack of readily accessible, cheap, and accurate tests that allow detection and differentiation of infecting species. Early diagnosis is critical to patient survival but, unlike *Aspergillus* diseases, where a number of CE-marked or FDA-approved biomarker tests are now available for clinical diagnosis, similar tests for fusariosis, scedosporiosis and mucormycosis remain experimental, with detection reliant on insensitive and slow culture of pathogens from invasive bronchoalveolar lavage fluid, tissue biopsy, or from blood. This review examines the ecology, epidemiology, and contemporary methods of detection of these mold pathogens, and the obstacles to diagnostic test development and translation of novel biomarkers to the clinical setting.

Abbreviations

AFB	acid-fast bacillus
AIDS	acquired immune deficiency syndrome
ABPA	allergic bronchopulmonary aspergillosis
ABPM	allergic bronchopulmonary mycosis
AML	acute myeloid leukemia
AB	*Aspergillus* bronchitis
AS	*Aspergillus* sensitization
BALf	bronchoalveolar lavage fluid
CCPA	chronic cavitary pulmonary aspergillosis
CF	cystic fibrosis
CFPA	chronic fibrosing pulmonary aspergillosis
CGD	chronic granulomatous disease
CNPA	chronic necrotizing pulmonary aspergillosis
CNS	central nervous system
CPA	chronic pulmonary aspergillosis
COPD	chronic obstructive pulmonary disease
CT	computed tomography
ELISA	enzyme-linked immunosorbent assay
EORTC/MSG	European Organisation for Research and Treatment of Cancer/ Mycology Study Group
EPS	extracellular polysaccharide
FDSC	*Fusarium dimerum* species complex
FOSC	*Fusarium oxysporum* species complex
FSSC	*Fusarium solani* species complex
GM	galactomannan

HSCT	hematopoietic stem-cell transplant
IPA	invasive pulmonary aspergillosis
ISHAM	International Society for Human and Animal Mycology
LFA	lateral-flow assay
LFD	lateral-flow device
MRI	magnetic resonance imaging
MAb	monoclonal antibody
MDS	myelodysplastic syndrome
NPV	negative predictive value
NTM	non-tuberculosis mycobacteriosis
PCR	polymerase chain reaction
PET	positron emission tomography
POCT	point-of-care test
PPV	positive predictive value
PSC	*Pseudallescheria/Scedosporium* species complex
PTB	pulmonary tuberculosis
SAIA	sub-acute invasive pulmonary aspergillosis
TAFC	triacetylfusarinine C

1. Introduction

In 1991, it was estimated that there were 1.5 million species of fungi on the planet (Hawksworth, 1991). With the advent of high throughput sequencing, that estimate was up-graded to as many as 5.1 million species (Blackwell, 2011). Despite this enormous number, in which fungi outnumber plants by at least 6–1, only ~700 species have been reported as causing infections in humans (Brown, Denning, & Levitz, 2012), while the number that are regularly reported as disease-causing agents in the clinical setting is limited to an even smaller number of species. Nevertheless, that handful of species, which includes yeasts, yeast-like fungi, and molds, collectively cause enormous but largely unrecognized damage to human health and well-being in the form of superficial infections of the mucosa, skin and nails (Schwartz, 2004), deep infections of the bones and joints (Taj-Aldeen et al., 2015), asthma and allergies (Denning et al., 2014), and life-threatening disseminated infections in already seriously ill patients, amounting to billions of affected people each year (Brown, Denning, Gow, et al., 2012; Brown, Denning, & Levitz, 2012).

Fungi that cause fatal disseminated infections are typically opportunistic, taking advantage of immune systems weakened caused by primary immunodeficiencies (Lanternier et al., 2013) or AIDS (Limper, Adenis, Le, & Harrsion, 2017), hematological malignancies (Pergam, 2017), or the use of potent immuno-suppressive therapies that deplete microbial immunity (Gundacker & Baddley, 2015). The fungi regarded as the most important medically are *Aspergillus*, *Candida*, *Coccidioides*, *Cryptococcus*, *Histoplasma*, *Paracoccidioides*, *Pneumocystis*, and *Talaromyces*. Only one of these, *Aspergillus*, is a true filamentous fungus or mold, the remainder being yeasts, yeast-like, or thermally dimorphic fungi. It is estimated that *Aspergillus* diseases, caused principally by the single species *Aspergillus fumigatus*, collectively affect around 9 million people worldwide, with a further 10 million people estimated to be at risk of *Aspergillus* infection. However, because of under-diagnosis and under-reporting of *Aspergillus* diseases, these figures are likely to be under-estimates (Denning, Pleuvry, & Cole, 2013). Also under-diagnosed or under-reported, are the diseases caused by other mold pathogens (*Fusarium* spp., species in the *Pseudallescheria/Scedosporium* complex (PSC), *Lomentospora prolificans*, and the mucormycetes comprising, amongst others, *Mucor*, *Rhizomucor*, and *Rhizopus* spp.), despite their increasing frequency as infectious etiologies in the ever-expanding populations of immunocompromised and diabetic patients, and their involvement in polymicrobial diseases following traumatic injury.

The aim of this review is to examine the ecology, epidemiology, and methods of detection of these under-reported molds in the context of human disease. It aims to highlight the paucity of diagnostic tests for these pathogens that, despite repeated calls by learned societies, opinion leaders, and action groups for better access to cheap diagnostics, remain pitifully inadequate, adding to the unacceptably high rates of mordity and mortality caused by these fungi, and the added financial burden to healthcare providers consequent of delayed or incorrect diagnosis. The reasons for our repeated failure to develop cheap, accurate, and user-friendly diagnostic tests are discussed, as are the steps that might be taken to 'release the brakes' on bench-to-bedside translation of diagnostics to the clinical setting, enabling access to point-of-care tests, particularly in poor and developing countries that lack access to well-resourced and well-equipped diagnostic facilities.

2. *Aspergillus* species pathogenic to humans

The genus *Aspergillus* comprises between 260 and 837 species of ascomycete fungi, making it one of the most abundant groups of fungi on the planet (Hawksworth, 2011). Despite their abundance, only a small number of species are known to cause diseases in humans, with a single anamorphic species *Aspergillus fumigatus* (teleomorph *Neosartorya fischeri*) in the section *Fumigati* responsible for >80% of recorded infections in humans (Latgé, 1999). Other well-established species capable of causing infections include *A. flavus* (section *Flavi*), *A. terreus* (section *Terrei*), *A. niger* (section *Nigri*) and *A. nidulans* (section *Nidulante*) (Sugui, Kwon-Chung, Juvvadi, Latge, & Steinbach, 2015), while a number of sibling species of *A. fumigatus* have emerged as human pathogens over recent years. These rarer species in the section *Fumigati* bare morphological similarities to *A. fumigatus*, but worryingly also exhibit intrinsic resistance to anti-fungal drugs (Sugui et al., 2015; Van der Linden, Warris, & Verweij, 2011), further complicating detection and treatment with the advent of azole-resistance in clinical strains of *A. fumigatus* (Abdolrasouli et al., 2018). Of these emerging species, *A. udagawae* (*Neosartorya udagawae*), *A. lentulus*, and *A. pseudofischeri* (*Neosartorya pseudofischeri*) are the most frequently reported as causing disease in humans. Notwithstanding these, *A. fumigatus* remains the most important pathogen in the *Aspergillus* genus in terms of prevalence and capacity to cause disease in humans. It is almost the most important mold pathogen of humans compared to all other mold species capable of causing life-threatening disseminated infections.

2.1 *Aspergillus* diseases

While *Aspergillus* spp. have recently emerged as the cause of deep-seated tissue infections following combat-related injuries in military personnel (Paolino, Henry, Hospenthal, Wortmann, & Hartzell, 2012; Tribble & Rodriguez, 2014; Warkentien et al., 2012), they are best known as opportunistic pathogens which cause life-threatening infections in immunocompromised individuals whose impaired immunity is unable to eliminate the infectious propagules (air-borne spores) inhaled into the lungs. In patients with defective cell-mediated immunity due to cytotoxic chemotherapy or T-cell dysfunction due to corticosteroid or other immunosuppressive therapy, *Aspergillus* spores germinate to produce invasive hyphae which proliferate, leading to the often-fatal disseminated disease invasive pulmonary aspergillosis (IPA). These fungi can also cause allergic and chronic respiratory diseases as a result of an over-zealous host response to saprotrophic colonization of the lung following inhalation of air-borne spores. The clinical consequences of this are most severe in the context of underlying respiratory disorders such as chronic obstructive pulmonary disease (COPD), post pulmonary tuberculosis (PTB), non-tuberculosis mycobacteriosis (NTM), bronchiectasis, and cystic fibrosis (CF). The role of *Aspergillus* spp. in bronchiectasis in the non-CF lung is less well understood, but *Aspergillus* sensitization and/or allergic bronchopulmonary aspergillosis (ABPA) is a cause and also a consequence of bronchiectasis (Chotirmall & Martin-Gomez, 2018; Chotirmall & McElvaney, 2014). The sequel to saprotrophic colonization is infection and the development of chronic pulmonary aspergillosis (CPA), a slowly progressive lung disease, with poor clinical outcome (Denning et al., 2013; Denning, Pleuvry, & Cole, 2011; Takazono & Izumikawa, 2018).

The global burden of *Aspergillus* lung diseases is vast (Bongomin, Gago, Oladele, & Denning, 2017). There are an estimated 300,000 cases of IPA per year with approximately 10 million at risk annually (Bongomin et al., 2017; Brown, Denning, Gow, et al., 2012). The global burden of ABPA in asthma is approximately 4.8 million, ABPA in cystic fibrosis approximately 6675, CPA approximately 3 million (including 1.6 million cases post-tuberculosis), with CPA complicating ABPA approximately 400,000 (Denning et al., 2013). In the United Kingdom, there are estimated to be around 4000 cases of IPA each year, 178,000 cases of ABPA in people with asthma, and 3600 patients with CPA, based on burden estimates post PTB and in sarcoidosis (Pegorie, Denning, & Welfare, 2017).

2.2 Invasive pulmonary aspergillosis

A fatality due to disseminated invasive pulmonary aspergillosis was first described in 1953 following autopsy of a patient with agranulocytosis disease (Rankin, 1953). At that time, IPA was a rare, albeit poorly recognized, disease of immunocompromised patients. It is perhaps ironic that modern advances in the treatment of hematological malignancies, especially remission-induction therapy for acute myeloid leukemia (AML) or myelodysplastic syndrome (MDS), and the use of highly-effective immune modulating drugs (Kyi, Hellmann, Wolchok, Chapman, & Postow, 2014; Reinwald, Boch, Hofmann, & Buchheidt, 2016), have led to a dramatic increase in the incidence of IPA in immunodeficient patients, such that the mortality rate of IPA exceeds 50% in neutropenic patients, and reaches 90% in hematopoietic stem-cell transplant (HSCT) recipients (Kousha, Tadi, & Soubani, 2011; Segal, 2009). The major risk factors for the disease include neutropenia, HSCT and solid organ transplantation, high-dose corticosteroids, hematological malignancy, cytotoxic therapy, advanced AIDS, and chronic granulomatous disease (CGD) (Dutkiewicz & Hage, 2010; Latgé, 1999).

2.3 Allergic bronchopulmonary aspergillosis

Of the different yeasts and filamentous fungi able to cause allergic bronchopulmonary mycosis (ABPM) in humans (Chowdhary et al., 2014), *A. fumigatus* and allergic bronchopulmonary aspergillosis (ABPA) are the most extensively studied. Allergic bronchopulmonary aspergillosis is one of the most severe *Aspergillus*-related diseases, with the potential to progress into pleuropulmonary fibrosis and respiratory failure (Carsin et al., 2017). It occurs almost exclusively in cystic fibrosis or asthmatic patients, and is characterized by a hypersensitivity reaction to *Aspergillus* spp., with diverse clinical and radiological manifestations.

2.4 Bronchiectasis

Bronchiectasis is a chronic respiratory disease characterized by progressive and irreversible dilation of the bronchial lumen, with impaired mucociliary clearance, recurrent infection, and chronic inflammation. *Aspergillus* species are the most common filamentous fungi cultured from respiratory secretions of CF, non-CF, and bronchiectasis patients (Aogáin et al., 2018; Máiz et al., 2015; Máiz, Nieto, Cantón, de la Pedrosa, & Martinez-García, 2018). *Aspergillus fumigatus* is the most frequently recovered species and the species most likely to cause infections, followed by *A. flavus*, *A. niger* and *A. terreus* which tend to colonize, rather than infect, the lung (Máiz et al., 2018). These species also differ in their geographical distributions, with a higher prevalence of *A. niger* and *A. terreus* in Japan, and *A. flavus* in India and China (Máiz et al., 2018).

2.5 Chronic pulmonary aspergillosis

Chronic pulmonary aspergillosis (CPA) is a semi-invasive disease (Chan et al., 2016), lying somewhere in between allergic and invasive forms of aspergillosis (Sehgal et al., 2018). It comprises a number of lung diseases that are long-term sequelae to pulmonary tuberculosis (PTB), namely *Aspergillus* nodule, simple pulmonary aspergilloma, chronic cavitary pulmonary aspergillosis (CCPA; presence of one or more cavities with or without aspergilloma, accompanied by pulmonary and systemic symptoms), chronic fibrosing pulmonary aspergillosis (CFPA; cavitation and fibrosis involving two or more lobes), and sub-acute invasive pulmonary aspergillosis (SAIA) (Denning et al., 2016) also called chronic necrotizing pulmonary aspergillosis (CNPA) (Kaymaz, Ergün, Candemir, & Çicek, 2016) leading to progressive destruction of the lung (Ohba et al., 2012). CCPA is the most common form of CPA (Denning, Riniotis, Dobrashian, & Sambatakou, 2003), with most cases of CPA occurring in South-East Asia, Western Pacific, and Africa, where prevalence of PTB is the highest. In high-income settings, COPD appears to be the most important pre-disposing condition for the disease (Smith & Denning, 2011). Unlike IPA, CPA occurs in immunocompetent patients after treatment for TB, leading to weight loss, intense fatigue, severe shortness of breath and life-threatening hemoptysis. *Aspergillus fumigatus* is the most frequently identified species in negative acid-fast bacillus (AFB) sputum smears from sub-Saharan Africans following treatment for PTB. Indeed, CPA is an important differential diagnosis for what appears to be smear-negative tuberculosis (Denning et al., 2011).

3. Detection of *Aspergillus* diseases

Detection of *Aspergillus* lung diseases remains a significant clinical challenge. Clinical manifestations of *Aspergillus* lung infection or colonization are non-specific, as are radiological abnormalities seen on X-rays or on computed tomograms of the chest (chest-CT). Despite this, anomalies on a chest-CT are used in some centers to initiate antifungal drug treatment in at-risk patients. However, the burgeoning costs of antifungal treatments (Schumock et al., 2016) and prolonged hospitalization for patients with suspected, but often unproven, pulmonary mycosis, has turned the spotlight on diagnostics and the need for improved detection methods that enable the implementation of antifungal stewardship programs which have the potential deliver significant savings in antifungal drug expenditure (Nwankwo et al., 2018).

Diagnostic-driven approaches to antifungal treatment have been shown to be more effective than empirical treatment, driving down costs, and improving outcomes for patients (Barnes, 2013). The emergence of triazole resistance in clinical strains of *A. fumigatus* and other *Aspergillus* spp. (Abdolrasouli et al., 2018; Rivero-Menendez, Alastruey-Izquidero, Mellado, & Cuenca-Estrella, 2016; Zoran et al., 2018), driven by a multiplicity of factors including widespread use of azole fungicides in agriculture (Fisher, Hawkins, Sanglard, & Gurr, 2018; Mortensen et al., 2010; Verweij, Snelders, Kema, Mellado, & Melchers, 2009), anti-fungal prophylaxis (Fisher et al., 2018), evolution of resistance in infecting strains in the lung during or after long-term (sometimes life-long) therapy with the same antifungal drug (Howard et al., 2009; Verweij et al., 2016), and sub-therapeutic dosing of patients (Van der Elst et al., 2015), is a serious cause for concern. Up to 12% of *Aspergillus* infections are estimated to be resistant to antifungal drugs (Rivero-Menendez et al., 2016), with up to 7% of *Aspergillus* strains from stem cell and solid organ transplant patients found to be resistant (Baddley et al., 2009; Kontoyiannis et al., 2010; Pappas et al., 2010).

Diagnostics has an important part to play in guiding patient treatment, by restricting the use of antifungal drugs to those that truly need them. It is important to note that a combination approach to diagnostics is more likely to be effective than a single test approach, despite pressures to cut healthcare costs. No single test fits all, especially where an organism such as *A. fumigatus* is capable of causing a number of different diseases of the lung. For this reason, abnormalities in a chest-CT that raise the suspicion of a fungal lung infection in an at-risk patient should be accompanied by complementary biomarker tests that add additional layers of proof. Biomarker tests such as 'pan-fungal' assays for fungal (1→3)-β-D-glucan, and the Bio-Rad *Platelia*® ELISA for *Aspergillus* galactomannan (GM-ELISA) in serum are FDA-approved and are included in the EORTC/MSG and ESCMID-ECMM-ERS guidelines for disease detection (De Pauw et al., 2008; Ullmann et al., 2018), while others such as the point-of-care *Aspergillus* lateral-flow device (*Asp*LFD) have been recently CE-marked and are now available commercially for IPA detection. These assays, and other commercial or experimental tests currently in development, are discussed in the context of allergic (ABPA), semi-invasive (CPA), and invasive (IPA) *Aspergillus* lung diseases.

3.1 Invasive pulmonary aspergillosis

3.1.1 Invasive diagnostic procedures

3.1.1.1 Bronchoscopy and culture

Diagnosis of IPA at the onset of disease is of paramount importance as prognosis worsens significantly over time. Detection of circulating biomarkers in blood such as *Aspergillus* galactomannan, fungal (1→3)-β-D-glucan, and *Aspergillus* DNA would indicate that the disease is already well progressed. Early detection of *Aspergillus* lung infection is only achievable by using tissue biopsy or bronchoalveolar lavage fluid (BALf) recovered during invasive bronchoscopy, which allow culture of the pathogen or detection of biomarkers of infection, respectively. The 'gold standard' test for diagnosis of IPA is culture of *Aspergillus* from a sterile lung biopsy, with associated necrosis of tissues due to presence of hyphal elements. There are a number of problems with culture, including the need for invasive tissue sampling, slow growth or no growth of isolates from cultured material, and difficulties in differentiating *Aspergillus* spp. from other agents of hyalohyphomycoses (e.g., other hyaline septate molds such as *Fusarium* and *Scedosporium* species) in the absence of characteristic spore-bearing structures. This leads to lengthy turnaround times, and overall poor sensitivity of culture-based detection compared to biomarker tests (Prattes et al., 2014). A major benefit of culture is the recovery of isolates for antifungal susceptibility testing and molecular detection of resistance mutations (Postina et al., 2018), especially given the increasing prevalence of triazole resistance in *A. fumigatus* isolates (Abdolrasouli et al., 2018).

3.1.1.2 Biomarker detection in BALf

3.1.1.2.1 Lateral-flow technology The advantage of assaying BALf for *Aspergillus* biomarkers, is that is allows increased sensitivity and speed of detection compared to culture. The *Aspergillus* lateral-flow device (*Asp*LFD), a point-of-care test (POCT) that incorporates a monoclonal antibody (mAb), JF5, which detects a biomarker of active growth (Thornton, 2008; Thornton, 2014), allows rapid near-patient diagnosis of IPA, with confirmatory detection of disease using laboratory-based GM-ELISA, $(1 \rightarrow 3)$-β-D-glucan, or polymerase chain reaction (PCR) tests of BALf samples. Combination biomarker testing adds considerable power to IPA diagnosis, with a number of studies demonstrating improvements in diagnostic sensitivity when these assays are used in combination (Eigl et al., 2017; Hoenigl et al., 2014; Johnson et al., 2015). There have been numerous studies conducted with the prototype *Asp*LFD test, with two meta-analysis studies (Pan et al., 2015; Zhang et al., 2019) evaluating its diagnostic performance in combination with GM-ELISA or $(1 \rightarrow 3)$-β-D-glucan tests. The more recent study by Zhang et al. (2019), which undertook a meta-analysis of 13 previously-published studies of the prototype *Asp*LFD, showed that a positive GM-ELISA and a positive $(1 \rightarrow 3)$-β-D-glucan or *Asp*LFD test provided a confirmatory diagnosis of IPA.

The *Asp*LFD was CE-marked in May 2018 for use with both serum and BALf, with Mercier, Schauwvlieghe, et al. (2019) reporting the performance of the CE-marked test in a retrospective evaluation of BALf samples from hematology patients across four centers in the Netherlands and Belgium. In this multicenter study, the CE-marked test had a good performance for the diagnosis of proven IPA versus controls (no IPA), with a specificity of 86% and sensitivity of 82%. Two follow-up studies (Jenks et al., 2019; Mercier, Dunbar, et al., 2019) have compared the *Asp*LFD with the newly developed sōna galactomannan lateral-flow assay (GM-LFA) produced by IMMY. In the study by Jenks et al. (2019), both the *Asp*LFD and GM-LFA were highly sensitive and specific for probable/proven IPA versus no IPA in patients with hematological malignancies, with specificities of close to 90% for both the *Asp*LFD and GM-LFA when read at 15 min, and with 100% sensitivity for the *Asp*LFD when read at 25 min. In patients at risk for IPA but without neutropenia or underlying hematological malignancy, the GM-LFA and *Asp*LFD tests showed sensitivities ranging from 58% to 69%, with specificities between 68% and 75% (Jenks et al., 2018). When the two tests were combined, sensitivity increased to 81% while specificity remained around 60%, making this a reasonable approach for IPA diagnosis in non-neutropenic patients.

In the study by Mercier, Dunbar, et al. (2019), which retrospectively tested 235 BALf samples of adult hematology patients from four centers on behalf of the Dutch-Belgian Mycosis Study Group (DB-MSG), performance of the *Asp*LFD and GM-LFA for proven IPA were similar. In cases of proven and probable IPA, the tests had identical specificity, but a higher sensitivity and a better NPV for the GM-LFA. However, a very faint test line is discernible in the GM-LFA with BALf samples from patients with no IPA, and is also present with the sample running buffer (i.e., strictly negative tests). This means that a comparator is required in cases where a weak test line is evident in order to determine whether this is due to non-specific binding (NSB), or due to an actual but weak line. Because of this ambiguity, a reading card is included in the IMMY test against which test line intensities are compared. Any test line considered fainter than a line of semi-quantitative intensity 1 is considered to be negative. Alternatively, plain running buffer can be used to create a negative control LFA against which clinical samples can be compared. Any test line more intense than the negative control can be considered positive. However, this procedure introduces additional uncertainty, variability, and cost, since a negative control must be included for every sample run.

While the GM-LFA and *Asp*LFD are both relatively easy to use, the *Asp*LFD is somewhat simpler, and does not suffer from NSB unlike the IMMY test. For the *Asp*LFD, no sample pre-treatment is required with non-bloody, non-viscous BALf, which can be applied directly to the test, with results available within 15 min. In the case of bloody or viscous BALf, heating with a buffer (3 min) followed by centrifugation (5 min) is required, with a total test time of approximately 25 min. For the GM-LFA, all samples must be pre-treated with heating for 6–8 min and centrifugation for 5 min, bringing the total test time to roughly 45 min. Notwithstanding this,

Mercier and co-workers (Mercier, Schauwvlieghe, et al., 2019; Mercier, Dunbar, et al., 2019) conclude that both tests show a good performance for diagnosis of IPA in hematology patients using BALf. Both tests can be used for fast screening of patients, although confirmation using an adjunct test such as GM-ELISA, PCR or culture is merited.

3.1.1.2.2 Cytokines An important limitation of biomarker tests is that sensitivity may decrease during antifungal prophylaxis or empirical therapy of patients with hematological malignancies (Eigl et al., 2015; Prattes et al., 2014). To mitigate this loss of test sensitivity, alternative signature molecules in BALf have recently been explored which might act as adjunct biomarkers of the disease. Gonçalves et al. (2017) investigated the alveolar cytokine profiles of patients at high risk of developing IPA. The cytokines IL-1β, IL-6, IL-8, IL-17A, IL-23, and TNF-α were significantly increased among patients with the disease. However, IL-6, IL-8, and IL-23 best differentiated between cases of IPA and controls, whereas the remaining cytokines displayed an inconsistent contribution to discrimination. The cytokine IL-8 was the dominant discriminator, with alveolar levels of ≥904 pg/mL predicting IPA with a sensitivity of 90%, a specificity of 73%, and a negative predictive value of 88%.

3.1.1.2.3 Siderophores Iron is essential for the growth of *Aspergillus fumigatus*, and targeting its acquisition blocks infection by the pathogen (Leal et al., 2013). In iron-poor environments such as serum, the pathogen produces low-molecular-weight siderophores to scavenge ferric iron from the host (Haas, 2003; Haas, Petrik, & Descristoforo, 2015). One of these, the extracellular siderophore triacetylfusarinine C (TAFC), is produced following spore germination, and is detectable in serum from patients with proven or probable IPA (Carroll, Amankwa, Pinto, Fuller, & Moore, 2016). The siderophore is also detectable in BALf, and an evaluation by Orasch et al. (2017) of BALf samples from patients with hematological malignancies (of which 73% received antifungal prophylaxis or empirical treatment at the time of bronchoscopy) showed that diagnostic sensitivities of the GM-ELISA and *Asp*LFD tests could be increased when used in combination with TAFC detection, in line with previous BALf GM-ELISA/ *Asp*LFD combination testing (Hoenigl et al., 2014; Prattes et al., 2014).

3.1.1.2.4 Polymerase chain reaction *Aspergillus* polymerase chain reaction (PCR) has been shown for many years to be a promising diagnostic procedure for IPA detection, especially when combined with other diagnostic biomarker tests (Buchheidt, Reinwald, Hofmann, Bloch, & Seiss, 2017;

White, Parr, Thornton, & Barnes, 2013). Despite the widespread use of 'in-house' tests, PCR was not included in the 2008 EORTC/MSG diagnostic criteria for IPA (De Pauw et al., 2008) due to a lack of assay standardization and validation. However, PCR will be included in the latest iteration of the guidelines, following systematic review of the evidence for its clinical use in comparison with antigen testing (White et al., 2015). The availability of commercial PCR tests has aided assay standardization (Rath & Steinmann, 2018), with three PCR assays specifically tailored to *Aspergillus* detection, and which have been evaluated using BALf or respiratory samples. The MycAssay *Aspergillus*® assay is a real-time PCR which detects *A. fumigatus* and other *Aspergillus* spp., with a sensitivity of 80–94.1%, and a specificity of 87.6–98.6%, in non-hematology and mixed patient groups (Torelli et al., 2011; Guinea et al., 2013; Orsi et al., 2012). The AsperGenius® test comprises two PCR assays, one for the detection and differentiation of *Aspergillus* spp. (AsperGenius Species multiplex®), and the other (AsperGenius Resistance multiplex®) for the detection of four azole resistance markers within the Cyp51A gene of *A. fumigatus* (L98H, TR34, T289A, Y121F). The AsperGenius Species multiplex® assay contains a probe for the *A. fumigatus* complex (*A. fumigatus* and the emerging sibling pathogens *A. lentulus*, *A. udagawae*, and *A. viridinutans*), and a probe for *Aspergillus* spp. (the *A. fumigatus* complex and *A. flavus*, *A. terreus*, and *A. niger*). Using BALf, the AsperGenius® test has a sensitivity of 68.4–84%, and a specificity of 80% to >92% (Chong et al., 2016; Guegan et al., 2018). In a study of 91 patients with GM-ELISA positive BAlf samples, 79% were positive for *A. fumigatus* or *Aspergillus* spp. DNA, with the azole resistance mutation TR_{34}/L98H detected in 8 of the cases, and T289A/Y121F mutations in a further 3 cases (Schauwvlieghe et al., 2017). The MycoGenie® assay detects *A. fumigatus* and the TR34/L98H mutation, and has a test sensitivity of 71.1% and specificity of >92% when used with BALf samples (Guegan et al., 2018). Notwithstanding improvements in PCR standardization, the Infectious Diseases Society of America (ISDA) currently recommends that *Aspergillus* PCR be used in individual cases if combined with other diagnostic and clinical data (Lass-Flörl, 2019; Misch & Safdar, 2016; Patterson et al., 2016).

3.1.2 Semi-invasive and non-invasive diagnostic procedures

3.1.2.1 Biomarker detection in serum

3.1.2.1.1 ELISA, PCR, (1 → 3)-β-D-glucan, and LFD tests The literature on the relative specificities and sensitivities of fungal (1 → 3)-β-D-glucan tests, Bio-Rad *Platelia*® GM-ELISA, in-house PCR assays, and prototype *Asp*LFD test, for serum-based diagnosis of *Aspergillus* diseases is capacious, and has been reported extensively elsewhere (Held, Schmidt, Thornton, Kotter, & Bertz, 2012; White et al., 2013; Pan et al., 2015; Held & Hoenigl, 2017; Rath & Steinmann, 2018; Ruhnke et al., 2018; Patterson & Donnelly, 2019).

To date, there have been no studies evaluating the diagnostic performance of the CE-marked *Asp*LFD when used with serum samples. However, an ELISA (GP-ELISA) which employs the same mAb (JF5) used in the *Asp*LFD (and also in the experimental *Aspergillus* proximity ligation assay (Mercier, Guldentops, van Daele, & Maertens, 2018)) has recently been developed and CE-marked by Euroimmun Medizinische Labordiagnostika AG, and its performance compared to the *Platelia*® GM-ELISA for IPA diagnosis using serum samples from immunocompromised patients (Dichtl, Seybold, Ormanns, Horns, & Wagener, 2019). While the specificities of the GM-ELISA and GP-ELISA were 99% and 96%, respectively, both assays demonstrated low sensitivities. When tests were extended to all sera available in the time frame of 7 days before to 7 days after the day of proven diagnosis, 47% and 56% of the cases were detected by the GM-ELISA and GP-ELISA, respectively. The low sensitivities of both tests underline the need for serial tests when using patient serum.

3.1.2.2 Molecular imaging using mAb JF5 and siderophores

Abnormalities in a chest-CT as diagnostic indicators of IPA (defined as dense, well-delineated, nodular infiltrates with and without ground glass attenuation (the so-called 'halo sign')) (De Pauw et al., 2008) are not disease-defining, with other pathologies (non-*Aspergillus* lung infections, and neoplastic and inflammatory processes) giving similar anomalies. Furthermore, these radiological indicators are transient in neutropenic patients, or are rare in non-neutropenic patients (Prattes et al., 2014). While magnetic resonance imaging (MRI) provides unrivalled spatial resolution and soft tissue contrast, and can be used as a detection aid for *Aspergillus* cerebral and central nervous system infections (Marzolf et al., 2016; Starkey, Moritani, & Kirby, 2014), its utility as an imaging modality for lung infections is limited due to the lack of detectable protons in air-filled spaces and potential artefacts between air-tissue interfaces.

For these reasons, attempts have been made to improve the accuracy of radiology for diagnosis of IPA by combining positron emission tomography (PET) with CT and MRI (Thornton, 2018). The *Aspergillus*-specific mAb,

JF5, when conjugated to the radionuclide ^{64}Cu, has been used in immunoPET/MRI as a disease-specific tracer ([^{64}Cu]NODAGA-JF5), and allows accurate and sensitive non-invasive detection of *A. fumigatus* lung infections in vivo (Davies et al., 2017; Rolle et al., 2016). Furthermore, humanization of JF5 (Davies et al., 2017) has allowed translation of the imaging technology to the clinic. A first-in-human clinical trial commenced in 2018, with successful detection of IPA in a neutropenic hematology patient, corroborated by BALf detection using the *Asp*LFD, GM-ELISA, and culture (Thornton, pers. comm.).

Alternative approaches to the molecular imaging of IPA include the use of *Aspergillus* siderophores in PET. When coupled to ^{68}Ga, a positron emitter with complexing properties similar to those of Fe(III), the iron-chelating siderophore TAFC showed accumulation in the lungs of immunosuppressed rats with IPA, with uptake correlating with severity of *Aspergillus* infection (Petrik et al., 2010; Petrik, Fransenn, et al., 2012; Petrik, Haas, et al., 2012). While still at the pre-clinical stage of development, a [^{68}Ga]TAFC tracer for molecular imaging of *Aspergillus* lung infections in vivo, holds enormous diagnostic potential, alongside antibody-guided PET/MR imaging (immunoPET/MRI), for non-invasive detection of IPA in humans.

3.1.2.3 Biomarker detection in urine

3.1.2.3.1 GM antigenuria Evidence for the presence of *Aspergillus* carbohydrate biomarkers in urine was first shown in counter-immunodiffusion assays with urine from rabbits experimentally infected with *A. fumigatus* (Lehmann & Reiss, 1978). Antiserum used in the assay was prepared from an immunosuppressed and infected rabbit (rather than an animal immunized with an *A. fumigatus* antigen preparation), and showed that a carbohydrate antigen was detected in the urine of the experimentally infected rabbits. Detection of the antigen was not found in uninfected animals or in animals infected with *Candida albicans*. While human urinary antigen was not investigated, and the identity of the carbohydrate was not established, this early study was the first to demonstrate the excretion of *A. fumigatus* carbohydrate antigen(s) into the urine of animals with IPA.

The identity of a carbohydrate antigen in the urine of experimentally infected rabbits, and in the urine of patients with IPA, was subsequently demonstrated by Dupont and co-workers (Dupont, Huber, Kim, & Bennett, 1987). Galactomannan (GM) was detected in animal and human urine samples using a sandwich ELISA employing rabbit antiserum raised against purified GM. Using this assay, urinary GM was detectable throughout the course of lethal aspergillosis in all 16 experimentally infected animals, in concentrations of 24–1900 ng GM/mL urine, with antigen excretion roughly paralleling the extent of disease. Galactomannan was detected in the urine of 7 of 13 patients with IPA, with concentrations of 1–83 ng GM/mL urine. A contemporaneous study by Bennett, Friedman, and Dupont (1987) showed that within 24 h after intravenous injection, 35% of galactomannan was excreted into the urine of immunocompetent rabbits. Galactomannan detection in urine was also investigated as a means of diagnosing experimental and spontaneous aspergillosis in cattle (Jensen, Stynen, Sarfati, & Latge, 1993). An inhibition ELISA using a GM-specific mAb EB-A1 (Stynen et al., 1992), reacted positively with urine samples from normal cattle, and so could not be used to detect systemic bovine aspergillosis. The same mAb was used by Haynes, Latgé, and Rogers (1990) to detect GM in the urine of neutropenic patients undergoing bone marrow transplantation or remission induction therapy for leukemia. In immunoblotting studies, EB-A1 showed diffuse staining (indicative of carbohydrate) in urine samples from patients with IPA, but no staining with urine from patients without evidence of the disease.

The development of a commercial latex agglutination test for *Aspergillus* GM (Pastorex *Aspergillus*, Sanofi Diagnostics Pasteur, France) using another GM-reactive rat mAb, EB-A2, also developed by Stynen et al. (1992), enabled standardized testing of GM antigenemia in the serum of IPA patients as a means of diagnosing IPA. The test was also investigated as a diagnostic tool for IPA using GM antigenuria in bone marrow transplant recipients (Ansorg, von Heinegg, & Rath, 1994). After modification of the assay for urine testing, it had a detection limit in native urine of approximately 20 ng GM/mL. Antigen was found in 79 (36.4%) of 217 serial urine samples, compared to 40 (11.8%) of 340 serum samples. Overall, antigenuria preceded antigenemia, and was more persistent. The sensitivity, specificity, PPV and NPV of antigenuria for autopsy-proven aspergillosis and clinically-suspected *Aspergillus* infection were 57%, 53%, 31% and 77%, respectively, while those of antigenemia were 43%, 53%, 25% and 71%, respectively. It was therefore concluded that urine testing was more reliable than serum testing for the detection of *Aspergillus* GM.

Refinement of the latex agglutination test led to the development of a quantitative sandwich ELISA for GM detection in patient serum, and which forms the basis of the commercial Bio-Rad *Platelia*® ELISA for GM detection in patient serum and BALf samples. The ELISA, which detects <1 ng GM/mL sample, was also investigated as a urine test for GM in patients with IPA (Stynen, Goris, Sarfati, & Latgé, 1995). When GM was detected in urine, it was always lower than in matched serum samples. In addition, GM was always detected earlier in serum than in urine. The maximal number of days separating patient death and the first detection of GM in patient serum and urine were $\geq$39 and $\leq$12, respectively. Therefore, in contrast to a previous report (Rogers, Haynes, & Barnes, 1990), it was recommended that serum rather than urine be used for the detection of GM, because of the later diagnosis obtained with urine samples.

Since the pioneering work in the 1990s, there has been limited interest in GM antigenuria as a detection method for IPA due to reports of its limited utility and high rates of false positivity, which vary between 8% and 47% (Klont, Mennink-Kersten, & Verweij, 2004). However, since the introduction of the *Asp*LFD POCT, which has revolutionized point-of-care testing for the disease, there has been renewed interest in GM detection in urine, due to its non-invasive nature and the applicability of lateral-flow technology to urine biomarker testing. Dufresne et al. (2012) recently reported the development of a new GM-reactive mAb (mAb476), and demonstrated the feasibility of using the mAb in an immunochromatographic assay for detection of GM-like antigens in urine. Further development of the assay into a dip-stick format has demonstrated the potential of the test to detect proven or probable IPA in patients at high-risk for the disease (patients with hematological malignancy, and recipients of solid organ or HSCT). Per-patient sensitivity and specificity in the overall cohort was 80% and 92%, respectively, and the urine assay correlated with serum galactomannan indices (Marr et al., 2018). While the test uses non-invasive urine samples for GM detection, the requirement for sample pre-treatment means that it is unlikely to be a point-of-care test for IPA. Nevertheless, it demonstrates the applicability of lateral-flow technology to urine testing for *Aspergillus* diagnostic biomarkers, and for urine-based detection of other fungal pathogens of humans such as *Cryptococcus* (Drain et al., 2019) and *Histoplasma* (Libert, Procop, & Ansari, 2018).

3.1.2.3.2 Urine siderophores In addition to cell wall carbohydrates such as GM, *A. fumigatus* and other *Aspergillus* species produce proteins that are detectable in urine. These include the iron-scavenging siderophores TAFC and ferricrocin. In a rat model of experimental aspergillosis, TAFC and ferricrocin (FC) were quantified in urine and serum with matrix-assisted laser desorption ionization (MALDI) (Luptáková et al., 2017). The limits of detection of the ferri-forms of TAFC and FC in rat serum were 0.28 and

0.36 ng/mL, respectively, while in urine the limits of detection were 0.02 and 0.03 ng/mL, respectively. The mean concentrations of TAFC and FC in urine were 0.37 and 0.63 μg/mL, respectively. In a rat model of IPA, ferricrocin and TAFC detection in urine using mass spectrometry therefore appeared to provide a non-invasive means of detecting *Aspergillus* lung infections.

The siderophore TAFC has also been evaluated as a urine biomarker of IPA in humans using mass spectrometry (Hoenigl et al., 2019). Urine TAFC, normalized to creatinine, was determined in 44 samples from 24 patients with underlying hematological malignancies and probable, possible or no IPA, and compared to GM antigenuria using the Bio-Rad *Platelia*® ELISA. Matched blood samples were also tested for GM, and for (1→3)-β-D-glucan using the Fungitell test. The TAFC/crea index determination in urine samples showed a promising performance for diagnosis of IPA, with sensitivities and specificities comparable to those reported for GM in serum and BALf, and superior to GM antigenuria determinations. Compared to TAFC detection in human BALf and serum samples (Carroll et al., 2016; Orasch et al., 2017), the diagnostic performance of TAFC levels in urine seems to be superior, which may be due to accumulation in the bladder as shown in recent animal models (Petrik et al., 2014, 2015).

3.2 Allergic bronchopulmonary aspergillosis (ABPA)

Diagnosis of ABPA is extremely difficult, reflected in a number of different diagnostic definitions of the disease. To unify diagnosis and therapy of ABPA, the International Society for Human and Animal Mycology (ISHAM) produced consensus-based guidelines in 2013 (Agarwal et al., 2013). Isolation of *A. fumigatus* from sputum samples can sometimes be useful, but is not included in the guidelines due to its poor sensitivity and specificity. As with other *Aspergillus* lung diseases, consolidation seen in a chest-CT is non-pathognomonic and may indicate another pulmonary infection such as pulmonary tuberculosis. Nevertheless, radiographic pulmonary opacities consistent with ABPA (Agarwal et al., 2013) are included under 'other criteria', of which there should be at least two of three present in a patient with bronchial asthma or cystic fibrosis (the other two being presence of precipitating IgG antibodies against *A. fumigatus* in serum, or total eosinophil count >500 cells/μL in steroid naïve patients). These additional criteria should accompany both of the 'obligatory criteria' which are

(1) a positive Type I skin test (immediate cutaneous hypersensitivity to *Aspergillus* antigen) or elevated IgE levels against *A. fumigatus*, and (2) elevated total IgE levels (>1000 IU/mL).

The heavy reliance on serology to include or exclude ABPA has led to efforts to improve test reproducibility since production of IgG and IgE antibodies to *Aspergillus* spp. is usually determined using a crude antigen extract with inherent variability and cross-reactivity with other fungal antigens (Crameri, Zeller, Glaser, Vilhelmsson, & Rhyner, 2009; Reed, 1978). The cloning of genes that encode allergenic proteins of *A. fumigatus* enabled improvements in the reliability of serological diagnosis of ABPA using recombinant antigens (Crameri, 1998; Crameri, Hemmann, Ismail, Menz, & Blaser, 1998; Fricker-Hidalgo et al., 2010). A recent study by Alghamdi, Barton, Wilcox, and Peckham (2019) of CF patients with ABPA, *Aspergillus* sensitization (AS), or *Aspergillus* bronchitis (AB) showed that patients with ABPA had significantly greater IgE reactivity to the allergenic proteins Asp f1, f2, f3, and f4 compared to patients with AS. Patients with AB expressed higher IgG positivity to Asp f1 and Asp f2 compared with those with ABPA. There were very low IgE antibody levels against all recombinant antigens in AS patients. Asp f1 IgG reactivity in ABPA patients correlated with positive sputum culture. It was concluded that the use of multiple recombinant allergens may improve the diagnostic accuracy in CF complicated with ABPA or AB. Furthermore, Asp f1 reactivity might be a useful marker for guiding antifungal therapy in ABPA.

The role of fungi in bronchiectasis not due to CF is poorly defined. Aogáin et al. (2018) characterized for the first time the mycobiome in bronchiectasis, assessing its clinical relevance in two geographically distinct cohorts from the CAMEB study, an international multicenter cross-sectional Cohort of Asian and Matched European Bronchiectasis patients. The mycobiome was determined in 238 patients by targeted amplicon shotgun sequencing of the 18S-28S rRNA internal transcribed spacer regions ITS1 and ITS2. The bronchiectasis mycobiome profiles were dominated by *A. fumigatus* in Singapore/Kula Lumpur, and by *A. terreus* in Dundee, the latter associated with exacerbations. High frequencies of *Aspergillus*-associated disease including sensitization and ABPA were detected. This study showed that the mycobiome is of clinical relevance in bronchiectasis, and screening for *Aspergillus*-associated disease should be considered even in apparently stable patients.

3.3 Chronic pulmonary aspergillosis (CPA)

Symptoms of CPA (taken here as CCPA, CFPA, and SAIA/CNPA) are non-specific and, as with other *Aspergillus* lung diseases, diagnosis of the disease based on culture of fungi from respiratory specimens lacks sensitivity, with culture positivity as low as 11.8% (Takazono & Izumikawa, 2018). Radiological findings on a chest-CT are variable and overlapping and depend on the state and stage of the disease. Diagnosis of a simple pulmonary aspergilloma in an immunocompetent individual is based on detection of a single lung cavity with no progression over at least 3 months of observation, while an *Aspergillus* nodule is defined by the presence of one or more nodules without cavitation (Denning et al., 2016; Muldoon, Sharman, Page, Bishop, & Denning, 2016). The diseases CCPA and SAIA share the similar characteristics of one or more cavities with or without fungal ball, accompanied by radiological progression such as expanding thick-walled cavities and pericavitary infiltration (Denning et al., 2016). They differ in that SAIA also involves hyphal invasion of the lung parenchyma (Hope, Walsh, & Denning, 2005). This subtle difference creates a diagnostic dilemma, especially when histopathology is lacking. The time course of radiological progression (CCPA >3 months; SAIA 1–3 months) and process of cavitary formation (CCPA typically occurs in pre-existing cavities, while SAIA cavities are formed during *Aspergillus* infection) are therefore needed for disease differentiation, but are reliant on serial radiography. The syndrome CFPA is defined by cavitation and fibrosis involving two or more lobes leading to a major loss of lung function, and is generally regarded as the end result of untreated CCPA (Denning et al., 2003, 2016).

Galactomannan-ELISA of serum has limited diagnostic value for CPA detection, with low sensitivity and specificity even when the threshold index value for test positivity is reduced from ≥1.5 (the manufacturer's recommended threshold index value for serodiagnosis of IPA (Maertens et al., 2007)) to ≥0.5 (Kitasato et al., 2009). Galactomannan-ELISA of BALf showed relatively higher sensitivity (77.2%) and specificity (77.0%), when the index threshold value for test positivity was reduced further to ≥0.4 (Izumikawa et al., 2012). In a more recent study (Salzer et al., 2018), the GM-ELISA was evaluated alongside the *Asp*LFD and levels of cytokines for diagnosis of CPA (Salzer et al., 2018). Sensitivity and specificity of the GM-ELISA with a cut-off of ≥0.5 was 41% and 100%, respectively. When the cut-off was increased to ≥1.0, the sensitivity decreased to 30% while specificity remained at 100%. The *Asp*LFD was positive with only 7% of patients. Therefore, based on test sensitivities, both the GM-ELISA and *Asp*LFD showed insufficient performance for diagnosing CPA. However, their high specificities provide high positive predictive values, which may help identify semi-invasive or invasive disease. In this study, the cytokine profiles of CPA patients did not differ significantly from patients with other respiratory disorders, but showed significantly higher levels of IFN-γ, IL-1b, IL-6, IL-8, and TNF-α compared to healthy controls.

Further evaluation of the *Asp*LFD test using serum and BALf samples from CPA patients has shown improved diagnostic performance (Takazono et al., 2019). In this study, the diagnostic performance of the *Asp*LFD was compared to the GM-ELISA and (1$\rightarrow$3)-β-D-glucan tests using samples from patients with SAIA, simple pulmonary aspergilloma, and respiratory disease controls (PTB, NTM, COPD, interstitial pneumonia, prior lung surgery, lung cancer, and bacterial infection (pneumonia, lung abscess)). The sensitivity and specificity of the *Asp*LFD using serum were 62.0% and 67.7%, respectively, while serum GM-ELISA (cut-off index value of $\geq$1.5) showed a sensitivity of 22% and a specificity of 92.3%, respectively. Plasma (1$\rightarrow$3)-β-D-glucan tests (cut-off value of 19.3 pg/mL), had a sensitivity of 48% and a specificity of 90.8%, respectively. Using BALf samples, the sensitivity and specificity of the *Asp*LFD were 66.7% and 69.2%, respectively, while the GM-ELISA (cut-off index value of $\geq$0.6) had a sensitivity of 72.7% and specificity of 83.1%. *Aspergillus* precipitating antibody had a test sensitivity of 70%. Given these results, it was concluded that the performance of the *Asp*LFD serum test was acceptable for the diagnosis of CPA as a POCT.

As with ABPA, detection of *Aspergillus*-specific IgG plays and important role in the diagnosis of CPA (Richardson & Page, 2018). There are a number of *Aspergillus*-specific IgG tests available commercially (Page, Richardson, & Denning, 2016; Takazono et al., 2019). In a study comparing six of these assays (Page et al., 2016), the automated ImmunoCAP fluoroenzyme immunoassay and Immunolite ELISA variant (Page, Richardson, & Denning, 2015) were significantly superior to the other assays (Dynamiker, Genesis, and Serion) in diagnosing CPA, with identical specificities (both 96%) and specificities (both 98%). Precipitin testing performed poorly, as did a recently introduced quantitative serum *Aspergillus fumigatus*-specific IgM assay (Yao, Zhou, Shen, et al., 2018). Questions remain concerning the optimum cut-off values for IgG test positivities, with values provided by manufacturer's appearing to be sub-optimal for diagnosis of the disease (Page et al., 2016; Page, Richardson, & Denning, 2019; Sehgal et al., 2018). Despite this, it appears that serial serum testing for *Aspergillus*-specific IgG may provide a useful means of monitoring treatment response and outcome since serum IgG levels decrease during antifungal treatment and resolution (Yao, Zhou, Yang, et al., 2018).

The limitations of the commercial *Aspergillus*-specific IgG tests are their specificity for *A. fumigatus*, their expense, and their reliance on automation. Approximately 40% of CPA cases are caused by non-*fumigatus* spp. (e.g., *A. niger* and *A. versicolor*), which these tests would not detect (Tashiro et al., 2011). The cost of the tests and their reliance on automated procedures makes them unsuitable for resource-limited settings. While most CPA (and ABPA) patients are diagnosed in high-income countries, prevalence of these diseases is believed to be higher in low- to middle-income countries (Denning et al., 2011). Consequently, a test for CPA that requires minimal laboratory equipment, akin to the *Asp*LFD test for IPA, is needed. This need appears to have been met by a novel anti-*Aspergillus* antibody LFA manufactured by LDBIO Diagnostics, with other LFAs for *Aspergillus* immunoglobulins in development (Richardson & Page, 2018). The LDBIO Diagnostics LFA or immuno-chromatographic test (ICT) has recently been evaluated in a multicenter comparison of assays for CPA diagnosis (Piarroux et al., 2019). The sensitivity and specificity of the ICT were 88.9% and 96.3%, respectively, compared to 93.1% sensitivity and 94.3% specificity for an anti-*Aspergillus* IgG immunoblot assay. Since the ICT displays good diagnostic performance and complies with the ASSURED (Affordable, Sensitive, Specific, User-friendly, Equipment-free, and Delivered) criteria, it is appropriate for diagnosis of CPA in resource-limited settings. Marked improvements in IPA detection have been met through combination diagnostic testing. As advocated previously (Kobayashi & Thornton, 2014), a similar approach may prove beneficial for CPA diagnosis. A combination of serological and biomarker tests such as the LDBIO ICT and OLM *Asp*LFD might allow rapid, simple, cheap, but accurate diagnosis of the disease.

4. *Fusarium* species pathogenic to humans

The genus *Fusarium* comprises plant, human, and animal pathogenic species with worldwide distribution (Jain et al., 2011; Muraosa et al., 2017). As a group of organisms, they are a major constraint to food security, not only as pathogen of crop plants, but also as producer of mycotoxins contaminating the food chain (Thornton & Wills, 2015). In humans, *Fusarium* spp. cause a spectrum of diseases including nail infections (onychomycosis) (Arrese, Piérard-Franchimont, & Piérard, 1996; Westerberg & Voyack, 2013), bone and joint infections (Koehler, Tacke, & Cornely, 2014), infections of the skin and of burn wounds (Nucci & Anaissie, 2002; Latenser, 2003; Gurusidappa & Mamatha, 2011; Muhammed et al., 2013; Nucci et al., 2013; Van Diepeningen, Brankovics, Iltes, van der Lee, & Waalwijk, 2015), mycotic eye infections (fungal keratitis) (He et al., 2011; Jurkunas, Behlau, & Colby, 2009), and deep tissue infections following combat-related blast injury (Paolino et al., 2012; Tribble & Rodriguez, 2014; Warkentien et al., 2012). Up until the 1980s, most reported *Fusarium* infections were to the nails and eyes, or were locally invasive infections. Since then, there has been an increase in the number of cases of life-threatening disseminated disease (fusariosis) in patients with hematological malignancies (García-Ruiz et al., 2015), in HIV patients (Esnakula, Summers, & Naab, 2013), in patients with diabetes and liver cirrhosis (Chen, Chou, Lai, & Lin, 2017; Chen, Kondori, et al., 2017), in solid organ transplant recipients (Mohanty & Sahu, 2014; Muhammed et al., 2013), and most particularly in allogeneic HSCT recipients (Fanci, Pini, Bartolesi, & Pecile, 2013; Scheel et al., 2013; Al-Hatmi, Hagen, Menken, Meis, & de Hoog, 2016), reflecting their greater immunosuppression and profound and prolonged neutropenia. Fusariosis is now the second most common mold disease of humans after aspergillosis (Guarro, 2013), multidrug-resistant and refractory to treatment (Al-Hatmi et al., 2016; Fanci et al., 2013), with poor prognosis (Tortorano et al., 2014), and with mortality rates reaching 75%.

At least 70 species of *Fusarium* are able to cause infections in humans, and are grouped into species complexes (Guarro, 2013). Most infections in humans are caused by only four species; *F. petroliphilum* and *F. keratoplasticum* in the *Fusarium solani* species complex (FSSC) (Short et al., 2013), an unnamed species in the *Fusarium dimerum* species complex (FDSC), and a further unnamed species in the *Fusarium oxysporum* species complex (FOSC). Members of these species complexes are readily isolated from environmental samples, including soil and plant material (Thornton & Wills, 2015), air (Scheel et al., 2013), and water systems and plumbing fixtures (Dogget, 2000; Anaissie, Penzak, & Dignani, 2002; Mehl & Epstein, 2008; Anaissie et al., 2001; Short, O'Donnell, Zhang, Juba, & Geiser, 2011). Indeed, contaminated water systems appear to be an important environmental source of pathogenic fusaria in community-acquired and nosocomial outbreaks of fusariosis, with a recent study showing that hospital and communal sink biofilms are heavily colonized by pathogenic species in the FSSC, FDSC, and FOSC complexes (Al-Maqtoofi & Thornton, 2016), and thus a potential source of infectious propagules.

4.1 Detection of *Fusarium* diseases

4.1.1 Radiology, histology and culture

In countries with high rates of fungal keratitis, confocal microscopy of the eye has proved useful in the non-invasive detection of fungal structures (hyphae, pseudo-hyphae, and yeast-like structures) in the infected cornea (Brasnu et al., 2007). However, direct microscopy is non-specific, and so is unable to discriminate between the various different fungi (*Aspergillus*, *Candida*, *Fusarium*) capable of causing mycotic infections of the cornea.

Radiological findings in CT may be useful in raising the suspicion of a *Fusarium* lung infection, but are non-specific, with findings including alveolar and interstitial infiltrates, nodules and cavities that are typical of other mold infections (Marom et al., 2008). In chest CT, nodules and masses were the most common findings, with a halo sign being absent in 80% of cases.

The ESCMID and ECMM joint guidelines for diagnosis of hyalohyphomycosis strongly recommend histopathology for the diagnosis of fusariosis (Tortorano et al., 2014). However, in tissue, the hyphae of *Fusarium* typically resemble those of *Aspergillus*, with hyaline and septate filaments that dichotomize at acute and right angles (Nucci & Anaissie, 2007). Attempts have been made to improve the accuracy of in situ detection of *Fusarium* in tissues, but these remain experimental, relying on immunohistochemical procedures that employ cross-reactive antisera (Fukuzawa et al., 1995; Kaufman, Standard, Jalbert, & Kraft, 1997; Saito et al., 1999), or on molecular detection using in-house PCR assays (Salehi et al., 2016). Definitive diagnosis of invasive *Fusarium* infections therefore relies on culture of the fungus from skin biopsies and other infected tissues (sinuses, lungs) or from the blood. Fusariosis is perhaps unique amongst the invasive mold diseases, in that the pathogens' infectious propagules are spread hematogenously, and can be recovered from the bloodstream in up to 82% of cases (Grossman, Fox, Kovarik, & Rosenbach, 2012; Muhammed et al., 2013). In contrast, *Aspergillus* is rarely detected in the blood. Hematogenous spread of *Fusarium* spores accounts for the multiple necrotic skin lesions that develop in 75–90% of disseminated infections (Van Diepeningen, Brankovics, 2015), a clinical manifestation uncommon in disseminated aspergillosis. Culture from tissue and blood samples enables identification of *Fusarium* based on the crescent or banana-shaped macroconidia characteristic of this genus. However, identification to species level, which is essential for guiding clinical management (Muhammed, Coleman, Carneiro, & Mylonakis, 2011), is difficult and relies on nucleic acid-based tests of tissue specimens or axenic cultures.

4.1.2 Nucleic acid tests

Ribosomal RNA gene sequences, and sequencing of the internal transcribed spacer (ITS) region, a procedure widely used for the identification of other mold species, is of limited value for identifying and bar-coding *Fusarium* species due to the presence of duplicated divergent alleles in this region. For this reason, molecular identification of clinical isolates relies on three loci; *TEF-1α* (translation elongation factor-1α) (García-Ruiz et al., 2015; Zarrin, Ganj, & Faramarzi, 2016), *RPB1* (the largest subunit of RNA polymerase), and *RPB2* (the second largest subunit of RNA polymerase) (Al-Hatmi et al., 2016). Based on these loci, DNA sequence databases have been constructed, and can be interrogated at FUSARIUM-ID (http://isolate.fusariumdb.org/guide.php) and at CBS-KNAW (http://www.westerdijkinstitute.nl/fusarium/), for the identification of environmental and clinical isolates. Using a newly generated *Fusarium*-specific monoclonal antibody (ED7) that detects an extracellular antigen (a 200 kDa water-soluble carbohydrate), Al-Maqtoofi and Thornton (2016) were able to track pathogenic fusaria in hospital and communal sink biofilms, with FSSC sequence type (ST) 1-a, FOSC ST 33, and FDSC ST ET-gr isolates identified using *TEF-1α* sequencing of recovered strains. This study demonstrated the specificity of mAb ED7, which might find use in the detection of *Fusarium* infections in humans based on detection of the 200 kDa biomarker in patient samples. Multi-locus sequence typing (MLST) is currently regarded as the most robust method for differentiating Fusarium species complexes (Wang et al., 2011) or identifying isolates to species level (Tortorano et al., 2014), with *TEF-1α* and *RPB2* employed in many MLST diagnostic regimens (Van Diepeningen, Feng, et al., 2015).

There are currently no nucleic acid-based detection systems CE-marked or FDA-approved for the diagnosis of human fusariosis, but a number of experimental polymerase chain reaction (PCR) tests have been reported for the detection and quantification of *Fusarium* DNA in murine models of fusariosis, and in human tissue specimens (Bernal-Martínez, Buitrago,

Castelli, Rodríguez-Tudela, & Cuenca-Estrella, 2012; Muraosa et al., 2014; Salehi et al., 2016). A duplex RT-PCR using two specific molecular beacon probes targeting a highly conserved region of the rDNA gene was developed for *F. solani* and non-*Fusarium solani* DNA detection, and validated in two mouse models of invasive infection (Bernal-Martínez et al., 2012). Specificity of the assay was 100%. While sensitivity of the assay in a *F. solani* infection model was 93.9% for lung tissues and 86.7% for serum samples, its sensitivity in a *F. oxysporum* infection model was 87% for lung tissues and 42.8% for serum samples. A *Fusarium* genus-specific and FSSC-specific RT-PCR system has also been developed, which targets the 28s ribosomal RNA gene (Muraosa et al., 2014). To apply the RT-PCR system to the molecular diagnosis of fusariosis, performance was evaluated using plasma and whole blood samples from a mouse model of invasive *F. solani* infection. The sensitivity of the RT-PCR system was found to be 100% in plasma, but no amplification was detected in whole blood. In a retrospective multicenter study, Salehi et al. (2016) reported the discrimination of fusariosis, aspergillosis, mucormycosis, and scedosporiosis in formalin-fixed paraffin-embedded human tissue specimens using multiple real-time quantitative PCR assays. The qPCR assays, targeting the ITS2 region of ribosomal DNA using fluorescently labelled primers, was used to identify clinically important genera and species of *Fusarium*, *Aspergillus*, *Scedosporium*, and mucormycetes, and the molecular identification compared to results from histological examination. *Fusarium oxysporum* and *F. solani* DNA was amplified from five specimens from patients initially diagnosed by histopathology as having aspergillosis. This study demonstrates that histopathological features of molds may be easily confused in tissue sections, and that qPCR can be used for the rapid and accurate identification of fungal pathogens to the genus and species levels directly from FFPE tissues. Despite this, molecular tests should be used only to supplement conventional laboratory tests (Tortorano et al., 2014).

4.1.3 Antigen tests

At present, there are no antigen tests available commercially which are specific for *Fusarium* detection. While (1 → 3)-β-D-glucan detection can be used to diagnose fusariosis both in animal models (Khan, Ahmad, & Theyyathel, 2008) and in humans, its pan-fungal reactivity means that it lacks specificity. Disconcertingly, cross-reaction of the *Aspergillus* GM-ELISA with *Fusarium* spp. has been reported, with roughly half of patients with fusariosis giving a positive test result with the assay (Tortorano et al., 2012). The reason for this cross-reactivity is uncertain, but may be due to the presence of the β-1,5-linked Gal*f* moieties recognized by the GM-reactive mAb EB-A2 being present in polysaccharide antigens from fungi other than *Aspergillus*, including *Fusarium* spp. (Tortorano et al., 2012). Certainly, galactomannan has been found in *Fusarium* culture supernatants (Tortorano et al., 2012; Wiedemann et al., 2016). The Platelia *Candida* mannan ELISA has also been reported to cross-react with *F. verticillioides*, *F. solani* and *F. oxysporum* (Rimek, Singh, & Kappe, 2003). These findings call into question the validity of these tests for the specific diagnosis of aspergillosis and candidiasis, respectively. Despite this, the GM-ELISA has been evaluated in humans as a means of diagnosing fusariosis and as a monitoring tool for treatment response (Nucci et al., 2014), but there is little obvious legitimacy for using this assay for fusariosis detection in the clinical setting given the similar patient groups susceptible to candidiasis, aspergillosis and fusariosis.

5. *Pseudallescheria/Scedosporium* species complex and *Lomentospora prolificans*

Previously, *Pseudallescheria boydii* was considered synonymous with *Pseudallescheria angusta*, *Pseudallescheria ellipsoidea*, and *Pseudallescheria fusoidea* (Rainer, de Hoog, Wedde, Graser, & Gilges, 2000), with a subsequent taxonomic study suggesting that *P. boydii* is a complex comprising six species (*P. boydii*, *P. angusta*, *P. ellipsoidea*, *P. fusoidea*, *P. minutispora*, and *Scedosporium aurantiacum*) (Gilgado, Cano, Gené, & Guarro, 2005). *Scedosporium apiospermum* along with *Graphium* spp. (*G. capitatum*, *G. eumorphum*, and *G. penicillioides*) were considered to be synanomorphs of *P. boydii* (Rainer & de Hoog, 2006), with *S. apiospermum* then shown to be a species distinct from *P. boydii* (Gilgado, Cano, Gené, Sutton, & Guarro, 2008), challenging previous doctrine that this fungus was the asexual or anamorphic state of *P. boydii*. Currently, *Pseudallescheria* and *Scedosporium* are regarded as a composite of species, the *Pseudallescheria/Scedosporium* complex (PSC) (Luplertlop, 2018), with the genus *Scedosporium* now comprising *S. aurantiacum*, *S. desertorum*, *S. minutisporum*, *S. cereisporum*, and *S. dehoogii*, in addition to the *S. apiospermum* complex consisting of *S. angustum*, *S. apiospermum*, *S. boydii*, *S. ellipsoidea*, and *S. fusoideum* (Ramirez-Garcia et al., 2018). Prior to the advent of molecular taxonomy, the dematiaceous fungus *Scedosporium prolificans* (formerly *S. inflatum*) was believed to be a member of the genus *Scedosporium*, but has since been shown to be unrelated to *Scedosporium*, and has been re-classified as *Lomentospora prolificans* (Hennebert & Desai, 1974) with the genus *Lomentospora* being re-instated for this species (Lackner et al., 2014).

The species *S. apiospermum*, *S. aurantiacum*, and *S. boydii* are pathogenic to humans, causing eumycetoma, localized cutaneous infections, muscle, joint and bone infections, and fatal disseminated infections (Cortez et al., 2008; Kondo, Goto, & Yamanaka, 2018; Ramirez-Garcia et al., 2018; Taylor et al., 2014), and are consequently the most studied species in the genus. These fungi are soil saprotrophs that appear to have a proclivity for habitats impacted by human activity (Kalteis, Rainer, & de Hoog, 2009; Rougeron et al., 2015). However, the evidence for this association is questionable, since these organisms are also readily isolated from habitats such as brackish waters and estuarine muds (Thornton, 2009). Notwithstanding this, the numerous reports of these fungi as agents of disease in near-drowning events and following natural disasters such as tsunamis due to aspiration of soil-laden water (Garzoni et al., 2005; Leroy & Smismans, 2007; Linscott, 2007; Nakamura et al., 2011, 2013; Angelini, Drago, & Ruggieri, 2013; Nakamura et al., 2013), demonstrates their abundance in the natural environment, and their ability to cause life-threatening central nervous system (CNS) and cerebral infections following traumatic implant of infectious propagules in the lungs (Ramirez-Garcia et al., 2018). Less is known about the ecology of *Lomentospora prolificans*, but it is a soil- and compost-borne fungus (Grantina-Ievina, Andersone, Berkolde-PĪre, Nikolajeva, & Ievinsh, 2013; Hennebert & Desai, 1974), and is present, alongside members of the PSC, in estuarine muds (Thornton, Ryder, Le Cocq, & Soanes, 2015).

Pathogens in the PSC complex and *L. prolificans* are persistent colonizers of the lungs of cystic fibrosis (CF) patients, and are the second most frequent molds after *A. fumigatus* in the sputum of CF patients and those with underlying respiratory diseases (Blyth, Harun, et al., 2010; Blyth, Middleton, et al., 2010; Bouchara, Symoens, Schwartz, & Chaturvedi, 2018; Cimon et al., 2000; Cooley, Spelman, Thursky, & Slavin, 2007; Schwarz et al., 2017; Schwarz et al., 2018; Schwarz, Thronicke, Staab, & Tintelnot, 2015; Sedlacek et al., 2015; Tomazin & Matos, 2017). While chronic airway colonization of the lung by these fungi can lead to fatal outcomes in CF patients (Borghi, Iatta, Manca, Montagna, & Morace, 2010), the role of these fungi in CF exacerbation, other than causing acute obstruction (Padoan et al., 2016), is not fully understood. Notwithstanding this, a trait common to all strains isolated from the CF lung is their high resistance to antifungal drugs, with *L. prolificans* demonstrating multi-drug resistance (Borghi et al., 2010; Sedlacek et al., 2015), and successful treatment of lung infections requiring triple antifungal drug regimes in several cases

(Schwarz et al., 2018). Important risk factors for fatal disseminated infections (scedosporiosis and lomentosporiosis) include neutropenia and hematological malignancy (Maertens et al., 2000; Rodriguez-Tudela et al., 2009; Song, Lee, & Lee, 2009), advanced AIDS (Tammer et al., 2011), chronic granulomatous disease (Santos, Oleastra, Galicchio, & Zelazko, 2000), allogeneic stem cell transplantation (Tamaki et al., 2016), and solid organ transplantation (Husain et al., 2005). Malignancy, fungemia, central nervous system (CNS) and lung involvement are prognostic factors predicting worse outcome for scedosporiosis and lomentosporiosis (Seidel et al., 2019). Disseminated infections following lung transplantation are particularly problematic (Morio et al., 2010; Rolfe, Haddad, & Wills, 2013; Vagefi et al., 2005), with a *S. apiospermum* and L. *prolificans* mixed disseminated infection reported recently (Balandin et al., 2016).

5.1 Detection of *Scedosporium* and *Lomentospora* diseases

5.1.1 Radiology, histology and culture

Scedosporium and *Lomentospora* infections are often clinically indistinguishable from other invasive mold infections, with radiology based on chest CT providing little diagnostic worth due to non-specific indicators of fungal lung infection (Dabén et al., 2008; Holmes, Trevillyan, Kidd, & Leong, 2013; Nakamura et al., 2013; Schwarz et al., 2015). In contrast, brain MRI is an important diagnostic modality for detecting disseminated CNS infections in immunosuppressed stem cell and solid organ transplant recipients, and in victims of near-drowning, particularly in cases where antifungal treatment fails to resolve disease, or where a clear spreading focus is absent (Bhuta, Hsu, & Kwan, 2012; Holmes et al., 2013; Morio et al., 2010; Nakamura et al., 2013; Ramirez-Garcia et al., 2018; Sharma & Singh, 2015; Tamaki et al., 2016).

Histopathological examination of biopsy material has limited diagnostic utility as it is difficult to distinguish *Scedosporium*/*Lomentospora*-infected tissues from those infected with other hyaline fungi such as *Aspergillus* and *Fusarium* species (Ramirez-Garcia et al., 2018), although the dematiaceous (melanised) hyphae of *L. prolificans* may provide an opportunity for species differentiation. Immunohistological procedures using polyclonal fluorescent antibodies have attempted to improve the accuracy of in situ detection (Kaufman et al., 1997), but cross-reaction with antigens from other fungi such as *Aspergillus* hamper their usefulness. For these reasons, diagnosis of scedosporiosis and lomentosporosis relies on the recovery of the etiological agent from clinical specimens and identification using macroscopic and microscopic features (Luna-Rodríguez et al., 2019; Ramsperger, Duan, Sorrell, Meyer, & Chen, 2014) in vitro. As with *Fusarium* species, accurate differentiation of species is critical for guiding clinical management due to variable antifungal susceptibilities (Box et al., 2018; Cortez et al., 2008; Gilgado, Serena, Cano, Gené, & Guarro, 2006). While *Scedosporium* and *Lomentospora* species can be recovered from sterile sites such as blood, bone and tissue biopsy specimens using standard mycological media, for example Saboraud's dextrose agar (SDA), their frequent occurrence in sputum samples of CF patients (Chen, Chou, et al., 2017; Chen, Kondori, et al., 2017; Sedlacek et al., 2015) necessitated the development of selective media that prevent overgrowth by faster growing *Aspergillus*, *Candida* and *Fusarium* species in polymicrobial clinical samples (Rainer, Kalteis, de Hoog, & Summerbell, 2008; Blyth, Harun, et al., 2010; Blyth, Middleton, et al., 2010; Hong et al., 2017). One of the most effective media for selective isolation of *Scedosporium* species is *Scedosporium* Selective agar (SceSel+) (Blyth, Harun, et al., 2010; Blyth, Middleton, et al., 2010; Rainer et al., 2008) containing the antifungal agents benomyl and dichloran that limit the growth of *Aspergillus* and *Candida*. *Scedosporium* species and *L. prolificans* can then be further differentiated in culture with the inclusion of cycloheximide that is inhibitory to *L. prolificans* (Ramsperger et al., 2014).

5.1.2 Nucleic acid tests

Speciation of isolates within the *Pseudallescheria/Scedosporium* complex and their discrimination from *L. prolificans* based on microscopic features requires considerable expertise and can be problematic. For this reason, a number of molecular methods have been developed for the detection and differentiation of these fungi in axenic culture, sputum, and tissue samples including multiplex PCR (Harun et al., 2011); repetitive sequence-based PCR (rep-PCR)(Steinmann, Schmidt, Buer, & Rath, 2011); qPCR, loop-mediated isothermal amplification (LAMP) and PCR-based reverse line blot (PCR-RLB) (Lu, Gerrits van den Ende, Bakkers, et al., 2011; Lu, Gerrits van den Ende, de Hoog, et al., 2011); oligonucleotide arrays (Bouchara et al., 2009); DNA sequencing and restriction fragment length polymorphisms (RFLP) (Delhaes et al., 2008; Gilgado et al., 2005, 2008), and Rolling Circle Amplification (RCA) (Lackner, Najafzadeh, Sun, Lu, & de Hoog, 2012). These techniques and their different capabilities are comprehensively reviewed elsewhere (Ramirez-Garcia et al., 2018; Ramsperger et al., 2014). Despite these myriad methods for pathogen detection and speciation, no single technique has been universally adopted nor are any commercial tests currently recommended for this purpose. Current guidelines recommend that nucleic acid-based assays be used in combination with conventional laboratory tests (Tortorano et al., 2014).

5.1.3 Proteomics and antigen tests

Matrix-assisted laser desorption/ionization time-of-flight mass spectrometry (MALDI-TOF/MS) has been investigated as an alternative strategy for identification of fungi within the PSC (Coulibaly et al., 2011), and other pathogenic molds (Bader, 2013; Cassagne et al., 2011; Lau, Drake, Calhoun, Henderson, & Zelany, 2013; Santos, Paterson, Venancio, & Lima, 2010). This technique currently lacks the requisite standardization needed for clinical acceptability in medical mycology, but it nevertheless has the potential to replace gene-sequencing methods for mold identification, as it is accurate, comparatively inexpensive and relatively quick compared to sequencing methods. However, it does require robust online reference databases constructed using similar fungal growth and extraction procedures for interrogation of MS spectra from unknown species. Commercially available MALDI-TOF/MS identification solutions are apparently inadequate for *Scedosporium/Lomentospora* discrimination at present (Ramirez-Garcia et al., 2018).

Mass spectrometry (MS) has also recently used by Thornton et al. (2015) to identify the antigen bound by a mAb (CA4) specific to *L. prolificans*. This IgG1 mAb is highly specific, binding to an intracellular protein from *L. prolificans*, but not with proteins extracted from members of the PSC or from other unrelated molds. CA4-reactive peptides in western blots of cell extracts following two-dimensional electrophoresis were identified using MS as fragments of the melanin-biosynthetic enzyme tetrahydroxynaphthalene reductase (4HNR). This enzyme is one of a number of enzymes involved in DHN-melanin production of dematiaceous molds such as *L. prolificans* (Al-Laaeiby, Kershaw, Penn, & Thornton, 2016). Confirmation of the CA4 antigen as the enzyme 4HNR was established using genome sequencing and targeted disruption of the 4HNR-encoding gene. Mutants lacking the gene, and with abnormal pigmentation (orange-brown spores compared to the black spores of the wild type strain), were enzyme-deficient and no longer reactive with mAb CA4.

The study of Thornton et al. (2015) further demonstrates the power of hybridoma technology to develop mAbs specific to pathogenic molds, but the generation of a species-specific mAb to an intracellular antigen is unusual. The majority of genus or species-specific mAbs generated against cell extracts from pathogenic yeasts, yeast-like fungi, and molds, bind to carbohydrate moieties on extracellular carbohydrate or glycoprotein antigens (Thornton, Pitt, Wakley, & Talbot, 2002; Thornton, 2009; Davies & Thornton, 2014; Al-Maqtoofi & Thornton, 2016; Davies et al., 2017; Morad et al., 2018). This is the case of mAbs specific to members of the PSC. Thornton (2009) reported the development of IgG1 (clone HG12) and IgM (clone GA3) mAbs specific to a heat-stable, water-soluble, 120 kDa carbohydrate present on the surface of conidia and hyphae of *S. apiospermum*, *S. aurantiacum*, and *S. boydii*. The antigen is also readily detectable as an extracellular antigen that is produced in abundance during active growth of these pathogens. The mAbs do not react with the related fungus *S. dehoogii*, with *L. prolificans*, or other unrelated human pathogenic fungi. The copious production of the 120 kDa antigen and the availability of highly specific mAbs for antigen detection, make this an ideal system for the development of a diagnostic test for members of the PSC. To this end, we have developed a lateral-flow assay (*Sced*-LFA) for PSC detection that comprises the IgG1 mAb HG12 (Fig. 1), and which demonstrates the feasibility of developing a POCT for rapid detection of pathogens in the PSC. In addition, we have developed a quantitative sandwich ELISA that combines both mAbs HG12 and GA3. This ELISA was recently used by Box et al. (2018) to determine the pharmacodynamics of voriconazole in in vitro models of invasive pulmonary scedosporiosis. Given the recent CE-marking and commercial availability of the *Asp*LFD (OLM Diagnostics) and *Aspergillus* GP-ELISA (Euroimmun AG) which both employ the *Aspergillus*-specific mAb JF5 (Thornton, 2009), the development of complementary qualitative LFA and quantitative ELISA tests for the diagnosis of scedosporiosis, based on detection of the 120 kDa in human BALf and serum samples, is technically feasible.

6. Mucormycete species pathogenic to humans

Mucormycetes are thermotolerant, fast-growing saprotrophic molds of the phylum Zygomycota order Mucorales, which are abundant in a wide range of ecological niches including soil and compost (Richardson, 2009), food (e.g., cheese where they cause spoilage (Hymery et al., 2014)), edible crops where they cause post-harvest storage rots (Saito, Michailides, & Xiao, 2016), and household dust (Al-Humiany, 2010). The order Mucorales includes species in the genus *Apophysomyces*, *Cunninghamella*, *Lichtheimia*, *Mucor*, *Rhizomucor*, *Rhizopus*, and *Saksenaea*, which are separated from the

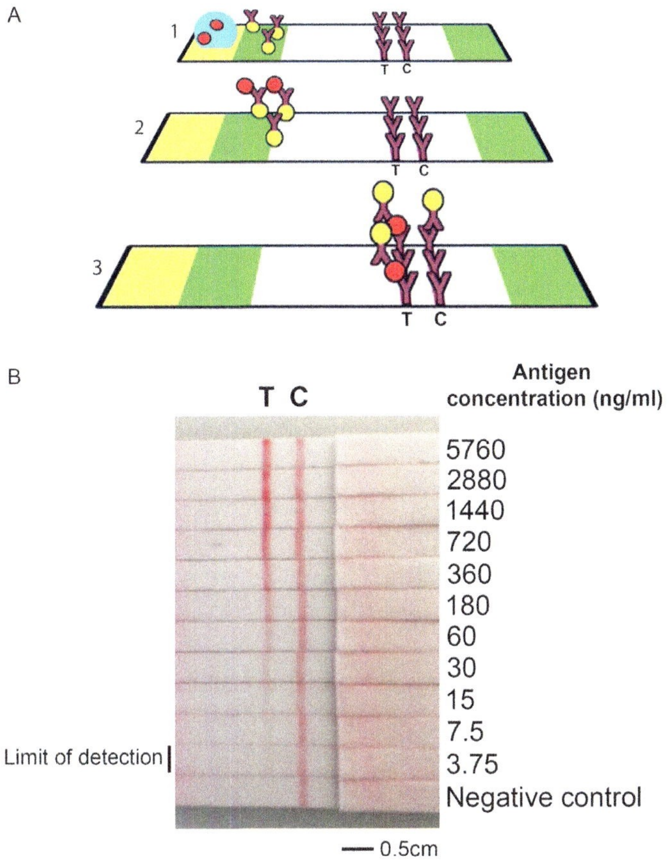

Fig. 1 Demonstration lateral-flow assay (*Sced*-LFA) for the detection of pathogens within the *Pseudallescheria/Scedosporium* species complex. (A) Schematic diagram showing the internal workings of the *Sced*-LFA test. The monoclonal antibody HG12 (Thornton, 2009) is immobilized in the test zone (T) on a porous nitrocellulose membrane. Anti-mouse immunoglobulin immobilized to the membrane in a separate zone serves as an internal control (C). On addition of 100 μL of analyte (for example bronchoalveolar lavage) to the release pad (1), mAb HG12 which is conjugated to NanoAct cellulose nanobeads (CNB) in the release pad (Y-shaped structures with *yellow spheres*), binds to the diagnostic 120 kDa antigen (*red spheres*) and the antibody-CNB/antigen complex then moves along the porous membrane by capillary action (2). The same mAb immobilized in the test zone (T) binds to the complex (3), resulting in a positive test line. Any mAb-CNB conjugate that migrates unbound to the target antigen, binds to the internal control (C) indicating that the assay has run correctly. Where the 120 kDa diagnostic antigen is present in the analyte, 2 lines appear (T and C) indicating infection, but samples that are antigen negative (no infection) result in a single control line (C) only. (B) Test results using filtrates from an actively growing culture of *S. apisopermum*, and with mAb HG12 conjugated to red CNB particles. Positive tests results (T and C both present) are visible for antigen concentrations $\geq$3.75 ng/mL filtrate (the limit of detection). The negative control (antigen-negative culture medium) produces a single internal control line only. The single-step *Sced*-LFA test is highly specific, quick (taking only 10 min to perform from addition of sample to interpretation of result), and is stable for 18 months at room temperature, making it ideal for use in resource-limited settings. Scale bar = 0.5 cm. *Panel (B): Courtesy of the* Sced-*LFA contract manufacturer LateralDx.*

entomophthoromycetes *Basiodiobolus* and *Condiobolus*, zygomycete fungi that are also capable of causing disease in humans (Binder, Maurer, & Lass-Flörl, 2014; Ribes, Vanover-Sams, & Baker, 2000). While mucormycetes such as *Rhizopus oryzae* (syn. *Rhizopus arrhizus*) and *Lichtheimia corymbifera* can cause allergic reactions in humans (Rognon et al., 2016; Sircar et al., 2015), it is the ability of these fungi to cause osteoarticular mycosis (Taj-Aldeen et al., 2017), and rhino-orbital-cerebral (Arndt, Aschendorff, Echternach, Daemmrich, & Maier, 2009; Son, Lim, Lee, Yang, & Lee, 2016), oral (Hingad, Kumar, & Deshmukh, 2012), pulmonary (Hung et al., 2015; Iqbal, Irfan, Jabeen, Kazmi, & Tariq, 2017; Wang, Guo, Xue, & Chen, 2016), gastrointestinal (Adhikari, Gautam, Paudyal, Sigdel, & Basnyat, 2019), cutaneous (Arruebarrena, Romano, Sciarretta, Hopkins, & Davis, 2016; Li, Hwang, Zhou, Du, & Zhang, 2013; Zahoor, Kent, & Wall, 2016) or disseminated infections (Walsh, Gamaletsou, McGinnis, Hayden, & Kontoyiannis, 2012) in pre-disposed adult and pediatric patients (Däbritz et al., 2011) that is of the greatest concern, with a single species, *Rhizopus oryzae* responsible for >80% of all life-threatening infections (Richardson, 2009; Richardson & Page, 2018; Raghunath, Priyanka, Kiran, & Deepthi, 2018). This fungus and other pathogenic Mucorales (predominantly *Lichtheimia, Mucor,* and *Rhizomucor*) differ from other opportunistic molds in their ability to infect a broader, more varied population of human hosts (Gomes, Lewis, & Kontoyiannis, 2011; Richardson, 2009).

Cutaneous and deep tissue infections typically occur by direct inoculation during trauma, such as following combat-related blast injury (Kronen et al., 2017; Paolino et al., 2012; Tribble & Rodriguez, 2014; Warkentien et al., 2012), natural disasters (Andresen et al., 2005; Fanfair et al., 2012), or traffic accidents (Lelievre et al., 2014), but are also reported in patients with diabetes or hematological malignancies (Arnáiz-García et al., 2009). Reports of disseminated mucormycosis, while rare, are increasing in incidence particularly in patients with uncontrolled diabetes, iron overload, and ketoacidosis (Binder et al., 2014; Rammaert, Lanternier, Poirée, Kania, & Lortholary, 2012), and also in patients with hematological malignancies and allogeneic stem cell transplants (Petrikkos et al., 2012; Sugui, Christensen, Bennett, Zelazny, & Kwon-Chung, 2011).

6.1 Detection of mucormycosis

6.1.1 Radiology, histology and culture

The aggressive and rapid course of mucormycete infections makes early detection critically important for the management both of localized cutaneous, deep-seated, and disseminated infections (Castrejón-Pérez, Miranda, Welsh, Welsh, & Ocampo-Candiani, 2017). This is especially important in the context of their intrinsic resistance to short-tailed mold-active azole drugs (voriconazole (VCZ) and fluconazole) due to an evolutionary conserved amino acid substitution in the enzyme lanosterol 14α-demethylase responsible for ergosterol biosynthesis (Caramalho et al., 2017). This has led to an increased incidence of breakthrough infections amongst immunocompromised and high-risk patients receiving VCZ prophylaxis, with the suggestion that VCZ prophylaxis should be thought of as one of the risk factors for mucormycosis (Ustun, Farrow, DeRemer, Fain, & Jillela, 2007).

As with aspergillosis, fusariosis, and scedosporiosis, abnormalities observed in a chest-CT consistent with angio-invasive mold infections (nodules, halo signs, cavities, wedge-shaped infiltrates, and pleurar effusions) are not pathognomonic of pulmonary mucormycosis, and so can only raise the suspicion of mucormycosis in the context of a patient's clinical history. Despite this, the reversed halo sign in neutropenic patients with hematological malignancies appears to have high predictive value for mucormycosis (Legouge et al., 2014; Walsh et al., 2014). Computed tomography has also proven useful in the differential diagnosis of rhino-orbital-cerebral mucormycosis (ROCM) and bacterial orbital cellulitis, where the early symptoms of both diseases are similar (Son et al., 2016). However, it is unable to discriminate between ROCM and aspergillosis in cases of invasive fungal rhinosinusitis (Arndt et al., 2009). Culture of mucormycetes from biopsy samples (Hong & Park, 2017) is problematic due to the fragile nature of mucormycete hyphae that are aseptate or pauci-septate (Skiada et al., 2013). Growth of mucormycetes in culture is rapid, but identification to the level of species and even genus requires considerable mycological expertise (Ziaee, Zia, Bayat, & Hashemi, 2016). As with other pathogenic molds, MALDI-TOF/MS can be used to differentiate species once in axenic culture (Cornely et al., 2014; Cassagne et al., 2011; De Carolis et al., 2012; Schrödl et al., 2012), but more data are needed to authenticate this technique, and the robustness of reference databases needs to be fully evaluated.

In situ detection of mucormycetes in tissue samples relies on the microscopic observation of characteristic broad ribbon-like hyphae that can be stained with optical brighteners such as calcofluor white or with Grocott-Gomori methamine silver (Antonov, Tang, & Grossman, 2015; Lass-Flörl, 2009; Liao, Karanjawala, Baba, Khalil, & Irani, 2015). Sputum and BALf samples can also show the characteristic broad aseptate hyphae

(Glazer et al., 2000), which is a first indicator of mucormycosis (Al-Abbadi, Russo, & Wilkinson, 1997). However, in one case study, only 25% of sputum or BALf samples were positive pre-mortem (Kontoyiannis, Wessel, Bodey, & Rolston, 2000). Attempts have been made to improve the specificity and sensitivity of microscopy by using mucormycete-reactive antibodies in immunohistochemistry. The production of rabbit hyper-immune antisera and mAbs to immuno-dominant intracellular antigens of the Mucorales was reported by Jensen, Aalbæk, and Schønheyder (1994) and Jensen, Aalbæk, Lind, and Krogh (1996). Jensen et al. (1994) demonstrated that heterologous absorption rendered antisera, raised against *Rhizopus oryzae* somatic antigens, monospecific by indirect immunofluorescence, whereas heterologous absorption of *Lichtheimia corymbifera* antiserum did not abolish reactivity with *R. oryzae*. The reactivity of the heterologously absorbed antisera, and a murine IgG1 mAb (1A7B4) raised against *L. corymbifera*, enabled mucormycetes within bovine lesions to be identified by immunohistochemistry. In a subsequent study, Jensen et al. (1996) showed that a mAb (WSSA-RA-1), raised against water-soluble somatic antigens from *R. oryzae*, similarly bound to intracellular homologous antigens of between 14kDa and 110kDa. The high degree of specificity exhibited by mAb WSSA-RA-1 allowed its use in immunohistochemistry to detect the pathogen in situ in placenta, lymph node and gastrointestinal tract samples of cows with suspected systemic bovine zygomycosis. In a subsequent study (Jensen, Salonen, & Ekfors, 1997), the same mAb (WSSA-RA-1) was used in immunohistochemistry to impove the sensitivity and specificity of mucormycosis detection in patients with hematological malignancies.

6.1.2 Nucleic acid tests

Identification of mucoralean fungi in culture based on morphological characteristics is notoriously difficult, particularly with strains that fail to sporulate. For this reason, nucleic acid-based procedures have been developed that enable discrimination of Mucorales at both genus- and species-level, although evidence that identification beyond genus-level aids treatment is lacking. Unlike *Fusarium*, sequencing of the internal transcribed spacer regions ITS1 and ITS2 of ribosomal RNA gene (rDNA) has been shown to be a reliable method for species differentiation in the mucormycetes (Gade, Hurst, Balajee, Lockhart, & Litvintseva, 2017), with ITS sequencing recommended by ISHAM as a first-line method of identification (Balajee et al., 2009). Numerous in-house molecular assays have been developed for direct detection of mucoralean DNA in clinical samples (Skiada et al., 2017), including BALf (Lengerova et al., 2014), serum (Millon et al., 2016), and biopsy (skin, lung, stomach mediastine, nose, parotid gland) samples (Bernal-Martínez et al., 2012). However, their current lack of standardization means that they are only moderately supported in the clinical guidelines for the diagnosis of mucormycosis (Cornely et al., 2014; Skiada et al., 2017). Despite this, PCR has proved useful in the detection of mucormycete infections, especially in infections that are culture-negative (Hammond et al., 2011). Lengerova et al. (2014) used PCR followed by high-resolution melt analysis (PCR/HRMA) to detect *Rhizopus* spp., *Rhizomucor pusillus*, *Lichtheimia corymbifera*, and *Mucor* spp. in BALf samples from immunocompromised patients who were at risk of invasive fungal disease. Real-time qPCR assays specific to the species identified in the PCR/HMRA test were used to validate the test. Of the 99 BALF samples collected from 86 patients with pulmonary abnormalities, 91% were negative and 9% were positive in the PCR/HMRA test, with a sensitivity and specificity 100% and 93%, respectively. By combining the positive PCR/HRMA results with positive real-time qPCR results, the specificity was increase to 98%. Due to its high NPV of 99%, the PCR/HRMA test is a fast and reliable technique for differential diagnosis of pulmonary mucormycosis in immunocompromised patients caused by the four most clinically important mucormycetes.

Because of the difficulty in obtaining fungal DNA from human fluids, the possibility of detecting mucormycete DNA in paraffin-embedded and fresh tissues has been investigated in both an experimental model of disseminated mucormycosis (Dannaoui et al., 2010), and also in humans (Gade et al., 2017). The universal primers ITS1 and ITS2 were used in PCR to amplify the ITS1 region from mucormycete DNA in formalin-fixed paraffin-embedded kidney and brain tissues from mice infected with one of five species *(Rhizopus oryzae, Rhizopus microsporus, Lichtheimia corymbifera, Rhizomucor pusillus*, and *Mucor circinelloides*). Amplicons were then sequenced for species identification. Using this procedure, identification of major mucormycete species from formalin-fixed paraffin-embedded could reach 100%, provided sufficient starting material was available for DNA extraction. Gade et al. (2017) investigated the utility of a PCR targeting a 200–300 bp region of the extended 28S region of rDNA for molecular identification of DNA from mucormycetes and other fungi in formalin-fixed paraffin-embedded and fresh tissues. They demonstrated this region could be used to identify all genera and some species of clinically relevant mucormycetes. They also demonstrated that PCR amplification and direct sequencing of the extended 28S region of rDNA was more sensitive compared to targeting the ITS2 region, since they were able to detect and identify mucormycetes and other fungal pathogens in tissues from patients with histopathological and/or culture evidence of fungal infections that were negative in PCR using ITS-specific primers.

6.1.3 Antibody and antigen tests

While there are no routine serological tests for diagnosis of disseminated mucormycosis, ELISA has been used to study the relationship between antibody levels and exposure to *Rhizopus microsporus* in sawmill workers with allergic alveolitis (Sandven & Eduard, 1992), while recombinant antigens have been developed for the serodiagnosis of farmer's lung caused by *Lichtheimia corymbifera* (Rognon et al., 2016). Detection of disseminated mucormycosis is not possible using tests for $(1\rightarrow3)$-β-D-glucan since mucormycetes lack this carbohydrate in their cell walls (Cornely et al., 2014), but it can be used to rule out IPA, the most frequent differential diagnosis, or combined *Aspergillus* and Mucorales infections (Skiada et al., 2013). Early attempts to detect disseminated disease serologically were based on immunodiffusion (Jones & Kaufman, 1978), electrophoresis and immunoblotting (Wysong & Waldorf, 1987), and ELISA (Kaufman, Turner, & McLaughlin, 1989) of homogenate antigens challenged with sera from patients with mucormycete infections. However, non-specific reactivity with the antigen preparations was observed with sera from patients with aspergillosis and candidiasis also, which meant that the tests were unable to generically or specifically identify the etiologic agents of mucormycosis.

The immunological determinant of mucoralean intracellular glycans has been shown to comprise linear $(1\rightarrow6)$-α-linked mannopyrasonyl residues (Miyazaki, Hayashi, Oshima, & Yadomae, 1979), with the cell walls of yeast and hyphal phases of mucoralean fungi containing fucomannopeptide and mannoprotein (Yamada, Hiraiwa, & Miyazaki, 1983; Yamada, Ohshima, & Miyazaki, 1982). While these antigens have provided useful targets for direct visualization of mucoralean hyphae in tissue samples using immunohistochemistry, their intracellular nature means that they have limited applicability as circulating biomarkers of mucormycosis in BALf and serum samples. Approaches have been used to identify extracellular antigens specific to the Mucorales (Hessian & Smith, 1982), which might circulate in the bloodstream and therefore be suitable targets for detection by, for example, ELISA. Extracellular polysaccharides (EPS) of the Mucorales that appear, using polyclonal IgG antibodies, to be specific to this group of fungi, have been isolated and characterized (De Ruiter et al., 1991; De Ruiter, Josso, et al., 1992; De Ruiter, van Bruggen-van der Lugt, et al., 1992; De Ruiter et al., 1993; De Ruiter et al., 1994; Notermans & Soentoro, 1986), and their immunodominant residues shown to comprise 2-O-methyl-D-mannose (De Ruiter et al., 1994).

Despite these extensive early studies, and despite our improved understanding of the polysaccharide cell wall architecture of mucoralean fungi (Lecointe, Cornu, Leroy, Coulon, & Sendid, 2019), no genus- or species-specific mAbs are currently available that target immunogenic EPS, and which might prove suitable for the development of immunoassays for diagnosis of mucormycosis in humans. Notwithstanding this, somatic antigens diagnostic of systemic *L. corymbifera* infections have been detected in the urine of cattle (Jensen, Frandsen, & Schønheyder, 1993), indicating that biomarkers of mucormycosis circulate in the bloodstream and are excreted. This raises the possibility of developing a non-invasive test for mucormycosis based on mucormycete antigenuria, provided that suitable targets can be identified and specific mAbs can be generated.

7. Conclusions and future prospects

One of the main priorities of the Global Action Fund for Fungal Infections (GAFFI) is availability and access of cheap and effective diagnostic tests for all, particularly those in poorly-resourced countries. A commendable priority, but one littered with obstacles. Technically, as this review has shown, it is possible to develop highly specific and sensitive assays for the detection of some of the most problematic of human pathogens, the filamentous fungi or molds. Furthermore, it is possible to demonstrate the clinical utility of these tests in retrospective or semi-prospective studies. However, the vast majority of these tests remain at the experimental stage in the laboratory, with little prospect of translation to the clinic as commercially-available CE-marked or FDA-approved assays. One of the biggest obstacles to bench-to-bedside translation is the escalating costs for the regulatory approvals required at each of the different technical and commercial phases of a test's development, and during its registration as an in vitro medical device (IVD).

As Fig. 1 and the accompanying text demonstrates, the generation of an immunoassay, a lateral-flow assay (LFA), that encompasses many of the desirable qualities of a near-patient point-of-care test for a human pathogenic fungus (short-time to result; inexpensive equipment/instrumentation; suitable for use by personnel with limited training; minimal number of steps; uncomplicated interpretation (Kozel & Wickes, 2014; Wickes & Wiederhold, 2018)) is relatively simple given the availability of a high quality monoclonal antibody (in this case a well-characterized mAb that detects a water-soluble, heat-stable, carbohydrate antigen specific to pathogens in the *Pseudallescheria/Scedosporium* complex (Thornton, 2009)), and the technical know-how of how to apply it to LFA technology. Production of this demonstration test took a month to complete, including antibody purification and conjugation, LFA formatting, and determination of specificity and analytical sensitivity. The limit of detection of purified antigen is 3.75 ng/mL (Fig. 1), which is well within the level of analytical sensitivity required for clinical detection of a circulating antigen in human samples. Despite this, the demand for a PSC-specific test is insufficient to warrant further commercial development of the *Sced*-LFA, despite the importance of this group of fungi in human disease.

To take a test such as the *Sced*-LFA from the demonstration phase through to prototyping and then into test validation and verification with an FDA-approved manufacturer, requires investment beyond the capabilities of many diagnostic companies, certainly beyond a small to mid-size enterprise (SME), without a cast-iron guarantee of financial return and profit within a short period of time post CE-marking or FDA approval. This guarantee is unlikely to be met with tests for emerging or rare pathogens such as *Fusarium*, *Lomentospora*, *Scedosporium*, or the mucormycetes, especially given the investment and time required for CE-marking and FDA approval of an IVD, and subsequent integration of a novel test into hospital diagnostic workstreams. With every hospital or regional healthcare authority demanding in-house validation of a test's performance compared to its own established diagnostic procedures, it can typically take several years before a test is widely accepted, and is purchased in sufficient numbers so as to justify the developer's costs for R&D, manufacture, marketing, sales and distribution of the test. These same principles apply to any diagnostic test, whether it be an antigen test such as the *Sced*-LFA, or a molecular diagnostic test for fungal DNA (Wickes & Wiederhold, 2018).

If GAFFI's desire for cheap and effective diagnostic tests for all is to be met, there needs to be a major re-think in the way that we, as a community of medical mycologists, diagnosticians, and medical practitioners, work together to better support the diagnostics sector to facilitate the development and translation of novel fungal diagnostics to the clinical setting. Without this community support, and without much-needed financial assistance from government funding agencies who have traditionally paid insufficient attention to human mycoses and their detection, the availability of cheap, easy-to-use, and effective diagnostic tests will be limited to those pathogens which infect sufficiently large number of individuals, such as *Aspergillus*, *Cryptococcus*, and *Pneumocystis*, to warrant diagnostic test development. An exemplar of where community and funding agency support worked to the patient's advantage was the Aspergillus Technology Consortium (AsTeC) for aspergillosis clinical laboratory diagnostics. A similar approach needs to be adopted for other invasive mold diseases.

Acknowledgments

This work was carried out, in part, during tenure of a grant from Innovate UK (reference number 105440). The author would like to thank Richard Campbell of LateralDx for supply of the image in Fig. 1B.

References

Abdolrasouli, A., Scourfield, A., Rhodes, J., Shah, A., Elborn, J. S., Fisher, M. C., et al. (2018). High prevalence of triazole resistance in *Aspergillus fumigatus* isolates in a specialist cardiothoracic centre. *International Journal of Antimicrobial Agents*, *52*, 637–642.

Adhikari, S., Gautam, A. R., Paudyal, B., Sigdel, K. R., & Basnyat, B. (2019). Case report: Gastric mucormycosis—A rare but important differential diagnosis of upper gastrointestinal bleeding in an area of *Helicobacter pylori* endemicity. *Wellcome Open Research*, *4*, 5.

Agarwal, R., Chakrabati, A., Shah, A., Gupta, D., Meis, J. F., Guleria, R., et al. (2013). Allergic bronchopulmonary aspergillosis: Review of literature and proposal of new diagnostic and classification criteria. *Clinical and Experimental Allergy*, *43*, 850–873.

Al-Abbadi, M. A., Russo, K., & Wilkinson, E. J. (1997). Pulmonary mucormycosis diagnosed by bronchoalveolar lavage: A case report and review of the literature. *Pediatric Pulmonology*, *23*, 222–225.

Alghamdi, N. S., Barton, R., Wilcox, M., & Peckham, D. (2019). Serum IgE and IgG reactivity to *Aspergillus* recombinant antigens in patients with cystic fibrosis. *Journal of Medical Microbiology*, *68*, 924–929.

Al-Hatmi, A. M. S., Hagen, F., Menken, S. B. J., Meis, J. F., & de Hoog, G. S. (2016). Global molecular epidemiology and genetic diversity of *Fusarium*, a significant emerging group of human opportunists from 1958 to 2015. *Emerging Microbes & Infections*, *5*, e124.

Al-Humiany, A. A. (2010). Opportunistic pathogenic fungi of the house dust in Turubah, Kingdom of Saudia Arabia. *Australian Journal of Basic and Applied Sciences*, *4*, 122–126.

Al-Laaeiby, A., Kershaw, M. J., Penn, T. J., & Thornton, C. R. (2016). Targeted disruption of melanin biosynthesis genes in the human pathogenic fungus *Lomentospora prolificans* and its consequences for pathogen survival. *International Journal of Molecular Sciences*, *17*, 444.

Al-Maqtoofi, M., & Thornton, C. R. (2016). Detection of human pathogenic *Fusarium* species in hospital and communal sink biofilms by using a highly specific monoclonal antibody. *Environmental Microbiology*, *18*, 3620–3634.

Anaissie, E. J., Kuchar, R. T., Rex, J. H., Francesconi, A., Kasai, M., Müller, F.-M. C., et al. (2001). Fusariosis associated with pathogenic *Fusarium* species colonisation of a hospital water system: A new paradigm for the epidemiology of opportunistic mold infections. *Clinical Infectious Diseases*, *33*, 1871–1878.

Anaissie, E. J., Penzak, S. R., & Dignani, M. C. (2002). The hospital water supply as a source of nosocomial infections: A plea for action. *Archives of Internal Medicine*, *162*, 1483–1492.

Andresen, D., Donaldson, A., Choo, L., Knox, A., Klaasen, M., Ursic, C., et al. (2005). Multifocal cutaneous mucormycosis complicating polymicrobial wound infection in a tsunami survivor from Sri Lanka. *Lancet*, *365*, 876–878.

Angelini, A., Drago, G., & Ruggieri, P. (2013). Post-tsunami primary *Scedosporium apiospermum* osteomyelitis of the knee in an immunocompetent patient. *International Journal of Infectious Diseases*, *17*, e646–e649.

Ansorg, R., von Heinegg, E. H., & Rath, P. M. (1994). *Aspergillus* antigenuria compared to antigenemia in bome marrow transplant recipients. *European Journal of Clinical Microbiology & Infectious Diseases*, *13*, 582–589.

Antonov, N. K., Tang, R., & Grossman, M. E. (2015). Utility of touch preparation for rapid diagnosis of cutaneous mucormycosis. *JAAD Case Reports*, *1*, 175–177.

Aogáin, M. M., Chandrasekaran, R., Lim, A. Y. H., Low, T. B., Tan, G. L., Hassan, T., et al. (2018). Immunological corollary of the pulmonary microbiome in bronchiectasis: The CAMEB study. *The European Respiratory Journal*, *52*, 1800766.

Arnáiz-García, M. E., Alonso-Peña, D., del Carmen González-Vela, M., García-Palomo, J. D., Sanz-Giménez-Rico, J. R., & Arnáiz-García, A. M. (2009). Cutaneous mucormycosis: Report of five cases and review of the literature. *Journal of Plastic, Reconstructive & Aesthetic Surgery*, *62*, e434–e441.

Arndt, S., Aschendorff, A., Echternach, M., Daemmrich, T. D., & Maier, W. (2009). Rhino-orbital-cerebral mucormycosis and aspergillosis: Differential diagnosis and treatment. *European Archives of Oto-Rhino-Laryngology*, *266*, 71–76.

Arrese, J. E., Piérard-Franchimont, C., & Piérard, G. E. (1996). Fatal hyalohyphomycosis following *Fusarium* onychomycosis in an immunocompromised patient. *The American Journal of Dermatopathology*, *18*, 196–198.

Arruebarrena, G. A., Romano, A. E., Sciarretta, J., Hopkins, M. A., & Davis, J. M. (2016). Cutaneous mucormycosis in a trauma patient with a new diagnosis of diabetes mellitus. *Surgical Infection Case Reports*, *1*, 4–7.

Baddley, J. W., Marr, K. A., Andes, D. R., Walsh, T. J., Kauffman, C. A., Kontoyiannis, D. P., et al. (2009). Patterns of susceptibility of *Aspergillus* isolates recovered from patients enrolled in the Transplant-Associated Infection Surveillance Network. *Journal of Clinical Microbiology*, *47*, 3271–3275.

Bader, O. (2013). MALDI-TOF-MS-based species identification and typing approaches in medical mycology. *Proteomics*, *13*, 788–799.

Balajee, S. A., Borman, A. M., Brandt, M. E., Cano, J., Cuenca-Estrella, M., Dannaoui, E., et al. (2009). Sequence-based identification of *Aspergillus*, *Fusarium*, and mucorales species in the clinical mycology laboratory: Where are we and where should we go from here? *Journal of Clinical Microbiology*, *47*, 877–884.

Balandin, B., Aguilar, M., Sánchez, I., Monzón, A., Rivera, I., Salas, C., et al. (2016). *Scedosporium apiospermum* and *S. prolificans* mixed disseminated infection in a lung transplant recipient: An unusual case of long-term survival with combined systemic and local antifungal therapy in intensive care unit. *Medical Mycology Case Reports*, *11*, 53–56.

Barnes, R. A. (2013). Directed therapy for fungal infections: focus on aspergillosis. *The Journal of Antimicrobial Chemotherapy*, *68*, 2431–2434.

Bennett, J. E., Friedman, M. M., & Dupont, B. (1987). Receptor-mediated clearance of *Aspergillus* galactomannan. *The Journal of Infectious Diseases*, *155*, 1005–1010.

Bernal-Martínez, L., Buitrago, M. J., Castelli, M. V., Rodríguez-Tudela, J. L., & Cuenca-Estrella, M. (2012). Detection of invasive infection caused by *Fusarium solani* and non-*Fusarium solani* species using a duplex quantitative PCR-based assay in a murine model of fusariosis. *Medical Mycology*, *50*, 270–275.

Bhuta, S., Hsu, C. C.-T., & Kwan, G. N. C. (2012). *Scedosporium apiospermum* endophthalmitis: Diffusion-weighted imaging in detecting subchoroidal abscess. *Clinical Ophthalmology*, *6*, 1921–1924.

Binder, U., Maurer, E., & Lass-Flörl, C. (2014). Mucormycosis—From the pathogens to the disease. *Clinical Microbiology and Infection*, *20*, 60–66.

Blackwell, M. (2011). The fungi: 1, 2, 3 ... 5.1 million species? *American Journal of Botany*, *98*, 426–438.

Blyth, C. C., Harun, A., Middleton, P. G., Sleiman, S., Lee, O., Sorrell, T. C., et al. (2010). Detection of occult *Scedosporium* species in respiratory tract specimens from patients with cystic fibrosis by use of selective media. *Journal of Clinical Microbiology*, *48*, 314–316.

Blyth, C. C., Middleton, P. G., Harun, A., Sorrel, T. C., Meyer, W., & Chen, S. C.-A. (2010). Clinical associations and prevalence of *Scedosporium* spp. in Australian cystic fibrosis patients: Identification of novel risk factors? *Medical Mycology*, *48*, S37–S44.

Bongomin, F., Gago, S., Oladele, R. O., & Denning, D. W. (2017). Global and multi-national prevalence of fungal diseases—Estimate precision. *Journal of Fungi*, *3*, 57.

Borghi, E., Iatta, R., Manca, A., Montagna, M. T., & Morace, G. (2010). Chronic airway colonisation by *Scedosporium apiospermum* with a fatal outcome in a patient with cystic fibrosis. *Medical Mycology*, *48*, S108–S113.

Bouchara, J.-P., Hsieh, H. Y., Croquefer, S., Barton, R., Marchais, V., Pihet, M., et al. (2009). Development of an oligonucleotide array for direct detection of fungi in sputum samples from patients with cystic fibrosis. *Journal of Clinical Microbiology*, *47*, 142–152.

Bouchara, J.-P., Symoens, F., Schwartz, C., & Chaturvedi, V. (2018). Fungal respiratory infections in cystic fibrosis (CF): Recent progress and future research agenda. *Mycopathologia*, *183*, 1–5.

Box, H., Negri, C., Livermore, J., Whalley, S., Johnson, A., McEntee, L., et al. (2018). Pharmacodynamics of voriconazole for invasive pulmonary scedosporiosis. *Antimicrobial Agents and Chemotherapy*, *62* e02516-17.

Brasnu, E., Bourcier, T., Dupas, B., Degorge, S., Rodallec, T., Laroche, L., et al. (2007). *In vivo* confocal microscopy in fungal keratitis. *The British Journal of Ophthalmology*, *91*, 588–591.

Brown, G. D., Denning, D. W., Gow, N. A., Levitz, S. M., Netea, M. G., & White, T. C. (2012). Hidden killers: Human fungal infections. *Science Translational Medicine*, *4*, 165rv13.

Brown, G. D., Denning, D. W., & Levitz, S. M. (2012). Tackling human fungal infections. *Science*, *336*, 647.

Buchheidt, D., Reinwald, M., Hofmann, W.-K., Bloch, T., & Seiss, B. (2017). Evaluating the use of PCR for diagnosing invasive aspergillosis. *Expert Review of Molecular Diagnostics*, *17*, 603–610.

Caramalho, R., Tyndall, J. D. A., Monk, B. C., Larentis, T., Lass-Flörl, C., & Lackner, M. (2017). Intrinsic short-tailed azole resistance in mucormycetes is due to an evolutionary conserved amino acid substitution of the lanosterol 14α-demethylase. *Scientific Reports*, *7*, 15898.

Carroll, C. S., Amankwa, K. N., Pinto, L. J., Fuller, J. D., & Moore, M. M. (2016). Detection of serum siderophores by LC-MS/MS as a potential biomarker of invasive aspergillosis. *PLoS ONE*, *11*, e0151260.

Carsin, A., Romain, T., Ranque, S., Reynaud-Gaubert, M., Dubus, J.-C., Mège, J.-L., et al. (2017). *Aspergillus fumigatus* in cystic fibrosis: An update on immune interactions and molecular diagnostics in allergic bronchopulmonary aspergillosis. *Allergy*, *72*, 1632–1642.

Cassagne, C., Ranque, S., Normand, A. C., Forquet, P., Thiebault, S., Planard, C., et al. (2011). Mould routine identification in the clinical laboratory by matrix-assisted laser desorption ionisation time-of-flight mass spectrometry. *PLoS ONE*, *6*, e28425.

Castrejón-Pérez, A. D., Miranda, I., Welsh, O., Welsh, E. C., & Ocampo-Candiani, J. (2017). Cutaneous mucormycosis. *Anais Brasileiros de Dermatologia, 92*, 304–311.

Chan, J. F.-W., Lau, S. K.-P., Wong, S. C.-Y., To, K. K.-W., So, S. Y.-C., Leung, S. S.-M., et al. (2016). A 10-year study reveals clinical and laboratory evidence for the 'semi-invasive' properties of chronic pulmonary aspergillosis. *Emerging Microbes & Infections, 5*, e37.

Chen, Y.-J., Chou, C.-L., Lai, K.-J., & Lin, Y.-J. (2017). *Fusarium* brain abscess in a patient with diabetes mellitus and liver cirrhosis. *Acta Neurologica Taiwanica, 26*, 128–132.

Chen, M., Kondori, N., Deng, S., Gerrits van den Ende, A. H. G., Lackner, M., Liao, W., et al. (2017). Direct detection of *Exophiala* and *Scedosporium* species in sputa of patients with cystic fibrosis. *Medical Mycology, 56*, 695–702.

Chong, G. M., van der Beek, M. T., von dem Borne, A., Boelens, J., Steel, E., Kampinga, G. A., et al. (2016). PCR-based detection of *Aspergillus fumigatus* Cyp51A mutations on bronchoalveolar lavage: A multicentre validation of the AsperGenius® assay in 201 patients with haematological disease suspected for invasive aspergillosis. *The Journal of Antimicrobial Chemotherapy, 71*, 3528–3535.

Chotirmall, S. H., & Martin-Gomez, M. T. (2018). *Aspergillus* species in bronchiectasis: Challenges in the cystic fibrosis and non-cystic fibrosis airways. *Mycopathologia, 183*, 45–59.

Chotirmall, S. H., & McElvaney, N. G. (2014). Fungi in the cystic fibrosis lung: Bystanders or pathogens? *The International Journal of Biochemistry & Cell Biology, 52*, 161–173.

Chowdhary, A., Agarwal, K., Kathuria, S., Gaur, S. N., Randhawa, H. S., & Meis, J. F. (2014). Allergic bronchopulmonary mycosis due to fungi other than *Aspergillus*: A global overview. *Critical Reviews in Microbiology, 40*, 30–48.

Cimon, B., Carrere, J., Vinatier, J. F., Chazalette, J. P., Chabasse, D., & Bouchara, J. P. (2000). Clinical significance of *Scedosporium apiospermum* in patients with cystic fibrosis. *European Journal of Clinical Microbiology & Infectious Diseases, 19*, 53–56.

Cooley, L., Spelman, D., Thursky, K., & Slavin, M. (2007). Infection with *Scedosporium apiospermum* and *S. prolificans*, Australia. *Emerging Infectious Diseases, 13*, 1170–1177.

Cornely, O. A., Arikan-Akdagli, S., Dannaoui, E., Groll, A. H., Lagrou, K., Chakrabarti, A., et al. (2014). ESCMID and ECMM joint clinical guidelines for the diagnosis and management of mucormycosis 2013. *Clinical Microbiology and Infection, 20*, 5–26.

Cortez, K. J., Roilides, E., Quiroz-Telles, F., Meletiadis, J., Antachopoulos, C., Knudsen, T., et al. (2008). Infections caused by *Scedosporium* spp. *Clinical Microbiology Reviews, 21*, 157–197.

Coulibaly, O., Marinach-Patrice, C., Cassagne, C., Piarroux, R., Mazier, D., & Ranque, S. (2011). *Pseudallescheria/Scedosporium* complex species identification by matrix-assisted laser desorption ionisation-time of flight mass spectrometry. *Medical Mycology, 49*, 621–626.

Crameri, R. (1998). Recombinant *Aspergillus fumigatus* allergens: From the nucleotide sequences to clinical applications. *International Archives of Allergy and Immunology, 115*, 99–114.

Crameri, R., Hemmann, S., Ismail, C., Menz, G., & Blaser, K. (1998). Disease-specific recombinant antigens for the diagnosis of allergic bronchopulmonary aspergillosis. *International Immunology, 10*, 1211–1216.

Crameri, R., Zeller, S., Glaser, A. G., Vilhelmsson, M., & Rhyner, C. (2009). Cross-reactivity among fungal antigens: a clinically relevant phenomenon? *Mycoses, 52*, 99–106.

Dabén, R. P., de Lucas, E. M., Cuesta, L. M., Velasco, T. P., Garica, J. A., Landeras, R., et al. (2008). Imaging findings of pulmonary infection casued by *Scedosporium prolificans* in a deep immunocompromised patient. *Emergency Radiology, 15*, 47–49.

Däbritz, J., Attarbaschi, A., Tintelnot, K., Kollmar, N., Kremens, B., von Loewenich, F. D., et al. (2011). Mucormycosis in paediatric patients: Demographics, risk factors and outcome of 12 contemporary cases. *Mycoses*, *54*, e785–e788.

Dannaoui, E., Schwartz, P., Slany, M., Loeffler, J., Jorde, A. T., Cuenca-Estrella, M., et al. (2010). Molecular detection and identification of zygomycete species from paraffin-embedded tissues in a murine model of disseminated zygomycosis: A collaborative European Society of Clinical Microbiology and Infectious Diseases (ESCMID) Fungal Infection Group (EFISG) evaluation. *Journal of Clinical Microbiology*, *48*, 2043–2046.

Davies, G., Rolle, A.-M., Maurer, A., Spycher, P. R., Schillinger, C., Solouk-Saran, D., et al. (2017). Towards translational immunoPET/MR imaging of invasive pulmonary aspergillosis: The humanised monoclonal antibody JF5 detects *Aspergillus* lung infections *in vivo*. *Theranostics*, 7, 3398–3414.

Davies, G., & Thornton, C. R. (2014). Differentiation of the emerging human pathogens *Trichosporon asahii* and *Trichosporon asteroides* from other pathogenic yeasts and moulds by using species-specific monoclonal antibodies. *PLoS ONE*, *9*, e84789.

De Carolis, E., Posteraro, B., Lass-Flörl, C., Vella, A., Florio, A. R., Torelli, R., et al. (2012). Species identification of *Aspergillus*, *Fusarium*, and Mucorales with direct surface analysis by matrix-assisted laser desorption ionisation time-of-flight mass spectrometry. *Clinical Microbiology and Infection*, *18*, 475–484.

De Pauw, B., Walsh, T. J., Donnelly, J. P., Stevens, D. A., Edwards, J. E., Calandra, T., et al. (2008). Revised definitions of invasive fungal disease from the European Organisation for Research and Treatment of Cancer/Invasive Fungal Infections Co-operative Group and the National Institute of Allergy and Infectious Diseases Mycoses Study Group (EORTC/MSG) Consensus Group. *Clinical Infectious Diseases*, *46*, 1813–1821.

De Ruiter, G. A., Josso, S. L., Colquhoun, I. J., Voragen, A. G. J., & Rombouts, F. M. (1992). Isolation and characterisation of β(1-4)-D-glucuronans from extracellular polysaccharides of moulds belonging to Mucorales. *Carbohydrate Polymers*, *18*, 1–7.

De Ruiter, G. A., van Bruggen-van der Lugt, A. W., Bos, W., Notermans, S. H. W., Rombouts, F. M., & Hofstra, H. (1993). The production and partial characterisation of a monoclonal IgG antibody specific for the moulds belonging to the order Mucorales. *Journal of General Microbiology*, *139*, 1557–1564.

De Ruiter, G. A., van Bruggen-van der Lugt, A. W., Mischnick, P., Smid, P., van Boom, J. H., Notermans, S. H., et al. (1994). 2-O-Methyl-D-mannose residues are immunodominant in extracellular polysaccharides of *Mucor racemosus* and related molds. *The Journal of Biological Chemistry*, *269*, 4299–4306.

De Ruiter, G. A., van Bruggen-van der Lugt, A. W., Nout, M. J. R., Middelhoven, W. J., Soentoro, P. S. S., Notermans, S. H. W., et al. (1992). Formation of antigenic extracellular polysaccharides by selected strains of *Mucor* spp., *Rhizopus* spp., *Rhizomucor* spp., *Absidia corymbifera* and *Syncephalastrum racemosum*. *Antonie Van Leeuwenhoek*, *62*, 189–199.

De Ruiter, G. A., van der Lugt, A. W., Voragen, A. G. J., & Rombouts, F. M. (1991). High-performance size-exclusion chromatography and ELISA detection of extracellular polysaccharides from Mucorales. *Carbohydrate Research*, *215*, 47–57.

Delhaes, L., Harun, A., Chen, S. C., Nguyen, Q., Slavin, M., Heath, C. H., et al. (2008). Molecular typing of Australian *Scedosporium* isolates showing genetic variability and numerous *S. aurantiacum*. *Emerging Infectious Diseases*, *14*, 282–290.

Denning, D. W., Cadranel, J., Beigelman-Aubry, C., Ader, F., Chakrabarti, A., Blot, S., et al. (2016). Chronic pulmonary aspergillosis: Rationale and clinical guidelines for diagnosis and management. *The European Respiratory Journal*, *47*, 45–68.

Denning, D. W., Pashley, C., Hartl, D., Wardlaw, A., Godet, C., Del Gaccio, S., et al. (2014). Fungal allergy in asthma—State of the art and research needs. *Clinical and Translational Allergy*, *4*, 14.

Denning, D. W., Pleuvry, A., & Cole, D. C. (2011). Global burden of chronic pulmonary aspergillosis as a sequel to pulmonary tuberculosis. *Bulletin of the World Health Organization, 89*, 864–872.

Denning, D. W., Pleuvry, A., & Cole, D. C. (2013). Global burden of allergic bronchopulmonary aspergillosis with asthma and its complication chronic pulmonary aspergillosis in adults. *Medical Mycology, 51*, 361–370.

Denning, D. W., Riniotis, K., Dobrashian, R., & Sambatakou, H. (2003). Chronic cavitary and fibrosing pulmonary and pleural aspergillosis: Case series, proposed nomenclature change, and review. *Clinical Infectious Diseases, 37*, S265–S280.

Dichtl, K., Seybold, U., Ormanns, S., Horns, H., & Wagener, J. (2019). Evaluation of a novel *Aspergillus* antigen ELISA. *Journal of Clinical Microbiology, 57*, e00136-19.

Dogget, M. S. (2000). Characterisation of fungal biofilms within a municipal water distribution system. *Applied and Environmental Microbiology, 66*, 1249–1251.

Drain, P. K., Hong, T., Krows, M., Govere, S., Thulare, H., Wallis, C. L., et al. (2019). Validation of clinic-based cryptococcal antigen lateral-flow assay screening in HIV-infected adults in South Africa. *Scientific Reports, 9*, 2687.

Dufresne, S. F., Datta, K., Li, X., Dadachova, E., Staab, J. F., Patterson, T. F., et al. (2012). Detection of urinary excreted fungal galactomannan-like antigens for diagnosis of invasive aspergillosis. *PLoS ONE, 7*, e42736.

Dupont, B., Huber, M., Kim, S. J., & Bennett, J. E. (1987). Galactomannan antigenemia and antigenuria in aspergillosis: Studies in patients and experimentally infected rabbits. *The Journal of Infectious Diseases, 155*, 1–11.

Dutkiewicz, R., & Hage, C. A. (2010). *Aspergillus* infections in the critically ill. *Proceedings of the American Thoracic Society, 7*, 204–209.

Eigl, S., Hoenigl, M., Spiess, B., Heldt, S., Prattes, J., Neumeister, P., et al. (2017). Galactomannan testing and *Aspergillus* PCR in same-day bronchoalveolar lavage and blood samples for diagnosis of invasive aspergillosis. *Medical Mycology, 55*, 528–534.

Eigl, S., Prattes, J., Reinwald, M., Thornton, C. R., Reischies, F., Spiess, B., et al. (2015). Influence of mould-active antifungal treatment on the performance of the *Aspergillus*-specific bronchoalveolar lavage fluid lateral-flow device test. *International Journal of Antimicrobial Agents, 46*, 401–405.

Esnakula, A. K., Summers, I., & Naab, T. J. (2013). Fatal disseminated Fusarium infection in a human immunodeficiency virus positive patient. *Case Reports in Infectious Diseases, 2013*, 379320.

Fanci, R., Pini, G., Bartolesi, A. M., & Pecile, P. (2013). Refactory disseminated fusariosis by *Fusarium verticillioides* in a patient with acute myeloid leukaemia relapsed after allogeneic hematopoietic stem cell transplantation: A case report and literature review. *Revista Iberoamericana de Micología, 30*, 51–53.

Fanfair, R. N., Benedict, K., Bos, J., Bennett, S. D., Lo, Y.-C., Adebanjo, T., et al. (2012). Necrotizing cutaneous mucormycosis after a tornado in Joplin, Missouri, in 2011. *The New England Journal of Medicine, 367*, 2214–2225.

Fisher, M. C., Hawkins, N. J., Sanglard, D., & Gurr, S. J. (2018). Worldwide emergence of resistance to antifungal drugs challenges human health and food security. *Science, 360*, 739–742.

Fricker-Hidalgo, H., Coltey, B., Llerena, C., Renversez, J.-C., Grillot, R., Pin, I., et al. (2010). Recombinant allergens combined with biological markers in the diagnosis of allergic bronchopulmonary aspergillosis in cystic fibrosis patients. *Clinical and Vaccine Immunology, 17*, 1330–1336.

Fukuzawa, M., Inaba, H., Hayama, M., Sakaguchi, N., Sano, K., Ito, M., et al. (1995). Improved detection of medically important fungi by immunoperoxidase staining with polyclonal antibodies. *Virchows Archiv, 427*, 407–414.

Gade, L., Hurst, S., Balajee, S. A., Lockhart, S. R., & Litvintseva, A. P. (2017). Detection of mucormycetes and other pathogenic fungi in formalin fixed paraffin embedded and fresh tissues using the extended region of 28S rDNA. *Medical Mycology*, *55*, 385–395.
García-Ruiz, J. C., Olazábal, I., Pedroso, R. M. A., López-Soria, L., Velaso-Benito, V., Sanchez-Aparicio, J. A., et al. (2015). Disseminated fusariosis and hematologic malignancies, a still devastating association. Report of three new cases. *Revista Iberoamericana de Micología*, *32*, 190–196.
Garzoni, C., Emonet, S., Legout, L., Benedict, R., Hoffmeyer, P., Bernard, L., et al. (2005). Atypical infections in tsunami survivors. *Emerging Infectious Diseases*, *11*, 1591–1593.
Gilgado, F., Cano, J., Gené, J., & Guarro, J. (2005). Molecular phylogeny of the *Pseudallescheria boydii* complex: Proposal of two new species. *Journal of Clinical Microbiology*, *43*, 4930–4942.
Gilgado, F., Cano, J., Gené, J., Sutton, D. A., & Guarro, J. (2008). Molecular and phenotypic data supporting distinct species statuses for *Scedosporium apiospermum* and *Pseudallescheria boydii* and the proposed new species *Scedosporium dehoogii*. *Journal of Clinical Microbiology*, *46*, 766–771.
Gilgado, F., Serena, C., Cano, J., Gené, J., & Guarro, J. (2006). Antifungal susceptibilities of of the species of the *Pseudallescheria boydii* complex. *Antimicrobial Agents and Chemotherapy*, *50*, 4211–4213.
Glazer, M., Nusair, S., Breuer, R., Lafair, J., Sherman, Y., & Berkman, N. (2000). The role of BAL in the diagnosis of pulmonary mucormycosis. *Chest Journal*, *117*, 279–282.
Gomes, M. Z. R., Lewis, R. E., & Kontoyiannis, D. P. (2011). Mucormycosis caused by unusual mucormycetes, non-*Rhizopus*, -*Mucor*, and -*Lichtheimia* species. *Clinical Microbiology Reviews*, *24*, 411–445.
Gonçalves, S. M., Lagrou, K., Rodrigues, C. S., Campos, C. F., Bernal-Martínez, L., Rodrigues, F., et al. (2017). Evaluation of bronchoalveolar lavage fluid cytokines as biomarkers for invasive pulmonary aspergillosis in at-risk patients. *Frontiers in Microbiology*, *8*, 2362.
Grantina-Ievina, L., Andersone, U., Berkolde-PĪre, D., Nikolajeva, V., & Ievinsh, G. (2013). Critical tests for determination of microbiological quality and biological activity in commercial vermicompost samples of different origins. *Applied Microbiology and Biotechnology*, *97*, 10541–10554.
Grossman, M. E., Fox, L. P., Kovarik, C., & Rosenbach, M. (2012). Hyalohyphomycosis. In *Cutaneous manifestations of infection in the immunocompromised host* (pp. 65–88). New York: Springer.
Guarro, J. (2013). Fusariosis, a complex infection caused by a high diversity of fungal species refractory to treatment. *European Journal of Clinical Microbiology & Infectious Diseases*, *32*, 1491–1500.
Guegan, H., Robert-Gangneux, F., Camus, C., Belaz, S., Marchand, T., Baldeyrou, M., et al. (2018). Improving the diagnosis of invasive aspergillosis by the detection of *Aspergillus* in bronchoalveolar lavage fluid: Comparison of non-culture-based assays. *The Journal of Infection*, *76*, 196–205.
Guinea, J., Padilla, C., Escribano, P., Muñoz, P., Padilla, B., Gijón, P., et al. (2013). Evaluation of MycAssay™ *Aspergillus* for diagnosis of invasive pulmonary aspergillosis in patients without haematological cancer. *PLoS ONE*, *8*, e61545.
Gundacker, N. D., & Baddley, J. W. (2015). Fungal infections in the era of biologic therapies. *Current Clinical Microbiology Reports*, *2*, 76–83.
Gurusidappa, S. B., & Mamatha, H. S. (2011). Fusarial skin lesion in immunocompromised. *Indian Journal of Cancer*, *48*, 116–117.
Haas, H. (2003). Molecular genetics of fungal siderophore biosynthesis and update: The role of siderophores in iron uptake and storage. *Applied Microbiology and Biotechnology*, *62*, 316–333.

Haas, H., Petrik, M., & Descristoforo, C. (2015). An iron-mimicking, Trojan horse-entering fungi—Has the time come for molecular imaging of fungal infections? *PLoS Pathogens*, *11*, e1004568.

Hammond, S. P., Bialek, R., Milner, D. A., Petschnigg, E. V., Baden, L. R., & Marty, F. M. (2011). Molecular methods to improve diagnosis and identification of mucormycosis. *Journal of Clinical Microbiology*, *49*, 2151–2153.

Harun, A., Blyth, C. C., Gilgado, F., Middleton, P., Chen, S. C.-A., & Meyer, W. (2011). Development and validation of a multiplex PCR for detection of *Scedosporium* spp. in respiratory tract specimens from patients with cystic fibrosis. *Journal of Clinical Microbiology*, *49*, 1508–1512.

Hawksworth, D. L. (1991). The fungal dimension of biodiversity: Magnitude, significance, and conservation. *Mycological Research*, *95*, 641–655.

Hawksworth, D. L. (2011). Naming *Aspergillus* species: Progress towards one name for each species. *Medical Mycology*, *49*, S70–S76.

Haynes, K. A., Latgé, J. P., & Rogers, T. R. (1990). Detection of *Aspergillus* antigens associated with invasive aspergillosis. *Journal of Clinical Microbiology*, *28*, 2040–2044.

He, D., Hao, J., Zhang, B., Yang, Y., Song, W., Zhang, Y., et al. (2011). Pathogenic spectrum of fungal keratitis and specific identification of *Fusarium solani*. *Investigative Ophthalmology & Visual Science*, *52*, 2804–2808.

Held, S., & Hoenigl, M. (2017). Lateral flow assays for the diagnosis of invasive aspergillosis: current status. *Current Fungal Infection Reports*, *11*, 45–51.

Held, J., Schmidt, T., Thornton, C. R., Kotter, E., & Bertz, H. (2012). Comparison of a novel *Aspergillus* lateral-flow device and the Platelia galactomannan assay for the diagnosis of invasive aspergillosis following haematopoietic stem cell transplantation. *Infection*, *41*, 1163–1169.

Hennebert, G. L., & Desai, B. G. (1974). *Lomentospora prolificans*, a new hyphomycete from greenhouse soil. *Mycotaxon*, *1*, 45–50.

Hessian, P., & Smith, J. M. B. (1982). Antigenic characterisation of some potentially pathogenic mucoraceous fungi. *Sabouraudia*, *20*, 209–216.

Hingad, N., Kumar, G., & Deshmukh, R. (2012). Oral mucormycosis causing necrotizing lesion in a diabetic patient: A case report. *International Journal of Oral & Maxillofacial Pathology*, *3*, 8–12.

Hoenigl, M., Orasch, T., Faserl, K., Prattes, J., Loeffler, J., Springer, J., et al. (2019). Triacetylfusarinine C: A urine biomarker for diagnosis of invasive aspergillosis. *The Journal of Infection*, *78*, 150–157.

Hoenigl, M., Prattes, J., Spiess, B., Wagner, J., Prueller, F., Raggam, R. B., et al. (2014). Performance of galactomannan, beta-d-glucan, *Aspergillus* lateral-flow device, conventional culture, and PCR tests with bronchoalveolar lavage fluid for diagnosis of invasive pulmonary aspergillosis. *Journal of Clinical Microbiology*, *52*, 2039–2045.

Holmes, N. E., Trevillyan, J., Kidd, S. E., & Leong, T. Y.-M. (2013). Locally extensive angio-invasive *Scedosporium prolificans* infection following resection for squamous cell lung carcinoma. *Medical Mycology Case Reports*, *2*, 98–102.

Hong, G., Miller, H. B., Allgood, S., Lee, R., Lechtzin, N., & Zhang, S. X. (2017). Use of selective fungal culture media increases rates of detection of fungi in the respiratory tract of cystic fibrosis patients. *Journal of Clinical Microbiology*, *55*, 1122–1130.

Hong, Y., & Park, J. (2017). The role of transbronchial lung biopsy in diagnosing pulmonary mucormycosis in a critical care unit. *Korean Journal of Critical Care Medicine*, *32*, 205–210.

Hope, W. W., Walsh, T. J., & Denning, D. W. (2005). The invasive and saprophytic syndromes due to *Aspergillus* spp. *Medical Mycology*, *43*, S207–S238.

Howard, S. J., Cerar, D., Anderson, M. J., Albarrag, A., Fisher, M. C., Pasqualotto, A. C., et al. (2009). Frequency and evolution of azole resistance in *Aspergillus fumigatus* associated with treatment failure. *Emerging Infectious Diseases*, *15*, 1068–1076.

Hung, H.-C., Shen, G.-Y., Chen, S.-C., Yeo, K.-J., Tsao, S.-M., Lee, M.-C., et al. (2015). Pulmonary mucormycosis in a patient with systemic lupus erythematosus: A diagnostic and treatment challenge. *Case Reports in Infectious Diseases, 2015*, 478789.

Husain, S., Munoz, P., Forrest, G., Alexander, B. D., Somani, J., Brennan, K., et al. (2005). Infections due to *Scedosporium apiospermum* and *Scedosporium prolificans* in transplant recipients: Clinical characteristics and impact of antifungal agent therapy on outcome. *Clinical Infectious Diseases, 40*, 89–99.

Hymery, N., Vasseur, V., Coton, M., Mounier, J., Jany, J.-L., Barbier, G., et al. (2014). Filamentous fungi and mycotoxins in cheese: A review. *Comprehensive Reviews in Food Science and Food Safety, 13*, 437–456.

Iqbal, N., Irfan, M., Jabeen, K., Kazmi, M. M., & Tariq, M. U. (2017). Chronic pulmonary mucormycosis: An emerging fungal infection in diabetes mellitus. *Journal of Thoracic Disease, 9*, E121–E125.

Izumikawa, K., Yamamoto, Y., Mihara, T., Takazono, T., Morinaga, Y., Kurihara, S., et al. (2012). Bronchoalveolar lavage galactomannan for the diagnosis of chronic pulmonary aspergillosis. *Medical Mycology, 50*, 811–817.

Jain, P. K., Gupta, V. K., Misra, A. K., Gaur, R., Bajpai, V., & Issar, S. (2011). Current status of *Fusarium* infection in human and animal. *Asian Journal of Animal and Veterinary, 6*, 201–227.

Jenks, J. D., Mehta, S. R., Taplitz, R., Aslam, S., Reed, S. L., & Hoenigl, M. (2018). Point-of-care diagnosis of invasive aspergillosis in non-neutropenic patients: *Aspergillus* galactomannan lateral flow assay versus *Aspergillus*-specific lateral flow device test in bronchoalveolar lavage. *Mycoses, 62*, 230–236.

Jenks, J. D., Mehta, S. R., Taplitz, R., Law, N., Reed, S. L., & Hoenigl, M. (2019). Bronchoalveolar lavage *Aspergillus* galactomannan lateral flow assay versus *Aspergillus*-specific lateral flow device test for diagnosis of invasive pulmonary aspergillosis in patients with hematological malignancies. *The Journal of Infection, 78*, 249–259.

Jensen, H. E., Aalbæk, B., Lind, P., & Krogh, H. V. (1996). Immunohistochemical diagnosis of systemic bovine zygomycosis by murine monoclonal antibodies. *Veterinary Pathology, 33*, 176–183.

Jensen, H. E., Aalbæk, B., & Schønheyder, H. (1994). Immunohistochemical identification of aetiological agents of systemic bovine zygomycosis. *Journal of Comparative Pathology, 110*, 65–77.

Jensen, H. E., Frandsen, P. L., & Schønheyder, H. (1993). Experimental systemic bovine zygomycosis with reference to pathology and secretion of antigen into urine. *Journal of Veterinary Medicine Series B, 40*, 55–65.

Jensen, H. E., Salonen, J., & Ekfors, T. O. (1997). The use of immunohistochemistry to improve sensitivity and specificity in the diagnosis of systemic mycoses in patients with haematological malignancies. *The Journal of Pathology, 181*, 100–105.

Jensen, H. E., Stynen, D., Sarfati, J., & Latge, J.-P. (1993). Detection of galactomannan and the 18kDa antigen from *Aspergillus fumigatus* in serum and urine from cattle with systemic aspergillosis. *Journal of Veterinary Medicine Series B, 40*, 397–408.

Johnson, G. L., Sarker, S. J., Nannini, F., Ferrini, A., Taylor, E., Lass-Flörl, C., et al. (2015). *Aspergillus*-specific lateral-flow device and real-time PCR testing of bronchoalveolar lavage fluid: A combination biomarker approach for clinical diagnosis of invasive pulmonary aspergillosis. *Journal of Clinical Microbiology, 53*, 2103–2108.

Jones, K. D., & Kaufman, L. (1978). Development and evaluation of an immunodiffusion test for diagnosis of systemic zygomycosis (mucormycosis): Preliminary report. *Journal of Clinical Microbiology*, 7, 97–103.

Jurkunas, U., Behlau, I., & Colby, K. (2009). Fungal keratitis: Changing pathogens and risk factors. *Cornea, 28*, 638–643.

Kalteis, J., Rainer, J., & de Hoog, G. S. (2009). Ecology of *Pseudallescheria* and *Scedosporium* species in human-dominated and natural environments and their distribution in clinical samples. *Medical Mycology*, *47*, 398–405.
Kaufman, L., Standard, P. G., Jalbert, M., & Kraft, D. E. (1997). Immunohistologic identification of *Aspergillus* spp. and other hyaline fungi by using polyclonal fluorescent antibodies. *Journal of Clinical Microbiology*, *35*, 2206–2209.
Kaufman, L., Turner, L. F., & McLaughlin, D. W. (1989). Indirect Enzyme-Linked Immunosorbent Assay for zygomycosis. *Journal of Clinical Microbiology*, *27*, 1979–1982.
Kaymaz, D., Ergün, P., Candemir, I., & Çicek, T. (2016). Chronic necrotising pulmonary aspergillosis presenting as transient thoracic mass: A diagnostic dilemma. *Respiratory Medicine Case Reports*, *19*, 140–142.
Khan, Z. U., Ahmad, S., & Theyyathel, A. M. (2008). Diagnostic value of DNA and $(1 \rightarrow 3)$-β-D-glucan detection in serum and bronchoalveolar lavage of mice experimentally infected with *Fusarium oxysporum*. *Journal of Medical Microbiology*, *57*, 36–42.
Kitasato, Y., Tao, Y., Hoshino, T., Tachibana, K., Inoshima, N., Yoshida, M., et al. (2009). Comparison of *Aspergillus* galactomannan antigen testing with a new cut-off index and *Aspergillus* precipitating antibody testing for the diagnosis of chronic pulmonary aspergillosis. *Respirology*, *14*, 701–708.
Klont, R. R., Mennink-Kersten, M. A., & Verweij, P. E. (2004). Utility of *Aspergillus* antigen detection in specimens other than serum specimens. *Clinical Infectious Diseases*, *39*, 1467–1474.
Kobayashi, M., & Thornton, C. R. (2014). Serological diagnosis of chronic pulmonary aspergillosis (CPA) and *Aspergillus* lateral-flow device (LFD). *The European Respiratory Journal*, *44*, P2501.
Koehler, P., Tacke, D., & Cornely, O. A. (2014). Bone and joint infections by Mucorales, Scedosporium, Fusarium, and even rarer fungi. *Critical Reviews in Microbiology*, *9*, 1–14.
Kondo, M., Goto, H., & Yamanaka, K. (2018). Case of *Scedosporium aurantiacum* infection in a subcutaneous abscess. *Medical Mycology Case Reports*, *20*, 26–27.
Kontoyiannis, D. P., Marr, K. A., Park, B. J., Alexander, B. D., Anaissie, E. J., Walsh, T. J., et al. (2010). Prospective surveillance for invasive fungal infections in hematopoietic stem cell transplant recipients, 2001-2006: Overview of the Transplant-Associated Infection Surveillance Network (TRANSNET) database. *Clinical Infectious Diseases*, *50*, 1091–1100.
Kontoyiannis, D. P., Wessel, V. C., Bodey, G. P., & Rolston, K. V. (2000). Zygomycosis in the 1990s in tertiary care cancer center. *Clinical Infectious Diseases*, *30*, 851–856.
Kousha, M., Tadi, R., & Soubani, A. O. (2011). Pulmonary aspergillosis: A clinical review. *European Respiratory Review*, *20*, 156–174.
Kozel, T. R., & Wickes, B. (2014). Fungal diagnostics. *Cold Spring Harbor Perspectives in Medicine*, *4*, a019299.
Kronen, R., Liang, S. Y., Bochicchio, G., Bochicchio, K., Powderly, W. G., & Spec, A. (2017). Invasive fungal infections secondary to traumatic injury. *International Journal of Infectious Diseases*, *62*, 102–111.
Kyi, C., Hellmann, M. D., Wolchok, J. D., Chapman, P. B., & Postow, M. A. (2014). Opportunistic infections in patients treated with immunotherapy for cancer. *Journal for Immunotherapy of Cancer*, *2*, 19.
Lackner, M., de Hoog, G. S., Yang, L., Moreno, L. F., Ahmed, S. A., Andreas, F., et al. (2014). Proposed nomenclature for *Pseudallescheria*, *Scedosporium* and related genera. *Fungal Diversity*, *67*, 1–10.
Lackner, M., Najafzadeh, M. J., Sun, J., Lu, Q., & de Hoog, G. S. (2012). Rapid identification of *Pseudallescheria* and *Scedosporium* strains by using rolling circle amplification. *Applied and Environmental Microbiology*, *78*, 126–133.

Lanternier, F., Cypowyj, S., Picard, C., Bustamante, J., Lortholary, O., Casanova, J.-L., et al. (2013). Primary immunodeficiencies underlying fungal infections. *Current Opinion in Pediatrics*, *25*, 736–747.

Lass-Flörl, C. (2009). Zygomycosis: Conventional laboratory diagnosis. *Clinical Microbiology and Infection*, *15*, 60–65.

Lass-Flörl, C. (2019). How to make a fast diagnosis in invasive aspergillosis. *Medical Mycology*, *57*, S155–S160.

Latenser, B. A. (2003). *Fusarium* infections in burn patients: A case report and review of the literature. *The Journal of Burn Care & Rehabilitation*, *24*, 285–288.

Latgé, J.-P. (1999). *Aspergillus fumigatus* and Aspergillosis. *Clinical Microbiology Reviews*, *12*, 310–350.

Lau, A. F., Drake, S. F., Calhoun, L. B., Henderson, C. M., & Zelany, A. M. (2013). Development of a clinically comprehensive database and a simple procedure for identification of moulds from solid media by matrix-assisted laser desorption ionisation-time of flight mass spectrometry. *Journal of Clinical Microbiology*, *51*, 1828–1834.

Leal, S. M., Roy, S., Vareechon, C., deJesus Carion, S., Clark, H., Lopez-Berges, M. S., et al. (2013). Targeting iron acquisition blocks infection with the fungal pathogens *Aspergillus fumigatus* and *Fusarium oxysporum*. *PLoS Pathogens*, *9* e1003436.

Lecointe, K., Cornu, M., Leroy, J., Coulon, P., & Sendid, B. (2019). Polysaccharides cell wall architecture of Mucorales. *Frontiers in Microbiology*, *10*, 469.

Legouge, C., Caillot, D., Chrétien, M. L., Lafon, I., Ferrant, E., Audia, S., et al. (2014). The reversed halo sign: Pathognomonic pattern of pulmonary mucormycosis in leukemic patients with neutropenia? *Clinical Infectious Diseases*, *58*, 672–678.

Lehmann, P. F., & Reiss, E. (1978). Invasive aspergillosis: Antiserum for circulating antigen produced after immunisation with serum from infected rabbits. *Infection and Immunity*, *20*, 570–572.

Lelievre, L., Garcia-Hermoso, D., Abdoul, H., Hivelin, M., Chouaki, T., Toubas, D., et al. (2014). Posttraumatic mucormycosis. *Medicine*, *93*, 395–404.

Lengerova, M., Racil, Z., Hrncirova, I., Kocmanova, I., Volfova, P., Ricna, D., et al. (2014). Rapid detection and identification of mucormycetes in bronchoalveolar lavage samples from immunocompromised patients with pulmonary infiltrates by use of high-resolution melt analysis. *Journal of Clinical Microbiology*, *52*, 2824–2828.

Leroy, P., & Smismans, A. (2007). Long-term risk of atypical fungal infection after near-drowning episodes: In reply. *Paediatrics*, *119*, 417–418.

Li, H., Hwang, S. K., Zhou, C., Du, J., & Zhang, J. (2013). Grangrenous cutaneous mucormycosis caused by *Rhizopus oryzae*: a case report and review of primary cutaneous mucormycosis in China over past 20 years. *Mycopathologia*, *176*, 123–128.

Liao, G. P., Karanjawala, B. E., Baba, S., Khalil, K. G., & Irani, A. D. (2015). A 23-year-old diabetic female with persistent pneumonia and chronic lung abscess. *Annals of the American Thoracic Society*, *12*, 1231–1234.

Libert, D., Procop, G. W., & Ansari, M. Q. (2018). *Histoplasma* urinary antigen testing obviates the need for coincident serum antigen testing. *American Journal of Clinical Pathology*, *149*, 362–368.

Limper, A. H., Adenis, A., Le, T., & Harrsion, T. S. (2017). Fungal infections in HIV/AIDS. *The Lancet Infectious Diseases*, *17*, e334–e443.

Linscott, A. J. (2007). Natural disasters—A microbe's paradise. *Clinical Microbiology Newsletter*, *29*, 57–62.

Lu, Q., Gerrits van den Ende, A. H. G., Bakkers, J. M. J. E., Sun, J., Lackner, M., Najafzadeh, M. J., et al. (2011). Identification of *Pseudallescheria* and *Scedosporium* species by three molecular methods. *Journal of Clinical Microbiology*, *49*, 960–967.

Lu, Q., Gerrits van den Ende, A. H. G., de Hoog, G. S., Li, R., Accoceberry, I., Durand-Joly, I., et al. (2011). Reverse line blot hybridisation screening of *Pseudallescheria/Scedosporium* species in patients with cystic fibrosis. *Mycoses*, *54*, 5–11.

Luna-Rodríguez, C. E., de Treviño-Rangel, R., Montoyas, A. M., Becerril-García, M. A., Andrade, A., & González, G. M. (2019). *Scedosporium* spp.: Chronicle of an emerging pathogen. *Medicina Universitaria*, *21*, 4–13.

Luplertlop, N. (2018). *Pseudallescheria/Scedosporium* complex species: From saprobic to pathogenic fungus. *Journal de Mycologie Médicale*, *28*, 249–256.

Luptáková, D., Pluháček, T., Petřik, M., Novák, J., Palyzová, A., Sokolová, L., et al. (2017). Non-invasive and invasive diagnoses of aspergillosis in a rat model by mass spectrometry. *Scientific Reports*, *7*, 16523.

Maertens, J. A., Klont, R., Masson, C., Theunissen, K., Meersseman, W., Lagrou, K., et al. (2007). Optimisation of the cutoff value for the *Aspergillus* double-sandwich enzyme immunoassay. *Clinical Infectious Diseases*, *44*, 1329–1336.

Maertens, J., Lagrou, K., Deweerdt, H., Surmont, I., Verhoef, G. E. G., Verhaegen, J., et al. (2000). Disseminated infection by *Scedosporium prolificans*: an emerging fatality among haematology patients. Case report and review. *Annals of Hematology*, *79*, 340–344.

Máiz, L., Girón, R., Olveira, C., Vendrell, M., Nieto, R., & Martínez-García, M. A. (2015). Prevalence and factors associated with isolation of *Aspergillus* and *Candida* from sputum in patients with non-cystic fibrosis bronchiectasis. *Respiration*, *89*, 396–403.

Máiz, L., Nieto, R., Cantón, R., de la Pedrosa, E. G. G., & Martínez-García, M. A. (2018). Fungi in bronchiectasis: A concise review. *International Journal of Molecular Sciences*, *19*, 142.

Marom, E. M., Holmes, A. M., Bruzzi, J. F., Truong, M. T., O'Sullivan, P. J., & Kontoyiannis, D. P. (2008). Imaging of pulmonary fusariosis in patients with hematologic malignancies. *American Journal of Roentgenology*, *190*, 1605–1609.

Marr, K. A., Datta, K., Mehta, S., Ostrander, D. B., Rock, M., Francis, J., et al. (2018). Urine antigen detection as an aid to diagnose invasive aspergillosis. *Clinical Infectious Diseases*, *67*, 1705–1711.

Marzolf, G., Sabou, M., Lannes, B., Cotton, F., Meyronet, D., Galanaud, D., et al. (2016). Magnetic resonance imaging of cerebral aspergillosis: Imaging and pathological correlations. *PLoS ONE*, *11*, e0152475.

Mehl, H. L., & Epstein, L. (2008). Sewage and community shower drains are environmental reservoirs of *Fusarium* species complex group 1, a human and plant pathogen. *Environmental Microbiology*, *10*, 219–227.

Mercier, T., Dunbar, A., de Kort, E., Schauwvlieghe, A., Reynders, M., Guldentops, E., et al. (2019). Lateral flow assays for diagnosing invasive pulmonary aspergillosis in adult hematology patients: A comparative multicenter study. *Medical Mycology*, 1–9.

Mercier, T., Guldentops, E., van Daele, R., & Maertens, J. (2018). Diagnosing invasive mold infections: What is next. *Current Fungal Infection Reports*, *12*, 161–169.

Mercier, T., Schauwvlieghe, A., de Kort, E., Dunbar, A., Reynders, M., Guldentops, E., et al. (2019). Diagnosing invasive pulmonary aspergillosis in hematology patients: A retrospective multicentre evaluation of a novel lateral flow device. *Journal of Clinical Microbiology*, *57*, e01913-18.

Millon, L., Herbrecht, R., Grenouillet, F., Morio, F., Alanio, A., Letscher-Bru, V., et al. (2016). Early diagnosis and monitoring of mucormycosis by detection of circulating DNA in serum: Retrospective analysis of 44 cases collected through the French Surveillance Network of Invasive Fungal Infections (RESSIF). *Clinical Microbiology and Infection*, *22*, 810.e1–810.e8.

Misch, E. A., & Safdar, N. (2016). Updated guidelines for the diagnosis and management of aspergillosis. *Journal of Thoracic Disease*, *8*, E1771–E1776.

Miyazaki, T., Hayashi, O., Oshima, Y., & Yadomae, T. (1979). Studies on fungal polysaccharides: The immunological determinant of the serologically active substances from *Absidia cylindrospora*, *Mucor hiemalis* and *Rhizopus nigricans*. *Journal of General Microbiology*, *111*, 417–422.

Mohanty, N. K., & Sahu, S. (2014). *Fusarium solani* infection in a kidney transplant recipient. *Indian Journal of Nephrology*, *24*, 312–314.

Morad, H. O. J., Wild, A.-M., Wiehr, S., Davies, G., Maurer, A., Pichler, B. J., et al. (2018). Pre-clinical imaging of invasive candidiasis using immunoPET/MR. *Frontiers in Microbiology*, *9*, 1996.

Morio, F., Horeau-Langlard, D., Gay-Andrieu, F., Talarmin, J.-P., Haloun, A., Treilhaud, M., et al. (2010). Disseminated *Scedosporium/Pseudallescheria* infection after double-lung transplantation in patients with cystic fibrosis. *Journal of Clinical Microbiology*, *48*, 1978–1982.

Mortensen, K. L., Mellado, E., Lass-Florl, C., Rodriguez-Tudela, J. L., Johansen, H. K., & Arendrup, M. C. (2010). Environmental study of azole-resistant *Aspergillus fumigatus* and other aspergilli in Austria, Denmark, and Spain. *Antimicrobial Agents and Chemotherapy*, *54*, 4545–4549.

Muhammed, M., Anagnostou, T., Desalermos, A., Kourkoumpetis, T. K., Carneiro, H. A., Glavis-Bloom, J., et al. (2013). *Fusarium* infection—Report of 26 cases and review of 97 cases from the literature. *Medicine*, *92*, 305–316.

Muhammed, M., Coleman, J. J., Carneiro, H. A., & Mylonakis, E. (2011). The challenge of managing fusariosis. *Virulence*, *2*, 91–96.

Muldoon, E. G., Sharman, A., Page, I., Bishop, P., & Denning, D. W. (2016). *Aspergillus* nodules; another presentation of chronic pulmonary aspergillosis. *BMC Pulmonary Medicine*, *16*, 123.

Muraosa, Y., Oguchi, M., Yahiro, M., Watanabe, A., Yaguchi, T., & Kamei, K. (2017). Epidemiological study of *Fusarium* species causing invasive and superficial fusariosis in Japan. *Medical Mycology Journal*, *58E*, E5–E13.

Muraosa, Y., Schreiber, A. Z., Trabasso, P., Matsuzawa, T., Taguchi, H., Moretti, M. L., et al. (2014). Development of cycling probe-based real-time PCR system to detect *Fusarium* species and *Fusarium solani* species complex (FSSC). *International Journal of Medical Microbiology*, *304*, 505–511.

Nakamura, Y., Suzuki, N., Nakajima, Y., Utsumi, Y., Murata, O., Nagashima, H., et al. (2013). *Scedosporium aurantiacum* brain abscess after near-drowning in a survivor of a tsunami in Japan. *Respiratory Investigation*, *51*, 207–211.

Nakamura, Y., Utsumi, Y., Suzuki, N., Nakajima, Y., Murata, O., Sasaki, N., et al. (2011). Multiple *Scedosporium apiospermum* abscesses in a woman survivor of a tsunami in northeastern Japan: A case report. *Journal of Medical Case Reports*, *25*, 526.

Notermans, S., & Soentoro, P. S. S. (1986). Immunological relationship of extracellular polysaccharide antigens produced by different mould species. *Antonie Van Leeuwenhoek*, *52*, 393–401.

Nucci, M., & Anaissie, E. (2002). Cutaneous infection by *Fusarium* species in healthy and immunocompromised hosts: Implications for diagnosis and management. *Clinical Infectious Diseases*, *35*, 909–920.

Nucci, M., & Anaissie, E. (2007). *Fusarium* infections in immunocompromised patients. *Clinical Microbiology Reviews*, *20*, 695–704.

Nucci, M., Carlesse, F., Cappellano, P., Varon, A. G., Seber, A., Garnica, M., et al. (2014). Earlier diagnosis of invasive fusariosis with *Aspergillus* serum galactomannan testing. *PLoS ONE*, *9* e87784.

Nucci, M., Varon, A. G., Garnica, M., Akiti, T., Barreiros, G., Trope, B. M., et al. (2013). Increased incidence of invasive fusariosis with cutaneous portal of entry, Brazil. *Emerging Infectious Diseases*, *19*, 1567–1572.

Nwankwo, L., Periselneris, J., Cheong, J., Thompson, K., Darby, P., Leaver, N., et al. (2018). A prospective real-world study of the impact of antifungal stewardship program in a tertiary respiratory-medicine setting. *Antimicrobial Agents and Chemotherapy*, *62*, e00402–e00418.

Ohba, H., Miwa, S., Shirai, M., Kanai, M., Eifuku, T., Suda, T., et al. (2012). Clinical characteristics and prognosis of chronic pulmonary aspergillosis. *Respiratory Medicine*, *106*, 724–729.

Orasch, T., Prattes, J., Faserl, K., Eigl, S., Düttmann, W., Lindner, H., et al. (2017). Bronchoalveolar lavage triacetylfusarinine C (TAFC) determination for diagnosis of invasive pulmonary aspergillosis in patients with hematological malignancies. *The Journal of Infection*, *75*, 370–373.

Orsi, C. F., Gennari, W., Venturelli, C., La Regina, A., Pecorari, M., Righi, E., et al. (2012). Performance of 2 commercial real-time polymerase chain reaction assays for the detection of *Aspergillus* and *Pneumocystis* DNA in bronchoalveolar lavage fluid samples from critical care patients. *Diagnostic Microbiology and Infectious Disease*, *73*, 138–143.

Padoan, R., Poli, P., Colombrita, D., Borghi, E., Timpano, S., & Berlucchi, M. (2016). Acute *Scedosporium apiospermum* endobronchial infection in cystic fibrosis. *The Pediatric Infectious Disease Journal*, *35*, 701–702.

Page, I. D., Richardson, M. D., & Denning, D. (2015). Antibody testing in aspergillosis—Quo vadis? *Medical Mycology*, *53*, 417–439.

Page, I. D., Richardson, M. D., & Denning, D. (2016). Comparison of six *Aspergillus*-specific IgG assays for the diagnosis of chronic pulmonary aspergillosis. *The Journal of Infection*, *72*, 240–249.

Page, I. D., Richardson, M. D., & Denning, D. (2019). Siemens Immulite *Aspergillus*-specific IgG assay for chronic pulmonary aspergillosis diagnosis. *Medical Mycology*, *57*, 300–307.

Pan, Z., Fu, M., Zhang, J., Zhou, H., Fu, Y., & Zhou, J. (2015). Diagnostic accuracy of a novel lateral-flow device in invasive aspergillosis: A meta-analysis. *Journal of Medical Microbiology*, *64*, 702–707.

Paolino, K. M., Henry, J. A., Hospenthal, D. R., Wortmann, G. W., & Hartzell, J. D. (2012). Invasive fungal infections following combat-related injury. *Military Medicine*, *177*, 681–685.

Pappas, P. G., Alexander, B. D., Andes, D. R., Hadley, S., Kauffman, C. A., Freifeld, A., et al. (2010). Invasive fungal infections among organ transplant recipients: Results of the Transplant-Associated Infection Surveillance Network (TRANSNET). *Clinical Infectious Diseases*, *50*, 1101–1111.

Patterson, T. F., & Donnelly, J. P. (2019). New concepts in diagnostics for invasive mycoses: Non-culture-based methodologies. *Journal of Fungi*, *5*, 9.

Patterson, T. F., Thompson, G. R., Denning, D. W., Fishman, J. A., Hadley, S., Herbrecht, R., et al. (2016). Practice guidelines for the diagnosis and management of aspergillosis: 2016 update by the Infectious Diseases Society of America. *Clinical Infectious Diseases*, *15*, e1–e60.

Pegorie, M., Denning, D. W., & Welfare, W. (2017). Estimating the burden of invasive and serious fungal disease in the United Kingdom. *The Journal of Infection*, *74*, 60–71.

Pergam, S. A. (2017). Fungal pneumonia in patients with hematologic malignancies and hematopoietic cell transplantation. *Clinics in Chest Medicine*, *38*, 279–294.

Petrik, M., Fransenn, G. M., Haas, H., Laverman, P., Hörtnagl, C., Schrettl, M., et al. (2012). Preclinical evaluation of two ^{68}Ga-siderophores as potential radiopharmaceuticals for *Aspergillus fumigatus* infection imaging. *European Journal of Nuclear Medicine and Molecular Imaging*, *39*, 1175–1183.

Petrik, M., Haas, H., Dobrozemsky, G., Lass-Flörl, C., Helbok, A., Blatzer, M., et al. (2010). ^{68}Ga-siderophores for PET imaging of invasive pulmonary aspergillosis: Proof of principle. *Journal of Nuclear Medicine*, *51*, 639–645.

Petrik, M., Haas, H., Laverman, P., Schrettl, M., Franssen, G. M., Blatzer, M., et al. (2014). ^{68}Ga-triacetylfusarinine C and ^{68}Ga-ferrioxamine E for *Aspergillus* infection imaging: Uptake specificity in various microrganisms. *Molecular Imaging and Biology*, *16*, 102–108.

Petrik, M., Haas, H., Schrettl, M., Helbok, A., Blatzer, M., & Decristiforo, C. (2012). *In vitro* and *in vivo* evaluation of selected (68)Ga-siderophores for infection imaging. *Nuclear Medicine and Biology*, *39*, 361–369.

Petrik, M., Vlckova, A., Novy, Z., Urbanek, L., Haas, H., & Decristoforo, C. (2015). Selected ^{68}Ga-siderophores versus ^{68}Ga-colloid and ^{68}Ga-citrate: Biodistribution and small animal imaging in mice. *Biomedical Papers of the Medical Faculty of the University Palacky, Olomouc, Czech Republic*, *159*, 60–66.

Petrikkos, G., Skiada, A., Lortholary, O., Roilides, E., Walsh, T., & Kontoyiannis, D. P. (2012). Epidemiology and clinical maifestations of mucormycosis. *Clinical Infectious Diseases*, *54*, S23–S34.

Piarroux, R. P., Romain, T., Martin, A., Vainqueur, D., Vitte, J., Lachaud, L., et al. (2019). Multicenter evaluation of a novel immunochromatographic test for anti-aspergillus IgG detection. *Frontiers in Cellular and Infection Microbiology*, *9*, 12.

Postina, P., Skladny, J., Boch, T., Cornely, O. A., Hamprecht, A., Rath, P.-M., et al. (2018). Comparison of two molecular assays for detection and characterisation of *Aspergillus fumigatus* triazole resistance and Cyp51A mutations in clinical isolates and primary clinical samples of immunocompromised patients. *Frontiers in Microbiology*, *9*, 555.

Prattes, J., Flick, H., Prüller, F., Koidl, C., Raggam, R. B., Palfner, M., et al. (2014). Novel tests for diagnosis of invasive aspergillosis in patients with underlying respiratory diseases. *American Journal of Respiratory and Critical Care Medicine*, *190*, 922–929.

Raghunath, V., Priyanka, K. H. R., Kiran, C. G., & Deepthi, R. M. (2018). Mucormycosis—Can the diagnosis be challenging at times? *SRM Journal of Research in Dental Sciences*, *9*, 191–196.

Rainer, J., & de Hoog, G. S. (2006). Molecular taxonomy and ecology of *Pseudallescheria*, *Petriella* and *Scedosporium prolificans* (Microascaceae) containing opportunist agents on humans. *Mycological Research*, *110*, 151–160.

Rainer, J., de Hoog, G. S., Wedde, M., Graser, Y., & Gilges, S. (2000). Molecular variability of *Pseudallescheria boydii*, a neurotropic opportunist. *Journal of Clinical Microbiology*, *38*, 3267–3273.

Rainer, J., Kalteis, J., de Hoog, S. G., & Summerbell, R. C. (2008). Efficacy of a selective isolation procedure for members of the *Pseudallescheria boydii* complex. *Antonie Van Leeuwenhoek*, *93*, 315–322.

Ramirez-Garcia, A., Pellon, A., Rementeria, A., Buldain, I., Barreto-Bertger, E., Rollin-Pinheiro, R., et al. (2018). *Scedosporium* and *Lomentospora*: An updated overview of underrated opportunists. *Medical Mycology*, *56*, S102–S125.

Rammaert, B., Lanternier, F., Poirée, S., Kania, R., & Lortholary, O. (2012). Diabetes and mucormycosis: A complex interplay. *Diabetes & Metabolism*, *38*, 193–204.

Ramsperger, M., Duan, S., Sorrell, T. C., Meyer, W., & Chen, S. C.-A. (2014). The genus *Scedosporium* and *Pseudallescheria*: Current challenges in laboratory diagnosis. *Current Clinical Microbiology Reports*, *1*, 27–36.

Rankin, N. E. (1953). Disseminated aspergillosis and moniliasis associated with agranulocytosis and antibiotic therapy. *British Medical Journal*, *1*, 918–919.

Rath, P.-M., & Steinmann, J. (2018). Overview of commercially available PCR assays for the detection of *Aspergillus* spp. DNA in patient samples. *Frontiers in Microbiology*, *9*, 740.

Reed, C. (1978). Variability in antigenicity of *Aspergillus fumigatus*. *The Journal of Allergy and Clinical Immunology*, *61*, 227–229.

Reinwald, M., Boch, T., Hofmann, W. K., & Buchheidt, D. (2016). Risk of infectious complications in hemato-oncological patients treated with kinase inhibitors. *Biomarker Insights*, *10*, 55–68.

Ribes, J. A., Vanover-Sams, C. L., & Baker, D. J. (2000). Zygomycetes in human disease. *Clinical Microbiology Reviews*, *13*, 236–301.
Richardson, M. (2009). The ecology of the Zygomycetes and its impact on environmental exposure. *Clinical Microbiology and Infection*, *15*, 2–9.
Richardson, M., & Page, I. (2018). Role of serological tests in the diagnosis of mold infections. *Current Fungal Infection Reports*, *12*, 127–136.
Rimek, D., Singh, J., & Kappe, R. (2003). Cross-reactivity of the PLATELIA *CANDIDA* antigen detection enzyme immunoassay with fungal extracts. *Journal of Clinical Microbiology*, *41*, 3395–3398.
Rivero-Menendez, O., Alastruey-Izquidero, A., Mellado, E., & Cuenca-Estrella, M. (2016). Triazole resistance in *Aspergillus* spp.: A worldwide problem? *Journal of Fungi*, *2*, 21.
Rodriguez-Tudela, J. L., Berenguer, J., Guarro, J., Kantarcioglu, A. S., Horre, R., de Hoog, G. S., et al. (2009). Epidemiology and outcome of *Scedosporium prolificans* infection, a review of 162 cases. *Medical Mycology*, *47*, 359–370.
Rogers, T. R., Haynes, K. A., & Barnes, R. A. (1990). Value of antigen detection in predicting invasive pulmonary aspergillosis. *Lancet*, *336*, 1210–1213.
Rognon, B., Barrera, C., Monod, M., Valot, B., Roussel, S., Quadroni, M., et al. (2016). Identification of antigenic proteins from *Lichtheimia corymbifera* for farmer's lung disease diagnosis. *PLoS ONE*, *11*, e0160888.
Rolfe, N. E., Haddad, T. J., & Wills, T. S. (2013). Management of *Scedosporium apiospermum* in a pre- and post-lung transplant patient with cystic fibrosis. *Medical Mycology Case Reports*, *2*, 37–39.
Rolle, A.-M., Hasenberg, M., Thornton, C. R., Solouk-Saran, D., Männ, L., Weski, J., et al. (2016). ImmunoPET/MR imaging allows specific detection of *Aspergillus* lung infection *in vivo*. *Proceedings of the National Academy of Sciences of the United States of America*, *113*, E1026–E1033.
Rougeron, A., Schullar, G., Leto, J., Sitterié, E., Landry, D., Bougnoux, M.-E., et al. (2015). Human-impacted areas of France are environmental reservoirs of the *Pseudallescheria boydii/Scedosporium apiospermum* species complex. *Environmental Microbiology*, *17*, 1039–1048.
Ruhnke, M., Behre, G., Buchheidt, D., Christopeit, M., Hamprecht, A., Heinz, W., et al. (2018). Diagnosis of invasive fungal diseases in haematology and oncology: 2018 update of the recommendations of the infectious diseases working party of the German society for hematology and medical oncology (AGIHO). *Mycoses*, *61*, 796–813.
Saito, T., Imaizumi, M., Kudo, K., Hotchi, M., Chikaoka, S., Yoshinari, M., et al. (1999). Disseminated *Fusarium* infection identified by the immunohistochemical staining in a patient with a refractory leukemia. *The Tohoku Journal of Experimental Medicine*, *187*, 71–77.
Saito, S., Michailides, T. J., & Xiao, C. L. (2016). Mucor rot—an emerging postharvest disease of mandarin fruit caused by *Mucor piriformis* and other *Mucor* spp. in California. *Plant Disease*, *100*, 1054–1063.
Salehi, E., Hedayati, M. T., Zoll, J., Rafati, H., Ghasemi, M., Doroudinia, A., et al. (2016). Discrimination of aspergillosis, mucormycosis, fusariosis, and scedosporiosis in formalin-fixed paraffin-embedded tissue specimens by use of multiple real-time quantitative PCR assays. *Journal of Clinical Microbiology*, *54*, 2798–2803.
Salzer, H. J. F., Prattes, J., Flick, H., Reimann, M., Heyckendorf, J., Kalsdorf, B., et al. (2018). Evaluation of galactomannan testing, the *Aspergillus*-specific lateral-flow device test and levels of cytokines in bronchoalveolar lavage fluid for diagnosis of chronic pulmonary aspergillosis. *Frontiers in Microbiology*, *9*, 2223.
Sandven, P. E. R., & Eduard, W. (1992). Detection and quantification of antibodies against *Rhizopus* by enzyme-linked immunosorbent assay. *APMIS*, *100*, 981–987.

Santos, P. E., Oleastra, M., Galicchio, M., & Zelazko, M. (2000). Fungal infections in paediatric patients with chronic granulomatous disease. *Revista Iberoamericana de Micología, 17*, 6–9.

Santos, C., Paterson, R. R. M., Venancio, A., & Lima, N. (2010). Filamentous fungal characterisation by matrix-assisted laser desorption/ionisation time of flight spectrometry. *Journal of Applied Microbiology, 108*, 378–385.

Schauwvlieghe, A. F. A. D., Vank, A. G., Buddingh, E. P., Hoe, R. A. S., Dalm, V. A., Klaassen, C. H. W., et al. (2017). Detection of azole-susceptible and azole-resistant *Aspergillus* coinfection by cyp51A PCR amplicon melting curve analysis. *The Journal of Antimicrobial Chemotherapy, 72*, 3047–3050.

Scheel, C. M., Hurst, S. F., Barreiros, G., Akiti, T., Nucci, M., & Balajee, S. A. (2013). Molecular analyses of *Fusarium* isolates recovered from a cluster of invasive mold infections in a Brazilian hospital. *BMC Infectious Diseases, 13*, 49.

Schrödl, W., Heydel, T., Schwartze, V. U., Hoffmann, K., Grosse-Herrenthey, A., Walther, G., et al. (2012). Direct analysis and identification of pathogenic *Lichtheimia* species by matrix-assisted laser desorption ionisation-time of flight analyser-mediated mass spectrometry. *Journal of Clinical Microbiology, 50*, 419–427.

Schumock, G. T., Li, E. C., Suda, K. J., Wiest, M. D., Stubbings, J., Mutusiak, L. M., et al. (2016). National trends in prospective drug expenditures and projections for 2016. *American Journal of Health-System Pharmacy, 73*, 1058–1075.

Schwartz, R. A. (2004). Superficial fungal infections. *Lancet, 364*, 1173–1182.

Schwarz, C., Brandt, C., Antweiler, E., Krannich, A., Staab, D., Schmitt-Grohé, S., et al. (2017). Prospective multicentre German study on pulmonary colonisation with *Scedosporium/Lomentospora* species in cystic fibrosis: Epidemiology and new association factors. *PLoS ONE, 12*, e0171485.

Schwarz, C., Brandt, C., Whitaker, P., Sutharsan, S., Skopnik, H., Gartner, S., et al. (2018). Invasive pulmonary fungal infections in cystic fibrosis. *Mycopathologia, 183*, 33–43.

Schwarz, C., Thronicke, A., Staab, D., & Tintelnot, K. (2015). *Scedosporium apiospermum*: a fungal pathogen in a patient with cystic fibrosis. *JMM Case Reports, 2*, 1–5.

Sedlacek, L., Graf, B., Schwarz, C., Albert, F., Peter, S., Würstl, B., et al. (2015). Prevalence of *Scedosporium* species and *Lomentospora prolificans* in patients with cystic fibrosis in a multicentre trial by use of a selective medium. *Journal of Cystic Fibrosis, 14*, 237–241.

Segal, B. H. (2009). Aspergillosis. *The New England Journal of Medicine, 360*, 1870–1884.

Sehgal, I. S., Choudhary, H., Dhooria, S., Aggrawal, A. N., Garg, M., Chakrabarti, A., et al. (2018). Diagnostic cut-off of *Aspergillus fumigatus*-specific IgG in the diagnosis of chronic pulmonary aspergillosis. *Mycoses, 61*, 770–776.

Seidel, D., Meißner, A., Lackner, M., Piepenbrock, E., Salmanton-García, J., Stecher, M., et al. (2019). Prognostic factors in 264 adults with invasive *Scedosporium* spp. and *Lomentospora prolificans* infection reported in the literature and FungiScope. *Critical Reviews in Microbiology, 45*, 1–21.

Sharma, A., & Singh, D. (2015). *Scedosporium apiospermum* causing brain abscess in a renal allograft recipient. *Saudi Journal of Kidney Diseases and Transplantation, 26*, 1253–1256.

Short, D. P. G., O'Donnell, K., Thrane, U., Nielsen, K. F., Zhang, N., Juba, J. H., et al. (2013). Phylogenetic relationships among members of the *Fusarium solani* species complex in human infections and the descriptions of *F. keratoplasticum* sp. nov. and *F. petroliphilum* stat. nov. *Fungal Genetics and Biology, 53*, 59–70.

Short, D. P. G., O'Donnell, K., Zhang, N., Juba, J. H., & Geiser, D. M. (2011). Widespread occurrence of diverse human pathogenic types of the fungus *Fusarium* detected in plumbing drains. *Journal of Clinical Microbiology, 49*, 4264–4272.

Sircar, G., Saha, B., Mandal, R. S., Pandey, N., Saha, S., & Bhattacharya, S. G. (2015). Purification, cloning and immuno-biochemical characterisation of a fungal aspartic protease allergen Rhi o 1 from the airborne mold *Rhizopus oryzae*. *PLoS ONE*, *10*, e0144547.
Skiada, A., Lanternier, F., Groll, A. H., Pagano, L., Zimmerli, S., Herbrecht, R., et al. (2013). Diagnosis and treatment of mucormycosis in patients with hematological malignancies: Guidelines from the 3rd European Conference on Infections in Leukemia (ECIL 3). *Haematologica*, *98*, 492–504.
Skiada, A., Lass-Floerl, C., Klimko, N., Ibrahim, A., Roilides, E., & Petrikkos, G. (2017). Challenges in the diagnosis and treatment of mucormycosis. *Medical Mycology*, *56*, S93–S101.
Smith, N. L., & Denning, D. W. (2011). Underlying conditions in chronic pulmonary aspergillosis including simple aspergilloma. *The European Respiratory Journal*, *37*, 865–872.
Son, J. H., Lim, H. B., Lee, S. H., Yang, J. W., & Lee, S. B. (2016). Early differential diagnosis of rhino-orbito-cerebral mucormycosis and bacterial orbital cellulitis: Based on computed tomography findings. *PLoS ONE*, *11*, e0160897.
Song, M. J., Lee, J. H., & Lee, N. Y. (2009). Fatal disseminated *Scedosporium prolificans* infection in a paediatric patient with acute lymphoblastic leukemia. *Mycoses*, *54*, 81–83.
Starkey, J., Moritani, T., & Kirby, P. (2014). MRI and CNS fungal infections: Review of aspergillosis to histoplasmosis and everything else. *Clinical Neuroradiology*, *24*, 217–230.
Steinmann, J., Schmidt, D., Buer, J., & Rath, P.-M. (2011). Discrimination of *Scedosporium prolificans* against *Pseudallescheria boydii* and *Scedosporium apiospermum* by semi-automated repetitive sequence-based PCR. *Medical Mycology*, *49*, 475–483.
Stynen, D., Goris, A., Sarfati, J., & Latgé, J.-P. (1995). A new sensitive sandwich enzyme-linked immunosorbent assay to detect galactofuran in patients with invasive aspergillosis. *Journal of Clinical Microbiology*, *33*, 497–500.
Stynen, D., Sarfati, J., Goris, A., Prévost, M.-C., Lesourd, M., Kamphuis, H., et al. (1992). Rat monoclonal antibodies against *Aspergillus* galactomannan. *Infection and Immunity*, *60*, 2237–2245.
Sugui, J. A., Christensen, J. A., Bennett, J. E., Zelazny, A. M., & Kwon-Chung, K. J. (2011). Hematogenously disseminated skin disease caused by *Mucor velutinosus* in a patient with acute myeloid leukemia. *Journal of Clinical Microbiology*, *49*, 2728–2732.
Sugui, J. A., Kwon-Chung, K. J., Juvvadi, P. R., Latge, J.-P., & Steinbach, W. J. (2015). *Aspergillus fumigatus* and related species. *Cold Spring Harbor Perspectives in Medicine*, *5*, a019786.
Taj-Aldeen, S. J., Gamaletsou, M. N., Rammaert, B., Sipsas, N. V., Zeller, V., Roilides, E., et al. (2017). Bone and joint infections caused by mucormycetes: A challenging osteoarticular mycosis of the twenty-first century. *Medical Mycology*, *55*, 691–704.
Taj-Aldeen, S. J., Rammaert, B., Gamaletsou, M. N., Sipsas, N. V., Zeller, V., Roilides, E., et al. (2015). Osteoarticular infections caused by non-*Aspergillus* filamentous fungi in adult and pediatric patients. *Medicine*, *94*, 1–13.
Takazono, T., Ito, Y., Tashiro, M., Nishimura, K., Saijo, T., Yamamoto, K., et al. (2019). Evaluation of *Aspergillus*-specific Lateral-flow device test using serum and bronchoalveolar lavage fluid for diagnosis of chronic pulmonary aspergillosis. *Journal of Clinical Microbiology*, *57*, e00095-19.
Takazono, T., & Izumikawa, K. (2018). Recent advances in diagnosing chronic pulmonary aspergillosis. *Frontiers in Microbiology*, *9*, 1810.
Tamaki, M., Nozaki, K., Onishi, M., Yamamoto, K., Ujiie, H., & Sugahara, H. (2016). Fungal meningitis caused by *Lomentospora prolificans* after allogeneic hematopoietic stem cell transplantation. *Transplant Infectious Disease*, *18*, 601–605.

Tammer, I., Tintelnot, K., Braun-Dullaeus, R. C., Mawrin, C., Scherlach, C., Schlüter, D., et al. (2011). Infections due to *Pseudallescheria/Scedosporium* species in patients with advanced HIV disease—A diagnostic and therapeutic challenge. *International Journal of Infectious Diseases*, *15*, e422–e429.

Tashiro, T., Izumikawa, K., Tashiro, M., Takazono, T., Morinaga, Y., Yamamoto, K., et al. (2011). Diagnostic significance of *Aspergillus* species isolated from respiratory samples in an adult pneumology ward. *Medical Mycology*, *49*, 581–587.

Taylor, A., Talbot, J., Bennett, P., Martin, P., Makara, M., & Barrs, V. R. (2014). Disseminated *Scedosporium prolificans* infection in a Labrador retriever with immune mediated haemolytic anaemia. *Medical Mycology Case Reports*, *6*, 66–69.

Thornton, C. R. (2008). Development of an immunochromatographic lateral-flow device for rapid serodiagnosis of invasive aspergillosis. *Clinical and Vaccine Immunology*, *15*, 1095–1105.

Thornton, C. R. (2009). Tracking the emerging human pathogen *Pseudallescheria boydii* by using highly specific monoclonal antibodies. *Clinical and Vaccine Immunology*, *16*, 756–764.

Thornton, C. R. (2014). Breaking the mould—Novel diagnostic and therapeutic strategies for invasive pulmonary aspergillosis in the immune deficient patient. *Expert Review of Clinical Immunology*, *6*, 771–780.

Thornton, C. R. (2018). Molecular imaging of invasive pulmonary aspergillosis using immunoPET/MRI: The future looks bright. *Frontiers in Microbiology*, *9*, 691.

Thornton, C. R., Pitt, D., Wakley, G. E., & Talbot, N. J. (2002). Production of a monoclonal antibody specific to the genus *Trichoderma* and closely related fungi, and its use to detect *Trichoderma* spp. in naturally infested composts. *Microbiology*, *148*, 1263–1279.

Thornton, C. R., Ryder, L. S., Le Cocq, K., & Soanes, D. M. (2015). Identifying the emerging human pathogen *Scedosporium prolificans* by using a species-specific monoclonal antibody that binds to the melanin biosynthetic enzyme tetrahydroxynaphthalene reductase. *Environmental Microbiology*, *17*, 1023–1038.

Thornton, C. R., & Wills, O. E. (2015). Immunodetection of fungal and oomycete pathogens: Established and emerging threats to human health, animal welfare and global food security. *Critical Reviews in Microbiology*, *41*, 27–51.

Tomazin, G., & Matos, T. (2017). Fungal infections in patients with cystic fibrosis. *Zdravstveni Vestnik*, *86*, 42–52.

Torelli, R., Sanguinetti, M., Moody, A., Pagano, L., Caira, M., De Carolis, E., et al. (2011). Diagnosis of invasive aspergillosis by a commercial real-time PCR assay for *Aspergillus* DNA in bronchoalveolar lavage fluid samples from high-risk patients compared to a galactomannan enzyme immunoassay. *Journal of Clinical Microbiology*, *49*, 4273–4278.

Tortorano, A. M., Esposto, M. C., Prigitano, A., Grancini, A., Ossi, C., Cavanna, C., et al. (2012). Cross-reactivity of *Fusarium* spp. in the *Aspergillus* galactomannan enzyme-linked immunosorbent assay. *Journal of Clinical Microbiology*, *50*, 1051–1053.

Tortorano, A. M., Richardson, M., Roilides, E., van Diepeningen, A., Caira, M., Munoz, P., et al. (2014). ESCMID and ECMM joint guidelines on diagnosis and management of hyalohyphomycosis: *Fusarium* spp., *Scedosporium* spp. and others. *Clinical Microbiology and Infection*, *20*, 27–46.

Tribble, D. R., & Rodriguez, C. J. (2014). Combat-related invasive fungal wound infections. *Current Fungal Infection Reports*, *8*, 277–286.

Ullmann, A. J., Aguado, J. M., Arikan-Akdagli, S., Denning, D. W., Groll, A. H., Lagrou, K., et al. (2018). Diagnosis and management of *Aspergillus* diseases: Executive summary of the 2017 ESCMID-ECMM-ERS guideline. *Clinical Microbiology and Infection*, *24*, e1–e38.

Ustun, C., Farrow, S., DeRemer, D., Fain, H., & Jillela, A. P. (2007). Early fatal *Rhizopus* infection on voriconazole prophylaxis following allogeneic stem cell transplantation. *Bone Marrow Transplantation, 39*, 807–808.

Vagefi, M. R., Kim, E. T., Alvarado, R. G., Duncan, J. L., Howes, E. L., & Crawford, J. B. (2005). Bilateral endogenous *Scedosporium prolificans* endophthalmitis after lung transplantation. *American Journal of Ophthalmology, 139*, 370–373.

Van der Elst, K. C., Brouwers, C. H., van den Heuvel, E. R., van Wanrooy, M. J., Uges, D. R., van der Werf, T. S., et al. (2015). Subtherapeutic posaconazole exposure and treatment outcome in patients with invasive fungal disease. *Therapeutic Drug Monitoring, 37*, 766–771.

Van der Linden, J. W. M., Warris, A., & Verweij, P. E. (2011). *Aspergillus* species intrinsically resistant to antifungal agents. *Medical Mycology, 49*, S82–S89.

Van Diepeningen, A. D., Brankovics, B., Iltes, J., van der Lee, T. A. J., & Waalwijk, C. (2015). Diagnosis of *Fusarium* infections: Approaches to identification by the clinical mycology laboratory. *Current Fungal Infection Reports, 9*, 135–143.

Van Diepeningen, A. D., Feng, P., Ahmed, S., Sudhadham, M., Bunyaratavej, S., & de Hoog, G. S. (2015). Spectrum of *Fusarium* infections in tropical dermatology evidenced by multilocus sequencing typing diagnostics. *Mycoses, 58*, 48–57.

Verweij, P. E., Snelders, E., Kema, G. H., Mellado, E., & Melchers, W. J. (2009). Azole resistance in *Aspergillus fumigatus*: A side-effect of environmental fungicide use? *The Lancet Infectious Diseases, 9*, 789–795.

Verweij, P. E., Zhang, J., Debets, A. J. M., Meis, J. F., van de Veerdonk, F., Schoustra, S. E., et al. (2016). In-host adaptation and acquired triazole resistance in *Aspergillus fumigatus*: A dilemma for clinical management. *The Lancet Infectious Diseases, 16*, e251–e260.

Walsh, T. J., Gamaletsou, M. N., McGinnis, M. R., Hayden, R. T., & Kontoyiannis, D. P. (2012). Early clinical and laboratory diagnosis of invasive pulmonary, extrapulmonary, and disseminated mucormycosis (zygomycosis). *Clinical Infectious Diseases, 54*, S55–S60.

Walsh, T. J., Skiada, A., Cornely, O. A., Roilides, E., Ibrahim, A., Zaoutis, T., et al. (2014). Development of new strategies for early diagnosis of mucormycosis from bench to bedside. *Mycoses, 57*, 2–7.

Wang, X.-M., Guo, L.-C., Xue, S.-L., & Chen, Y.-B. (2016). Pulmonary mucormycosis: A case report and review of the literature. *Oncology Letters, 11*, 3049–3053.

Wang, H., Xiao, M., Kong, F., Chen, S., Dou, H.-T., Sorrell, T., et al. (2011). Accurate and practical identification of 20 *Fusarium* species by seven-locus sequence analysis and reverse line blot hybridization, and an *in vitro* antifungal susceptibility study. *Journal of Clinical Microbiology, 49*, 1890–1898.

Warkentien, T., Rodriguez, C., Lloyd, B., Wells, J., Weintrob, A., Dunne, J. R., et al. (2012). Invasive mold infections following combat-related injuries. *Clinical Infectious Diseases, 55*, 1441–1449.

Westerberg, D. P., & Voyack, M. J. (2013). Onychomycosis: Current trends in diagnosis and treatment. *American Family Physician, 88*, 762–770.

White, P. L., Parr, C., Thornton, C. R., & Barnes, R. A. (2013). An evaluation of real-time PCR, galactomannan ELISA and a novel lateral-flow device for the diagnosis of invasive aspergillosis. *Journal of Clinical Microbiology, 51*, 1510–1516.

White, P. L., Wingard, J. R., Bretagne, S., Löffler, J., Patterson, T. F., Slavin, M. A., et al. (2015). *Aspergillus* polymerase chain reaction: Systematic review of evidence for clinical use in comparison with antigen testing. *Clinical Infectious Diseases, 61*, 1293–1303.

Wickes, B. L., & Wiederhold, N. P. (2018). Molecular diagnostics in medical mycology. *Nature Communications, 9*, 5135.

Wiedemann, A., Kakoschke, T. K., Speth, C., Rambach, G., Ensinger, C., Jensen, H. E., et al. (2016). Distinct galactofuranose antigens in the cell wall and culture supernatants as a means to differentiate *Fusarium* from *Aspergillus* species. *International Journal of Medical Microbiology*, *306*, 381–390.

Wysong, D. R., & Waldorf, A. R. (1987). Electrophoretic and immunoblot analyses of *Rhizopus arrhizus* antigens. *Journal of Clinical Microbiology*, *25*, 358–363.

Yamada, H., Hiraiwa, M., & Miyazaki, T. (1983). Characterisation of mannoproteins from yeast and mycelial forms of *Mucor rouxii*. *Carbohydrate Research*, *119*, 129–140.

Yamada, H., Ohshima, Y., & Miyazaki, T. (1982). Characteristion of fucomannopeptide and mannoprotein from *Absidia cylindrospora*. *Carbohydrate Research*, *110*, 113–126.

Yao, Y., Zhou, H., Shen, Y., Yang, Q., Ye, J., Fu, Y., et al. (2018). Evaluation of a quantitative serum *Aspergillus fumigatus*-specific IgM assay for diagnosis of chronic pulmonary aspergillosis. *The Clinical Respiratory Journal*, *12*, 2566–2572.

Yao, Y., Zhou, H., Yang, Q., Lu, G., Yu, Y., Shen, Y., et al. (2018). Serum *Aspergillus fumigatus*-specific IgG antibody decreases after antifungal treatment in chronic pulmonary aspergillosis patients. *The Clinical Respiratory Journal*, *12*, 1772–1774.

Zahoor, B., Kent, S., & Wall, D. (2016). Cutaneous mucormycosis secondary to penetrative trauma. *International Journal of the Care of the Injured*, *47*, 1383–1387.

Zarrin, M., Ganj, F., & Faramarzi, S. (2016). Development of a polymerase chain reaction-restriction fragment length polymorphism method for identification of the *Fusarium* genus using the transcription elongation factor-1α gene. *Biomedical Reports*, *5*, 705–708.

Zhang, L., Guo, Z., Xie, S., Zhou, J., Chen, G., Feng, J., et al. (2019). The performance of galactomannan in combination with 1,3-β-D-glucan or aspergillus-lateral flow device for the diagnosis of invasive aspergillosis: Evidences from 13 studies. *Diagnostic Microbiology and Infectious Disease*, *93*, 44–53.

Ziaee, A., Zia, M., Bayat, M., & Hashemi, J. (2016). Molecular identification of *Mucor* and *Lichtheimia* species in pure cultures of zygomycetes. *Jundishapur Journal of Microbiology*, *9*, e35237.

Zoran, T., Sartori, B., Sappl, L., Aigner, M., Sánchez-Reus, F., Rezusta, A., et al. (2018). Azole-resistance in *Aspergillus terreus* and related species: An emerging problem or rare phenomenon? *Frontiers in Microbiology*, *9*, 516.

CHAPTER TWO

Bacteroidetes bacteria in the soil: Glycan acquisition, enzyme secretion, and gliding motility

Johan Larsbrink[a,b], Lauren Sara McKee[a,c,*]
[a]Wallenberg Wood Science Center, Gothenburg and Stockholm, Sweden
[b]Division of Industrial Biotechnology, Department of Biology and Biological Engineering, Chalmers University of Technology, Gothenburg, Sweden
[c]Division of Glycoscience, Department of Chemistry, KTH Royal Institute of Technology, AlbaNova University Centre, Stockholm, Sweden
*Corresponding author: e-mail address: mckee@kth.se

Contents

Abstract

The secretion of extracellular enzymes by soil microbes is rate-limiting in the recycling of biomass. Fungi and bacteria compete and collaborate for nutrients in the soil, with wide ranging ecological impacts. Within soil microbiota, the Bacteroidetes tend to be a dominant phylum, just like in human and animal intestines. The Bacteroidetes thrive because of their ability to secrete diverse arrays of carbohydrate-active enzymes (CAZymes) that target the highly varied glycans in the soil. Bacteroidetes use an energy-saving system of genomic organization, whereby most of their CAZymes are grouped into Polysaccharide Utilization Loci (PULs). These loci enable high level production of specific CAZymes only when their substrate glycans are abundant in the local environment. This gives the Bacteroidetes a clear advantage over other species in the competitive soil environment, further enhanced by the phylum-specific Type IX Secretion System (T9SS). The T9SS is highly effective at secreting CAZymes and/or tethering them to the cell surface, and is tightly coupled to the ability to rapidly glide over solid surfaces,

Advances in Applied Microbiology, Volume 110
ISSN 0065-2164
https://doi.org/10.1016/bs.aambs.2019.11.001

a connection that promotes an active hunt for nutrition. Although the soil Bacteroidetes are less well studied than human gut symbionts, research is uncovering important biochemical and physiological phenomena. In this review, we summarize the state of the art on research into the CAZymes secreted by soil Bacteroidetes in the contexts of microbial soil ecology and the discovery of novel CAZymes for use in industrial biotechnology. We hope that this review will stimulate further investigations into the somewhat neglected enzymology of non-gut Bacteroidetes.

1. Microbial life in the soil

A healthy microbial ecosystem is one that is resistant and resilient to changes in environmental conditions or invasion by new organisms. This type of ecological resilience can ideally be achieved by a combination of a high number of cells, a high diversity of species, a high level of overall metabolic activity, and a high degree of functional enzymatic redundancy (Gonzalez et al., 2011; Ochoa-Hueso, 2017). A healthy microbiota in the soil is vital for a wide range of biogeochemical processes, including soil formation, soil fertility, and carbon storage. The recycling of organic matter in the soil—mostly achieved through microbial respiration—is vital to the carbon cycle and to aboveground biodiversity in macroorganisms (Bardgett & van der Putten, 2014), and is largely driven by the activity of enzymes secreted by soil microbes. Environments such as anaerobic peatlands are able to store enormous quantities of carbon for very long periods precisely because of the low level of microbial enzyme activity that can be detected there (Freeman, Ostle, & Kang, 2001).

Soil organic matter is a complex natural material that holds together soil particles and provides a micro-habitat for the soil microbiota. It comprises polysaccharides and phenolic polymers of mostly natural origin, and it is bound together in aggregate form by clay and water, as well as bacterial secretions, fungal mycelium, and microbial enzyme activities that cross-link natural and anthropogenic organic macromolecules (Richnow et al., 1997; Senesi, 1992). The bacteria found in soil are likely the most abundant and the most diverse community on earth (Bardgett & van der Putten, 2014; Tiedje, Asuming-Brempong, Nusslein, Marsh, & Flynn, 1999). This diversity is supported by the wide range of metabolic characteristics found among soil bacteria, the rapid evolution and adaptation achievable through horizontal gene transfer, the ability of certain species to enter dormant states, and the preference of different species for specific niche nutrients (Fierer & Lennon, 2011).

There is a great deal of microbial variation at the species level in distantly sampled soils (Bardgett & van der Putten, 2014; Fierer et al., 2012; Fulthorpe, Roesch, Riva, & Triplett, 2008; Ramirez et al., 2014), and most soil organisms appear to be restricted to a narrow range of physical locations and physiological conditions (Bates et al., 2013; Fierer, Strickland, Liptzin, Bradford, & Cleveland, 2009; Tedersoo et al., 2012). However, the same few genera typically always dominate the bacterial community of soils (Delgado-Baquerizo et al., 2018). In fact, a recent global survey has found that just 2% of bacterial taxa account for almost half of total bacterial cell numbers in soils around the world, and that local species diversity is largely predictable by analysis of habitat characteristics such as pH, climate, vegetation type, and nutrient availability (Delgado-Baquerizo et al., 2018). The Bacteroidetes are commonly among the most highly represented phyla in soils studied by metagenomic sequencing, 16S qPCR, or similar molecular biology techniques, and typically represent around 5% of the total soil community (Fierer et al., 2012). Numerically, the soil Bacteroidetes are dominated by the genera *Chitinophaga*, *Flavobacterium*, *Hymenobacter*, and *Pedobacter* (Janssen, 2006). In a survey of four distantly located sites in North and South America, although only 1.5% of species were definitively detected in every site (Fulthorpe et al., 2008), the most or second most abundant genus was *Chitinophaga* in every sample tested (Roesch et al., 2007; Sangkhobol & Skerman, 1981).

From an ecological perspective, the identities of the species present in the soil are less important than the overall functionality that the entire microbial community can offer (Louca et al., 2018), and a high overall functional microbial capacity can generally be correlated with higher species diversity (Wagg, Schlaeppi, Banerjee, Kuramae, & van der Heijden, 2019). Functional redundancy in the enzymatic capacity of taxa for biomass deconstruction means that in any given habitat there will be numerous species capable of broadly similar levels of deconstruction of the major biomass components. Certain species will be more highly active at specific ranges of temperature, water availability, and other biotic and abiotic factors, while others are less active (Jin & Kirk, 2018). As long as there is a high degree of functional redundancy in the community, there will always be a relatively stable rate of biomass turnover, which ensures that the carbon sequestered in plant biomass continues to be mineralized and nutrients continue to be mobilized, ensuring ongoing soil fertility. The large diversity of microorganisms inhabiting various soils is to a large extent governed by the types of nutrient entering these environments, and complex carbohydrates play a key role here. The complex glycans available in the soil—and the diverse microbial enzymes that are required to deconstruct them—will be discussed in detail in the coming sections.

In terms of the protein- and carbohydrate-rich biomass available for nutrition and the high level of competition for those nutrients, the soil is a surprisingly similar bacterial environment to the human, animal, and ruminant intestines (Thomas, Hehemann, Rebuffet, Czjzek, & Michel, 2011). In the human gut, for example, there is a clear biogeographical gradient through the intestines in terms of glycan nutrients, mucosal thickness, and microbial cell numbers, as both microbial cell density and diversity increase from the small intestine to the colon (Fig. 1) (Donaldson, Lee, & Mazmanian, 2015). Metagenomic studies have shown that it is possible to spatially profile species diversity and abundance in the gut, and to correlate these factors with nutrient profiles, pH, and the flow rate of transiting material (Donaldson et al., 2015; Sheth et al., 2019).

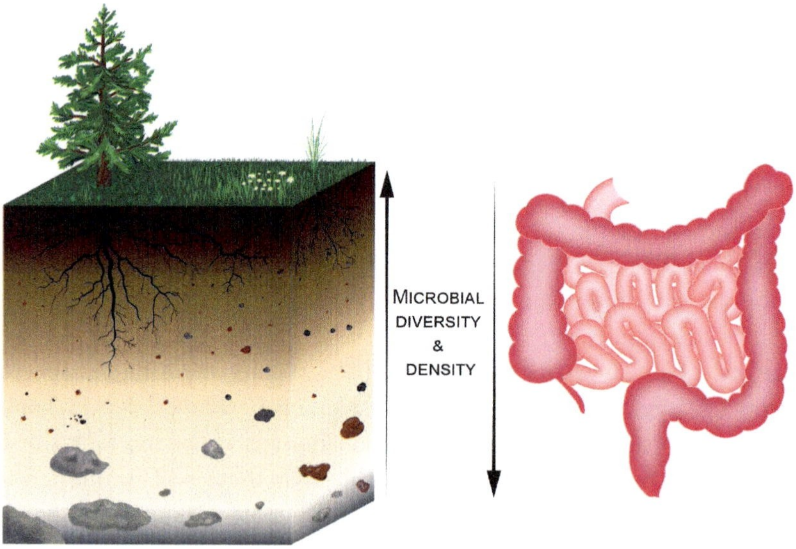

Fig. 1 Bacteroidetes bacteria are important and dominant carbohydrate degraders in two very different environments: the soil and the human gut. In both ecosystems, there is a spatial biogeographical gradient of nutrient glycan complexity, which correlates with both microbial cell densities and bacterial species diversity. In soils, the highest diversity is found in the litter and upper soil layer, where nutrients first enter the ecosystem and begin to be metabolized. Conversely, in the human digestive system, the highest diversity is found in the distal gut (colon) where the degradation of complex glycans inaccessible to the host occurs, and in turn where both cell densities and microbial diversity can be enormous. Bacteroidetes is a dominant phylum in both of these environments.

The microbial habitat in the gut is mostly anaerobic, and physically it can be described as a liquid environment, which means a relatively homogenous dispersal of soluble nutrients present at a particular stratum. Mastication (chewing) and rumination facilitate the fragmentation of food/feed particles, which makes them more uniform and more accessible to gut symbionts and their enzymes. Conversely, in the soil the physical environment is extremely heterogeneous, with micro- and macro-compartments providing distinct moisture and nutrient profiles (Brabcová, Nováková, Davidová, & Baldrian, 2016). There is also a higher proportion of non-organic components such as minerals and ash, substances that can exert a profound effect on the composition and function of the bacterial community (Carson, Campbell, Rooney, Clipson, & Gleeson, 2009; Carson, Rooney, Gleeson, & Clipson, 2007; Totsche et al., 2010). The microbiota in the upper litter layers and the deeper soil layers can be quite different, mostly due to differences in nutrient availability (Lopez-Mondejar, Voriskova, Vetrovsky, & Baldrian, 2015; Tlaskal, Voriskova, & Baldrian, 2016; Voriskova & Baldrian, 2013). Furthermore, soil particles within even small sampling sites vary by size, organic and mineral content, water content, surface area, surface charge, and hydrophobicity. In addition, while conditions such as pH and temperature in the gut are maintained by homeostatic mechanisms, in soil environments these are highly variable. In the soil, seasonal and diurnal changes in temperature and moisture availability, as well as nutrient availability, can cause significant alterations in bacterial species composition (Fulthorpe et al., 2008; Voriskova, Brabcova, Cajthaml, & Baldrian, 2014). All of these factors contributing to the variability inherent in the soil ecosystem mean that Bacteroidetes in this habitat have to employ different strategies for survival—such as using different modes of glycan acquisition—than their better known cousins in the human gut.

2. The soil provides a rich diet of diverse complex carbohydrates

Carbohydrates in the soil derive from a wide range of sources, such as decaying biomass of plant, microbial, and animal origin, and they show an equally diverse array of chemical structures (Fig. 2). The vast majority of glycans come from plants growing in the soil. The aboveground litter that consists of leaves, stems, and other newly dead plant tissue represents a layer not yet converted into soil; this upper horizon of complex biomass is an

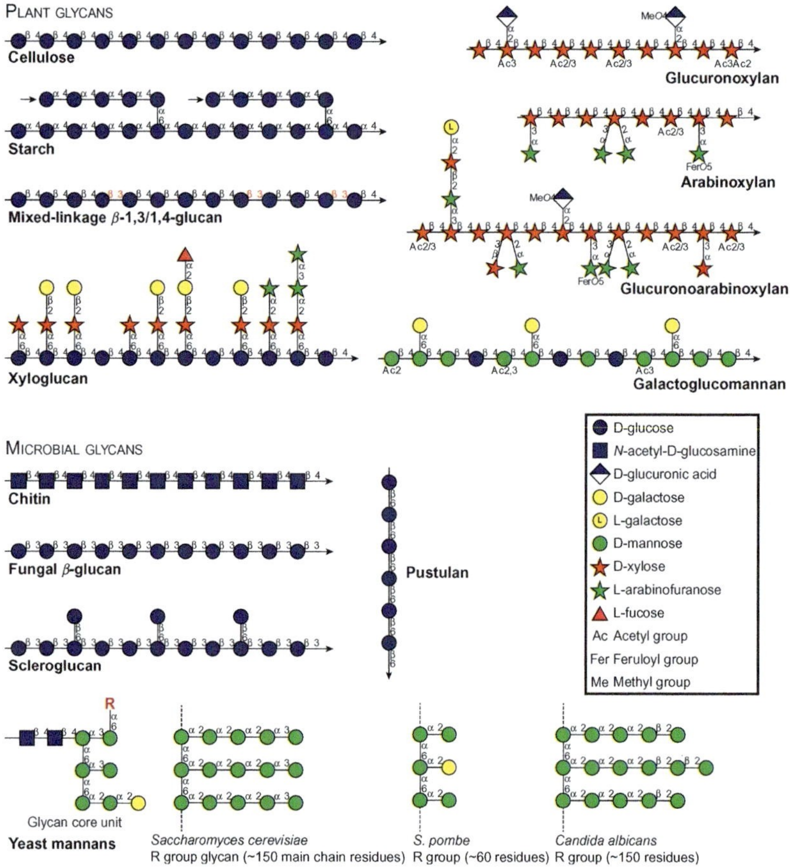

Fig. 2 Schematic representations of the major types of glycan (complex carbohydrate) found in plant and microbial cell walls, which contribute the bulk of carbohydrate biomass available for deconstruction by Bacteroidetes enzymes in the soil. These polysaccharides vary in length, from hundreds to hundreds of thousands of monosaccharides in length. Linkages (α or β), as well as monosaccharide types are indicated. Microbial glycans are highly diverse but less studied compared to their plant cell wall counterparts, and as such examples of the more structurally studied types are shown. The yeast mannans depicted represent some of the glycan moieties found in fungal cell wall mannoproteins.

important ecological niche. From this environment, many Bacteroidetes species with potentially key roles in carbohydrate turnover have been isolated, including *Chitinophaga pinensis*, the founding member of the abundant and widespread *Chitinophaga* genus (Sangkhobol & Skerman, 1981). Belowground, roots and root exudates contribute to the overall availability of glycans. Small soluble sugars and sugar alcohols leach from this plant material and can be directly metabolized by soil microbes in a short time-frame of hours to days. The bulk of plant biomass comprises cell wall material, which requires days, weeks, or even years to be totally deconstructed by the microbial community.

The plant cell wall is a complex inter-connected network of high molecular weight polysaccharides. The most abundant component, and the polymer that gives strength to the wall, is cellulose, a β-1,4-linked glucan that forms insoluble and crystalline microfibrils (Fig. 2). Despite being a homopolymer consisting only of glucose residues, this polysaccharide is highly recalcitrant to enzymatic attack, and full deconstruction into glucose requires multiple synergistic enzyme activities (Horn, Vaaje-Kolstad, Westereng, & Eijsink, 2012). In addition to cellulose, the major classes of cell wall polysaccharide are the various hemicelluloses and pectins. These are highly different from cellulose in terms of their sugar composition, linkage profiles, and degrees of substitution and crystallinity (Fig. 2). However, they share with cellulose the fact that multiple different enzyme activities are required for their complete saccharification to monosaccharides that can be metabolized for energy production and cellular growth.

Hemicelluloses are heteroglycans with β-1,4-linked backbones and varying degrees of substitution with monosaccharides, acetyl groups, ferulic acid groups, and others (Fig. 2) (Scheller & Ulvskov, 2010). Generally speaking, the most abundant hemicelluloses are xylan (which can be densely substituted with arabinose and/or glucuronic acid at specific linkages) (Busse-Wicher et al., 2016) and (Gluco)mannan (which is often heavily galactosylated) (Scheller & Ulvskov, 2010). The hemicelluloses can be thought of as cross-linkers in the cell wall, found surrounding and interconnecting the cellulose microfibrils (Martinez-Abad et al., 2017). Collectively, they represent a diverse source of monosaccharides available to the microbes that can deconstruct them, but the wide variety of sugars and sugar linkages (Fig. 2) mandates that large consortia of synergistic enzymes are used to achieve total deconstruction, either all produced by one organism, or by a collaborative microbial community.

Pectins are large, charged, and soluble polymers found in primary cell walls that again comprise a vast number of different sugars and different linkages (Mohnen, 2008; Ndeh et al., 2017). Most pectic polysaccharides have a high degree of branching, which gives them very high water solubility and also gel-forming properties. Glycans classified as pectins include homogalacturonan, rhamnogalacturonans I and II, arabinan, and (arabino) galactan, all high molecular weight and highly branched structures, found as a cross-linked network in the primary cell wall and middle lamella, surrounding the other carbohydrate components (Mohnen, 2008). The most complex pectic polysaccharide is rhamnogalacturonan II, but it has been shown that even this polymer, containing 21 distinct glycosidic linkages, can be almost completely deconstructed by a single Bacteroidetes species, the human gut symbiont *Bacteroides thetaiotaomicron* (Ndeh et al., 2017).

In addition to these carbohydrates, plant cell walls can contain large amounts of lignin. Lignin is a high molecular weight heteropolymer of phenolic groups connected by a range of different linkages, and its biosynthesis through radical polymerization makes its structure random (Ralph, Lapierre, & Boerjan, 2019). Deconstruction of lignin is mediated through the action of laccases, peroxidases, and other oxidoreductases, and while fungi have long been regarded as the main lignin decomposers in nature, recent discoveries have shown that bacteria also play a role in this process (Lee, Kang, Bae, Sohn, & Sung, 2019). Lignin is thought to be covalently connected to some of the carbohydrate components of plant cell walls (forming so-called lignin-carbohydrate complexes, or LCCs), further increasing the complexity of cell wall deconstruction (Giummarella, Balakshin, Koutaniemi, Kärkönen, & Lawoko, 2019; Martínez-Abad, Giummarella, Lawoko, & Vilaplana, 2018; Nishimura, Kamiya, Nagata, Katahira, & Watanabe, 2018; Takahashi & Koshijima, 1988; Yuan, Sun, Xu, & Sun, 2011).

As well as the plant material found in the soil, microbes themselves contribute to the overall glycan pool (Baldrian et al., 2013; Soudzilovskaia et al., 2015). Microbial necromass represents a rich source of cell wall carbohydrates, and is dominated by fungal cell walls that provide chitin (a polymer of β-1,4-linked *N*-acetylglucosamine), β-1,3- and β-1,6-linked glucans, and heavily mannosylated glycoproteins (Fig. 2) (Baldrian et al., 2013; Gow, Latge, & Munro, 2017; Mélida, Sandoval-Sierra, Diéguez-Uribeondo, & Bulone, 2013). Bacterial exopolysaccharides and fungal mycelial glycans play important roles in soil particle formation and adhesion of the leaf litter layer, and provide nutrition for multiple microorganisms (Costa, Raaijmakers, & Kuramae, 2018; Qurashi & Sabri, 2012).

The glycans available in the soil are not homogenously distributed throughout the whole environment. Certain glycans are bound within cell wall biomass, while other glycans may be adsorbed to soil particles. This leads to micro-habitats within the soil where different species can thrive, and where different enzyme activities dominate. For example, there are hotspots in forest soil where the decomposition of dead fungal mycelium is concentrated (Brabcová et al., 2016). In these micro-habitats, the Bacteroidetes phylum becomes increasingly dominant as polysaccharide deconstruction progresses, with the *Pedobacter* and *Chitinophaga* genera being especially abundant (Brabcová et al., 2016). Indeed, we have shown by integrated growth studies, biochemical, and proteomic investigations that the leaf litter-isolated *C. pinensis* is much more proficient at deconstructing fungal cell walls than plant cell walls (Larsbrink et al., 2017a, 2017b; McKee, Martinez-Abad, Ruthes, Vilaplana, & Brumer, 2019). The species is chitinolytic but not cellulolytic (Sangkhobol & Skerman, 1981), and its genome encodes several secreted chitinases and β-glucanases (Abe et al., 2017; Glavina Del Rio et al., 2010; Ramakrishna et al., 2018). The Bacteroidetes *Flavobacterium johnsoniae*, found in soil and freshwater, is likewise able to effectively metabolize chitin in a manner that intrinsically ties nutrient acquisition with gliding motility (McBride, Braun, & Brust, 2003), an intriguing feature of the phylum that will be discussed in further detail below.

3. Protein secretion mechanisms among the Bacteroidetes

We now know that there at least a dozen protein secretion systems functioning in the Bacterial kingdom as a whole (Costa et al., 2015; Green & Mecsas, 2016), with some being found throughout the kingdom, and others being restricted only to certain taxa. Several of these pathways are exploited by members of the Bacteroidetes for highly specialized functions. For example, there is evidence indicating that Bacteroidetes in the gut and soil ecosystems use a sub-type of the Type VI Secretion System (T6SS) to mediate antagonism between bacteria to reduce local competition for nutrients (Russell et al., 2014; Wexler & Goodman, 2017). This is achieved by transferring antibacterial proteins such as peptidoglycan hydrolases, phospholipases, and nucleases into the periplasm of competitor cells (Schwarz, Hood, & Mougous, 2010).

Of greater relevance to our discussion of glycan acquisition is the intricate Type IX Secretion System (T9SS), which is found exclusively in the Bacteroidetes phylum (McBride, 2019; McBride & Zhu, 2013). Extensive genetic studies of select Bacteroidetes species have enabled the elucidation of the components and mode of action of this secretion system, and its tight connection with gliding motility. The T9SS is critical for secretion of carbohydrate-active enzymes (CAZymes), gliding motility, and even pathogenesis in some species such as *Porphyromonas gingivalis* (de Diego et al., 2016; Lasica, Ksiazek, Madej, & Potempa, 2017). The system is found in most Bacteroidetes genera except the *Bacteroides* (Kulkarni, Zhu, Brendel, & McBride, 2017; McBride & Zhu, 2013; Wilson, Anderson, & Bernstein, 2015). Several excellent articles and reviews are available that describe the discovery, structure, and function of the T9SS (Johnston, Shrivastava, & McBride, 2018; Lasica et al., 2017; McBride, 2019; McBride & Nakane, 2015; McBride & Zhu, 2013), but we will summarize the details of the system here.

The T9SS is a noteworthy bacterial secretion pathway for a number of reasons. As well as an N-terminal signal peptide for targeting to the standard Sec system for export over the cell membrane (Tsirigotaki, De Geyter, Šoštaric´, Economou, & Karamanou, 2016), proteins secreted via the T9SS contain an additional C-terminal domain (CTD) that is cleaved after secretion (Veith et al., 2002). It is now known that there are two types of CTD, type A (TIGRFAM family annotation TIGR04183) and type B (TIGRFAM family annotation TIGR04131) (de Diego et al., 2016; Kulkarni, Johnston, Zhu, Hying, & McBride, 2019; Kulkarni et al., 2017; Lasica et al., 2016), and these may yet be revealed to process a different subset of protein substrates. The CTD was first identified by genetic experiments on the human pathogen *P. gingivalis* and was thought to tether some secreted proteins to the outer cell surface (Nelson, Glocka, Agarwal, Grimm, & McBride, 2007; Nguyen, Travis, & Potempa, 2007; Sato et al., 2005; Seers et al., 2006). Further knock-out experiments, microbiological observation, and genome comparisons revealed that orthologous genes were vital for protein secretion in *P. gingivalis* and for gliding motility in *F. johnsoniae* (Kharade & McBride, 2015; Kulkarni et al., 2019; McBride et al., 2003; Nelson et al., 2007; Sato et al., 2010; Shrivastava, Johnston, van Baaren, & McBride, 2013), suggesting that the T9SS is intimately involved in both of these apparently discrete functions.

The T9SSs of *P. gingivalis* and *F. johnsoniae* are the most well studied, and each consists of more than a dozen different protein components (Fig. 3) (McBride, 2019). The system is energized by the proton motive force via proteins spanning the inner membrane and the periplasm (GldL and GldM in *F. johnsoniae*) (Shrivastava et al., 2013; Vincent et al., 2017). Other T9SS components form a protein complex at the outer membrane, to which CTD-tagged proteins are directed after being transported over the inner membrane into the periplasm via Sec-mediated signal peptides. Crucial for secretion is a very large pore able to transport folded proteins

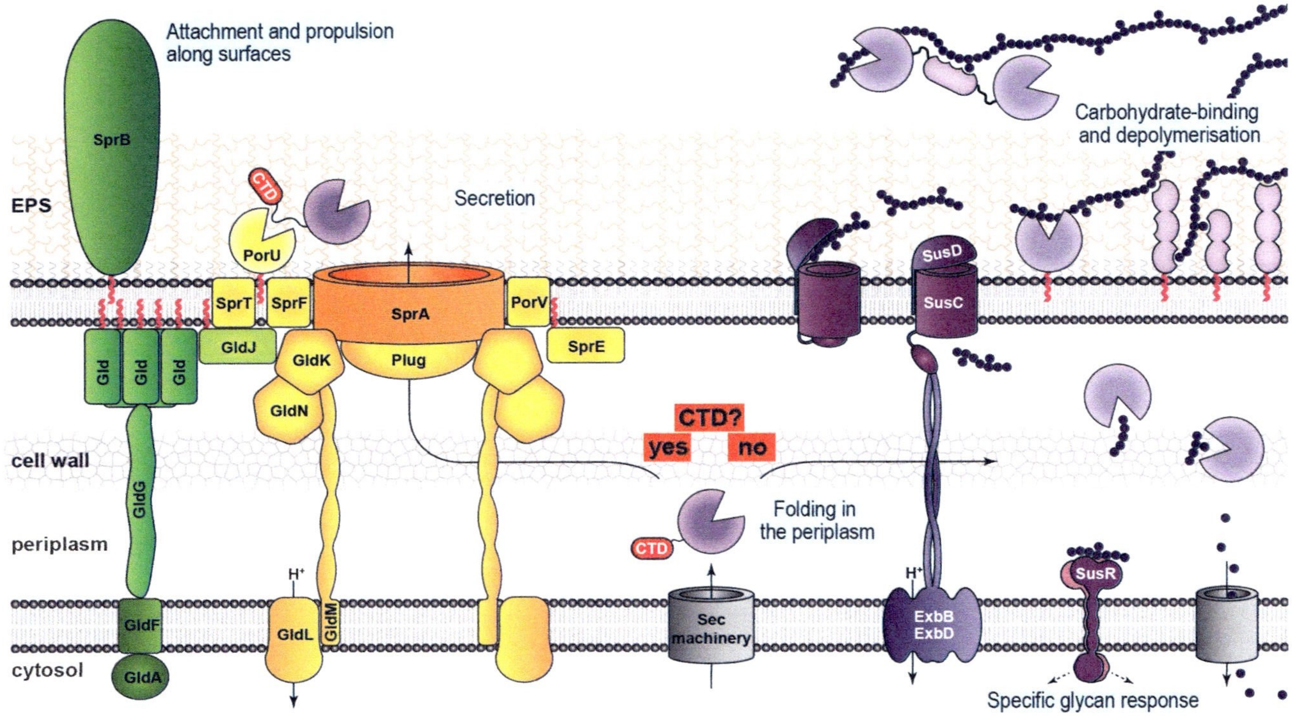

Fig. 3 Soil Bacteroidetes use a variety of systems to maximize the efficiency of CAZyme use and production. Polysaccharide Utilization Loci (purple) can restrict CAZyme production to times when a substrate is abundantly available, and co-produce CAZymes, sugar-binding proteins, and sugar importers to maximize the amount of metabolizable sugar taken in after the energetically expensive process of enzyme secretion. The Type IX Secretion System connects the active hunt for glycan nutrition in a difficult environment with the ability to glide rapidly over solid surfaces: it produces cellular machinery both for the secretion of proteins (yellow) and for attachment and gliding propulsion along solid surfaces (green). EPS—extracellular polymeric substance; pink wavy lines on proteins symbolize lipid anchors for lipoproteins.

(SprA in *F. johnsoniae*, ~70 Å in diameter) which is coupled to a Plug domain, most likely to prevent unspecific leakage through the membrane (Lauber, Deme, Lea, & Berks, 2018). After secretion, a peptidase (PorU in *F. johnsoniae*) (Glew et al., 2012; McBride, 2019) cleaves the CTD from the protein, which may then be released into the extracellular environment or become tethered to the outer membrane.

In addition to protein secretion, the T9SS is vital for the gliding motility of *F. johnsoniae* and *Cytophaga hutchinsonii* and likely other Bacteroidetes species (McBride, 2019; Zhu & McBride, 2014). The gliding motility exhibited by members of this phylum is driven by surface-attached adhesins that move rapidly over the cell surface to propel a cell at speeds of up to 2 μm per second (Lapidus & Berg, 1982; McBride, 2019). It is now well understood that the lectins and adhesins required for gliding motility by *Flavobacteria* and *Cytophaga* species are secreted via the T9SS (Shrivastava & Berg, 2015; Shrivastava et al., 2013; Shrivastava, Rhodes, Pochiraju, Nakane, & McBride, 2012), and that deletion of the T9SS causes defects in both gliding motility and the surface localization of key proteins, including important CAZymes (Burchard & Bloodgood, 1990; Rhodes et al., 2010).

Indeed, genetic experiments and physiological observations found that disruption of the T9SS leads to critical defects in the metabolism of certain glycans by Bacteroidetes. For instance, deletion of the SprP component in *Cytophaga hutchinsonii* crippled its cellulose utilization capabilities (Zhu & McBride, 2014). Several studies have since shown that the T9SS is vital for the secretion of specific CAZymes enabling polysaccharide metabolism by *F. johnsoniae* (Kharade & McBride, 2014; Larsbrink et al., 2016), *C. hutchninsonii*, and other soil-dwelling species (McKee et al., 2019; Taillefer, Arntzen, Henrissat, Pope, & Larsbrink, 2018; Zhu & McBride, 2014, 2017). Genome sequencing of the provisionally named *Candidatus Paraporphyromonas polyenzymogenes* has recently indicated that multi-modular cellulases used for biomass degradation are secreted via the T9SS (Naas et al., 2018).

As discussed earlier, the glycans found in the soil can be found in clumps of decaying biomass comprising bulky insoluble polymers, or as more soluble polymers, often in a dry or semi-dry state. As such, microbial scavenging for nutrients in the soil requires a great deal more motility than in the gut, where a Bacteroidetes cell can tether itself to the mucosal layer and catch nutritious glycans as they flow past. In the soil, cells are required to move to "hunt" for the glycans they need, which is likely the reason that gliding motility is so common among soil-dwelling species (Nan, 2017;

Nan, McBride, Chen, Zusman, & Oster, 2014). Perhaps the need to move through the soil to hunt for nutrition has driven the development of specific mechanisms among the Bacteroidetes for gliding over solid surfaces that are intrinsically tied to pathways for the secretion of CAZymes (McBride, 2019). This connection between motility and polysaccharide digestion is best characterized for *F. johnsoniae*, which can essentially be considered a model organism for the study of many physiological features common to the Bacteroidetes (McBride et al., 2009; Shrivastava & Berg, 2015). Multiple observations of how soil Bacteroidetes encode a large number of T9SS-secreted CAZymes suggest that similar tight connections between motility and metabolism of complex carbohydrates are common throughout the phylum.

4. Bacterial mechanisms of carbohydrate degradation in the soil

The Bacteroidetes are primary degraders of complex carbohydrate-based biomass and the genus is found ubiquitously in all ecosystems investigated to date, being particularly dominant in soils and in human and animal guts. Most sequenced Bacteroidetes genomes encode a large number of CAZymes (Table 1) (Lapébie, Lombard, Drula, Terrapon, & Henrissat, 2019), suggesting a strong metabolic focus on glycan scavenging (Thomas et al., 2011). CAZymes are grouped into classes and families in the carbohydrate-active enzymes database (http://www.cazy.org/; Lombard, Golaconda Ramulu, Drula, Coutinho, & Henrissat, 2014), where degradative enzymes are found in the glycoside hydrolase (GH), polysaccharide lyase (PL), carbohydrate esterase (CE), and auxiliary activities (AA) classes.

According to current literature on the topic, Bacteroidetes bacteria take an almost exclusively hydrolytic approach to biomass deconstruction. So far, and to the best of our knowledge, no examples of the relatively novel enzyme type called Lytic Polysaccharide Monooxygenase (LPMO) have been characterized that derive from a Bacteroidetes genome, although a few species do appear to encode LPMOs (Agostoni, Hangasky, & Marletta, 2017; Lombard et al., 2014). This suggests that the oxidative cleavage of polysaccharides is possible but rare in the phylum, which instead has a hydrolytic focus on glycan deconstruction (Bai, Eijsink, Kielak, van Veen, & de Boer, 2016). LPMOs are generally active on the crystalline substrates cellulose and chitin, although there are now many examples in the literature of LPMOs acting on (semi)soluble poly- and oligosaccharides

Table 1 The number of degradative CAZymes encoded by select Bacteroidetes species.

Species	Environment	Glycoside hydrolases	Carbohydrate esterases	Polysaccharide lyases	Predicted PULs
Bacteroides ovatus	Gut	352	30	35	112[a]
Bacteroides thetaiotaomicron	Gut	289	21	21	88[a]
Chitinophaga pinensis	Soil	201	12	11	32[b]
Flavobacterium johnsoniae	Soil	160	14	12	42[c]
Sporocytophaga myxococcoides	Soil	71	13	6	0[d]
Cytophaga hutchinsonii	Soil	52	14	3	0[e]

[a]Martens et al. (2011).
[b]This article (own observations of the published genome; Glavina Del Rio et al., 2010).
[c]McBride et al. (2009).
[d]Liu, Gao, Chen, and Wang (2014).
[e]Zhu et al. (2015).
The tally of GH, CE, and PL enzymes is taken from the CAZy database (http://www.cazy.org/). The number of predicted PULs for each organism is taken from the references indicated in the table.

(Agger et al., 2014; Couturier et al., 2018; Hüttner et al., 2019; Vu, Beeson, Span, Farquhar, & Marletta, 2014). Similar to LPMOs, the whole enzyme class of laccases, which can in certain circumstances deconstruct lignin, seems to be under-represented in Bacteroidetes genomes compared to other phyla (Janusz et al., 2017). Lignin deconstruction by Bacteroidetes bacteria appears to be very rare, although it has been demonstrated that certain *Sphingobacterium* species from environmental soil samples produce manganese superoxide dismutase enzymes identified by colorimetric assays—these enzymes can in some cases contribute to radical depolymerization of lignin (Ahmad et al., 2010; Rashid et al., 2015; Taylor et al., 2012; Wang, Lin, Du, Liang, & Ning, 2013).

There is extensive literature on the "typical" modes of microbial deconstruction of recalcitrant cell wall glycans such as cellulose (Artzi, Bayer, & Moraïs, 2016; Horn et al., 2012). Enzymatic cellulose deconstruction is generally believed to require LPMOs, *endo*-glucanases, and *exo*-acting cellobiohydrolases (CBHs) for efficient depolymerization, to introduce breaks in crystalline/amorphous regions and depolymerize individual chains from either the reducing- or non-reducing ends (Horn et al., 2012), followed by the action of β-glucosidases that cleave oligosaccharides into glucose. Some microorganisms freely secrete these enzymes to penetrate the plant cell wall material, while other species create cell surface-attached multi-enzyme complexes (Artzi et al., 2016; Horn et al., 2012). Ongoing investigations of Bacteroidetes enzymology reveal that this phylum may not always conform to the known enzymatic degradation models (Naas et al., 2014; Wang, Han, Chen, Zhang, & Liu, 2017; Zhu et al., 2015; Zhu & McBride, 2017).

As of yet there is still only a limited number of studies describing glycan conversion strategies used by the soil Bacteroidetes, but certain polysaccharide-degrading species have long been recognized for being abundant in soils and easy to isolate. Two species that have always been regarded as efficient metabolizers of cellulose are *Cytophaga hutchinsonii* and *Sporocytophaga myxococcoides* (Stanier, 1942). Both species secrete hemicellulose-degrading enzymes during growth, but are unable to grow on hemicelluloses as their sole carbon source (Sorensen, 1956; Taillefer et al., 2018). Their cellulolytic degradation is still not understood in any great detail despite a large number of enzymatic and genetic studies (Taillefer et al., 2018; Wang, Zhang, Zhou, Chen, & Liu, 2019; Zhu & McBride, 2017). In *C. hutchinsonii* and *S. myxococcoides*, neither LPMOs nor CBHs are encoded in the genomes, and exactly what mechanisms permit efficient cellulose conversion by these species remains an open question, although the enzymes and proteins they produce during growth have recently been mapped using proteomics on different cellular fractions (secreted soluble, outer membrane-bound, periplasmic, inner membrane-bound, total proteome) (Taillefer et al., 2018). Both species appear to rely on a highly redundant system of putative *endo*-glucanases and β-glucosidases, some of which have now been biochemically characterized, but the key determinants behind the exceptional ability of these species to metabolize cellulose remain enigmatic (Taillefer et al., 2018). Both species are known to require contact with their (solid) substrate for efficient growth, but the study illustrated that a large number of CAZymes were still freely secreted from the cells into growth medium. An intriguing aspect of these species is that during growth they secrete not only proteins but also a viscous slime layer that has not yet been structurally studied. *C. pinensis* likewise secretes a viscous yellow outer layer, which is difficult to remove and has stymied some of our own past research efforts. It could be speculated that such types of extracellular matrices have a role in sequestering secreted enzymes in close proximity to the enzyme-producing cell, as well as retaining the nutrients (oligosaccharides) these enzymes release during polysaccharide hydrolysis. This extracellular "slime" layer may even contribute to the structure and integrity of soil particles and as such help to shape the physical environment of these soil bacteria (Costa et al., 2018).

Another apparently unique degradative system found throughout the Bacteroidetes phylum are the so-called Polysaccharide Utilization Loci (PULs) that allow targeting of discrete carbohydrate types and present a sophisticated balance between enzyme production and carbohydrate capture. The features of these systems are described in detail in the following section, where we will compare what is currently known about PULs encoded by gut and soil species.

5. The high energy cost of enzyme secretion: Strategies for maximizing return on investment

The soil environment presents its native microbiota with a rich diet of carbohydrate- and protein-based biomass, much like in the human and animal guts. As such, and just like in those gut environments, there is a high level of competition for access to nutrition. To be successful in such a competitive environment, it can be advantageous for a species to produce a large amount of highly active enzymes to out-compete other species in the hunt for nutrients. This is a strategy, for example, favored by wood-degrading fungi, which produce extraordinary amounts of highly active and freely soluble cellulolytic enzymes, to obtain glucose for growth (Payne et al., 2015). Alternatively, a more balanced strategy is to only produce the right enzymes at the right time. This is the approach favored by the Bacteroidetes, using genetic systems to tie CAZyme production with substrate availability, CAZyme tethering to the cell surface, and the co-production of specific sugar transport proteins.

5.1 Glycan sensing and induced CAZyme secretion by Bacteroidetes bacteria

The efficient PULs that the Bacteroidetes have evolved are discrete loci in the genome that encode all necessary functions for degradation and import of specific carbohydrates: carbohydrate capture and transport over the membrane(s), degradative enzymes, and (hybrid) two-component carbohydrate sensing-gene regulation systems (Fig. 3). Generally speaking, each PUL is dedicated to the deconstruction and uptake of one particular polysaccharide (Lapébie et al., 2019). The first studied PUL was the Starch Utilization System (SUS) from the human gut symbiont *B. thetaiotaomicron* (Cho & Salyers, 2001; Foley, Cockburn, & Koropatkin, 2016; Reeves, Elia, Frias, & Salyers, 1996; Shipman, Cho, Siegel, & Salyers, 1999), and indeed the majority of PULs characterized to date derive from human or animal gut species. The archetypal SUS encodes surface-bound starch-binding proteins (SusD/E/F), an outer membrane transporter (SusC; TonB-dependent transporter), a surface-attached amylase (SusG), two periplasmic oligosaccharide-degrading enzymes (SusA/B), and an inner membrane-bound hybrid two-component system for sensing of starch-derived oligosaccharides and regulating the expression of the other Sus-encoded genes (SusR) (Fig. 3). The basic functional composition of the SUS is now known to be common to all PULs, and it is extremely highly conserved throughout the phylum. Certain species, such as the gut symbiont *Bacteroides ovatus*, encode over 100 different PULs (Table 1), which clearly reflects the importance of complex carbohydrates as nutrients for these bacteria (Bolam & Sonnenburg, 2011; Martens et al., 2011; Xu et al., 2003).

PULs invariably contain genes encoding SusC- and D-like proteins, used to recognize and import specific glycans and named for their similarity with their functional equivalents in the original SUS. This feature has been used to develop a PUL prediction algorithm and database, PULDB

(Terrapon et al., 2018; Terrapon, Lombard, Gilbert, & Henrissat, 2015). With some notable exceptions among the soil Bacteroidetes (Zhu et al., 2015), the SusC/D pair of outer membrane proteins are indispensable for glycan capture and acquisition (Martens, Koropatkin, Smith, & Gordon, 2009), and essentially form a SusC pore capped with the sugar-binding SusD (Glenwright et al., 2017). Since PULs confer highly beneficial traits, they are often shared between species through horizontal gene transfer (Hehemann et al., 2010; Hehemann, Kelly, Pudlo, Martens, & Boraston, 2012; Martens, Kelly, Tauzin, & Brumer, 2014; Martens et al., 2011; Thomas et al., 2011). Indeed, PULs can be hypothesized to be the major adaptation in nutrient acquisition that gives the Bacteroidetes a strong competitive advantage in biomass-degrading environments (Lapébie et al., 2019). The PUL system allows energy conservation by restricting high-level CAZyme secretion to times when those CAZymes are needed, and by using cell-tethered enzymes and carbohydrate binding proteins, ensuring that reaction products are kept close to the cell for maximum metabolic benefit. In this regard, PULs can be compared to the well-studied bacterial cellulosomes used by species from the *Clostridium* genus (Artzi et al., 2016). Cellulosomes are multi-enzyme complexes bound to the cell wall of Gram-positive bacteria, and some have recently been identified in anaerobic fungi (Haitjema et al., 2017). They are composed of scaffoldin proteins to which multiple CAZyme modules can attach through protein domains called cohesins and dockerins. The co-located enzyme activities act synergistically on complex biomass, and their physical attachment to the cell surface minimizes the "leakage" of released sugars to neighboring cells.

In recent years, PULs targeting a variety of glycans from different species have been studied, primarily in gut species. One ecologically interesting example is from *B. thetaiotaomicron* where, in addition to the SUS, a PUL targeting yeast mannan has been found that uses slow *endo*-acting enzymes on the surface of the cells, cleaving the glycans into large fragments that are then imported into the periplasm where rapid final degradation to monosaccharides takes place (Cuskin et al., 2015). The most complex PULs characterized to date are the rhamnogalacturonan II-degrading loci from *B. thetaiotaomicron*, which encompass over 20 different enzymes, reflecting the complexity of this highly branched type of pectin (Ndeh et al., 2017), and showing the impressively high degradative potential of the PUL system. From the close relative *B. ovatus*, PULs targeting the abundant hemicelluloses xyloglucan, xylan, and galactomannan have been characterized in detail using combined biochemical, microbiological, and crystallographic approaches (Bågenholm et al., 2017; Larsbrink et al., 2014; Rogowski et al., 2015; Tauzin et al., 2016). In addition, a probable cellulose-targeting metagenome-derived PUL from a rumen microbiota has been studied (Naas et al., 2014), highlighting that these systems have evolved to target all major plant cell wall polysaccharides, including the crystalline components that are typically considered to be recalcitrant to enzymatic deconstruction (Grondin, Tamura, Déjean, Abbott, & Brumer, 2017; Lapébie et al., 2019). Apart from yeast glycans, other microbial polysaccharides are also targeted by PUL-encoded systems, exemplified by the chitin utilization locus (ChiUL) from *F. johnsoniae* (Larsbrink et al., 2016), and the *N*-glycan-targeting Don locus of *Bacteroides fragilis* (Cao, Rocha, & Smith, 2014). PULs have also been identified that are specific for unusual marine glycans and algal polysaccharides such as porphyrin, alginate and laminarin (Hehemann et al., 2010; Kabisch et al., 2014). Of these studies, the majority have focused on gut microorganisms, though a vast number of PULs have also been identified in soil bacterial genomes, such as from the orders Flavobacteriia, Cytophagales, and Chitinophagia (Terrapon et al., 2018).

Although only a limited number of PULs have been studied to date, recent work highlights a potential difference in how the soil and gut Bacteroidetes make use of the PUL system. It is well established that a common strategy among gut bacterial PULs is to tether the key polysaccharide backbone-cleaving *endo*-type activities to the surface of the cell using outer membrane proteins (Cho & Salyers, 2001). The use of outer membrane-tethered enzymes and carbohydrate binding proteins (Cameron et al., 2012; Koropatkin, Martens, Gordon, & Smith, 2008; Shipman, Berleman, & Salyers, 2000; Tuson, Foley, Koropatkin, & Biteen, 2018) as well as periplasmic CAZymes for oligosaccharide deconstruction into monosaccharides (Anderson & Salyers, 1989) has long been proposed to offer gut Bacteroidetes an advantage over their competitors by taking higher molecular weight glycans out of the reach of other species before they are readily metabolizable (Cameron et al., 2014; Cuskin et al., 2015; Foley et al., 2016; Lapébie et al., 2019). This strategy is sometimes referred to as "selfish," as competing organisms are unable to benefit from the monosaccharide-liberating enzymes that are confined to the periplasmic space. This so-called selfishness is well illustrated by the human gut symbiont *B. thetaiotaomicron*, which as described above tethers slow acting enzymes to the cell surface, allowing relatively high molecular weight oligosaccharides to be brought into the periplasm early in the substrate degradation pathway, thereby hoarding the released sugars away from other competing species (Cuskin et al., 2015).

In contrast, the ChiUL from *F. johnsoniae* is dependent on the freely secreted soluble chitinase ChiA, which consists of two catalytic domains separated by a chitin-binding domain (Kharade & McBride, 2014; Larsbrink et al., 2016). When *F. johnsoniae* is grown in the lab, ChiA can be found only in cultivation medium. The presence of homologous large multidomain chitinases in syntenic PULs in related genomes is correlated with the ability of the encoding organisms to metabolize crystalline chitin, and as such these enzymes are believed to be an indispensable requirement for conversion of this recalcitrant polysaccharide. Several PUL-encoded enzymes were likewise detected in the secretomes of *C. pinensis* when grown on plant polysaccharides (Larsbrink et al., 2017a, 2017b), which suggests that the soil environment may favor PULs that rely on secreted enzymes, rather than cell-attached variants. More soil-derived PULs must of course be biochemically characterized in detail before such conclusions can be firmly drawn about the whole phylum.

A further departure from the nutrient acquisition approaches favored by gut Bacteroidetes may be a reduced reliance on the PUL system altogether. While *F. johnsoniae* encodes over 40 predicted PULs (McBride et al., 2009), other soil bacteria from the phylum encode far fewer and appear to use other mechanisms (Table 1). Intriguingly, while *C. pinensis* encodes over one hundred SusC/D pairs in its genome, only about a third of these appear to be located in PULs (Glavina Del Rio et al., 2010; McKee et al., 2019). The role(s) of the remaining isolated SusC/D pairs that lie scattered in the genome are currently unknown, but may be involved in capture and sensing of nutrients that are abundant in the soil, for instance from cross-feeding on mono- or oligosaccharides released by other organisms or by *C. pinensis* enzymes not found in PULs. Because of the relative lack of overall PUL organization, most CAZyme-encoding genes in the *C. pinensis* genome are found outside of such loci, and there is still no conclusive knowledge of when/whether these are expressed, and whether their expression is in any way tied to substrate sensing by the bacterium (McKee et al., 2019).

In an even more extreme case, the cellulose-metabolizing *C. hutchinsonii* encodes only two SusC/D pairs, and these genes were, upon disruption, found not to be involved in cellulose degradation (Zhu et al., 2015). Likewise, the provisionally named ruminant symbiont *Ca. P. polyenzymogenes* does not seem to encode any PULs at all, apparently relying instead on multi-modular CAZymes secreted via the T9SS, particularly cellulases that show activity on crystalline cellulose, amorphous cellulose, and mixed linkage β-glucans (Naas et al., 2018). Our own ongoing analysis of the genome and biochemistry of *C. pinensis* indicate a similarly high abundance of multi-modular CAZymes, predicted to be secreted through the T9SS and generally not found in PULs in the genome (Larsbrink et al., 2017a; McKee et al., 2019). Several of the *C. pinensis* enzymes we have identified seem to be secreted more or less constitutively, regardless of the carbon source provided (Larsbrink et al., 2017a, 2017b).

5.2 Community cross-feeding, or, the importance of reaping what you sow

To metabolize any complex glycan in the soil, the energy used in producing and secreting enzymes—often quite large multidomain proteins with catalytic and substrate-binding domains—across the outer membrane is extremely high, and it is vital that sufficient compensatory energy be garnered from the saccharides released by glycan deconstruction. Sometimes this demands that species work together for mutual benefit, and sometimes it causes one species to hoard resources.

The heterogeneous composition of the soil microbial community gives rise to limitless possible interactions between organisms. Cross-feeding between microorganisms is believed to be very common, and several scenarios for how species benefit each other have been proposed (D'Souza et al., 2018). It seems to be especially important in biomass recycling environments such as the litter layer of agricultural soils (Kramer et al., 2016). Species may simply take advantage of by-products from other species' metabolism (unidirectional relationship), benefit each other via mutual production and consumption of different metabolites (bidirectional relationship), or even require the presence of another species to thrive in an environment (obligate bidirectional/co-culture). Some species, like fungi, are known to be "leaky" in the sense that they freely secrete a multitude of CAZymes to depolymerize cell walls (or other macromolecules), where the resulting sugars and oligosaccharides can be exploited by other cells. In other cases, cells actively cooperate, such in the case of gut *Bifidobacteria* and *Eubacteria* growing on xylan, where one species metabolizes arabinose and the other xylose and their respective metabolic end products benefit the other organism (Riviere, Gagnon, Weckx, Roy, & De Vuyst, 2015), or *B. ovatus* which has been shown to secrete enzymes to degrade fructan polysaccharides into sugars it cannot metabolize itself (Rakoff-Nahoum, Foster, & Comstock, 2016). With the great complexity of sugars available in the soil, collaborative systems are likely to be very common, but they are difficult to study directly. Indirect indications of cooperativity and stable consortia can be gleaned from both co-cultivation and metagenomic studies (D'Onofrio et al., 2010; Zegeye et al., 2019).

Researchers have modeled competition and "cheating" behaviors in densely populated habitats such as the soil (Allison, 2005), where cheating organisms benefit from the products of polysaccharide degradation without contributing protein to the enzyme-catalyzed process (Velicer, 2003). Enzymes that are secreted freely into the environment are most likely to be exploited by cheating organisms, as the products of their reactions are free to diffuse away from the close proximity of the enzyme-producing cell (Allison, 2005). Such phenomena have certainly been observed among the human gut Bacteroidetes, members of which can display a more "selfish" or a more "sharing" mode of carbohydrate digestion, depending on the enzyme pathway utilized (Cuskin et al., 2015).

Most of the polysaccharides found in the soil are complex structures consisting of multiple types of monosaccharide connected through multiple different glycosidic linkages (Fig. 2). As such, an equally large number of enzyme activities are required for complete deconstruction of these carbohydrates. In many cases, single organisms possess all the enzymes needed to deconstruct a particular complex polysaccharide, while in other cases they are only able to target certain linkages. The PULs provide a system for Bacteroidetes to conserve energy by restricting high-level CAZyme production to times when the target substrate of a PUL is abundantly available, while also increasing the chances to collect all released sugars by co-producing specific sugar import systems (SusC/D pairs). Possibly, the typical mode of action of PULs, using surface-tethered enzymes to maximize the import of released sugars, is not as common among soil Bacteroidetes, as several observations indicate cross-feeding or unidirectional "leaky" enzyme systems for several species. As discussed above, both *F. johnsoniae* and *C. pinensis* freely secrete PUL-encoded soluble enzymes to deconstruct various glycans. The cellulolytic *C. hutchinsonii* and *S. myxococcoides* also secrete soluble enzymes during growth. Such cases may reflect a specialization where species focus on a select number of carbohydrates, but still need to remove other glycans to access their primary targets. In addition to cellulases, *C. hutchinsonii* and *S. myxococcoides* also secrete significant amounts of hemicellulose-degrading enzymes during growth, although neither species is able to metabolize all products resulting from this process, such as the xylose that is released from xylan depolymerization (Larsbrink et al., 2016; Sorensen, 1956; Taillefer et al., 2018). It is likely that their hemicellulolytic activities merely reflect a need to remove these polysaccharides from the underlying cellulose fibers. Released oligosaccharides that are not metabolized will instead constitute nutrients for neighboring species that can thereby take advantage of the activities of *C. hutchinsonii* and *S. myxococcoides*. In a similar case we have studied, *C. pinensis* has an extensive suite of enzymes for deconstructing β-glucans of various linkages. However, despite having the enzymatic potential to metabolize all types of β-glucan, we have shown that it only grows well on β-1,6-glucans and, to a lesser extent, β-1,3-glucans (McKee et al., 2019). This appears to be due to an inability to take up the products of enzymatic digestion of most β-glucans. The result of this failure of uptake is an accumulation of large amounts of oligosaccharides produced from β-1,4- and β-1,3-linked glucans (McKee et al., 2019). We have speculated that these observations may reflect a case of cross-feeding in the larger microbial community, where different species secrete complementary enzymes that act synergistically to produce simple metabolizable sugars from polymeric glycans. Further studies of co-cultured organisms or broad -omics studies are, however, required to better substantiate or refute these hypotheses.

The requirement for a high return on energy investment after synthesizing and secreting a CAZyme also requires that enzymes acting in the extracellular environment remain active for as long as possible, to maximize the amount of metabolizable sugar that can be collected by the cell. Due to the physical nature and limitations of protein molecules, each enzyme has an innate optimal pH and temperature where maximum activity can be achieved (sometimes depending on the presence of specific co-factors or metal ions). These optimal working conditions can be established quite easily in the laboratory, once an enzyme has been produced recombinantly. For most assays of glycoside hydrolase activity, it is relatively straight-forward to perform reactions at a range of temperatures and pH values. Such studies often reveal that a very narrow range of physical conditions is required for efficient turnover of an enzyme's substrate. This implies that there is a high likelihood of enzymes secreted into the soil showing low activity or losing their activity quite quickly. Analogous to how enzyme immobilization is exploited at industrial scale to increase the stability and reusability of these catalysts, adsorption to a cellular surface may limit exposure to otherwise denaturing conditions and extend the useful lifetime of the enzyme (Mohamad, Marzuki, Buang, Huyop, & Wahab, 2015). The tendency of the Bacteroidetes to tether many of their CAZymes to the extracellular cell surface, often achieved via the T9SS, may be a way to confine the enzyme activities to the cell surface, while also ensuring an optimal physicochemical environment for the enzymes themselves. Likewise, in soil, it is possible that the often observed extracellular matrices, or "slime" that species secrete, provide the proper environment not only for sequestration of solubilized sugars, but also for the prolongation of enzyme action.

6. Future perspectives

The abundance and diversity of carbohydrate-active enzymes produced by Bacteroidetes in the soil is almost certainly on at least the same order as for their gut symbiont cousins. But the soil environment remains underexplored as a source of novel enzymes for biotechnology. It is clear from the above discussion of characterized PULs that there has been a greater research effort on PUL discovery from gut symbiont Bacteroidetes than from their environmental cousins. Indeed, the methods of enzyme discovery from PUL sequences have been developed and to an extent formalized from study of human gut symbionts, principally *B. thetaiotaomicron* and *B. ovatus* (Luis & Martens, 2018), and is now quite routine in many laboratories around the world. The identification of CAZyme-encoding genes lying alongside homologs for *susC* and *susD* has proven a successful strategy for finding new combinations of enzymes that synergistically deconstruct a particular glycan. Many studies of this kind have uncovered new enzyme activities, new synergistic pathways, and even new enzyme families (Ndeh et al., 2017; Razeq et al., 2018). An indication of the likely substrate of a PUL can first be gleaned from the annotations of PUL-encoded genes. Following this, growth experiments and transcriptomic data can show which glycan is capable of inducing up-regulation of gene expression for a PUL identified from a genome sequence (Martens et al., 2011). The assumption of target substrate can be further strengthened by studying the carbohydrate binding of the SusD protein, produced recombinantly and analyzed by isothermal titration calorimetry and gel-based techniques such as affinity chromatography or pull-down assay (Larsbrink et al., 2014, 2016; Tauzin et al., 2016). These are techniques that can be applied to soil-derived CAZymes as well as to gut-derived CAZymes. Novel CAZymes can also be identified from bacterial secretomes by combining proteomic and biochemical techniques, ensuring that we do not miss any potentially exciting non-PUL-encoded activities (Larsbrink et al., 2017a; McKee & Brumer, 2015; Taillefer et al., 2018).

The functional capacity of the Bacteroidetes has only begun to be explored, and we hope that the near future of the field will see a greater focus on soil-derived PULs and enzymes. Nonetheless, we have described several important discoveries of enzymes and enzyme pathways from soil genera such as *Cytophaga*, *Flavobacteria*, and *Chitinophaga*. As described, in several of these organisms the main carbohydrate degradation pathways are not PUL-based, but apparently follow other schemes. This review highlights observations from recent studies that show how different the soil Bacteroidetes appear to be from their close relatives that dominate the gut environment. We hope that this summary will give crucial pointers for future research on bacteria from this fascinating phylum living in extremely diverse non-gut environments. Future studies may very well highlight further contrasts with the more studied gut-residing species and uncover more novel modes of carbohydrate deconstruction.

Beyond biotechnological tools that have been discovered in enzyme characterization studies, research in this area is also providing important insight into the microbial ecology of the soil environment. A healthy microbial ecosystem promotes soil particle formation, ensures soil fertility, and can even suppress disease in crop plants grown in that soil (Cretoiu, Korthals, Visser, & van Elsas, 2013). As we learn more about the enzymology of the Bacteroidetes phylum, and especially how enzyme production is induced in the soil, we may be able to devise evidence-based strategies for managing soil to use these species to promote plant growth and to enhance biodiversity (Breed et al., 2019). To briefly highlight just one exciting example, it has been shown that *C. pinensis* not only produces enzymes capable of fungal cell wall degradation (McKee et al., 2019), it also secretes anti-fungal proteins such as lantibiotics (Mohr et al., 2015), and so it may represent a potential biocontrol species. More work is needed to understand the rich microbial soil ecosystem and how it may be enriched and utilized to maintain ecological diversity and sustainable human agriculture and forestry.

Acknowledgments

J.L. is supported by a research project grant from the Swedish Energy Agency (Dnr 2015-009561). L.S.M. is supported by an award from the Swedish Research Council Vetenskapsrådet (project 2017-04906). Both J.L. and L.S.M. are supported by the Knut and Alice Wallenberg foundation via the Wallenberg Wood Science Centre.

References

Abe, K., Nakajima, M., Yamashita, T., Matsunaga, H., Kamisuki, S., Nihira, T., et al. (2017). Biochemical and structural analyses of a bacterial endo-β-1,2-glucanase reveal a new glycoside hydrolase family. *Journal of Biological Chemistry*, *292*, 7487–7506.

Agger, J. W., Isaksen, T., Varnai, A., Vidal-Melgosa, S., Willats, W. G., Ludwig, R., et al. (2014). Discovery of LPMO activity on hemicelluloses shows the importance of oxidative processes in plant cell wall degradation. *Proceedings of the National Academy of Sciences of the United States of America*, *111*(17), 6287–6292. https://doi.org/10.1073/pnas.1323629111.

Agostoni, M., Hangasky, J. A., & Marletta, M. A. (2017). Physiological and molecular understanding of bacterial polysaccharide monooxygenases. *Microbiology and Molecular Biology Reviews*, *81*(3), e00015–e00017. https://doi.org/10.1128/MMBR.00015-17.

Ahmad, M., Taylor, C. R., Pink, D., Burton, K., Eastwood, D., Bending, G. D., et al. (2010). Development of novel assays for lignin degradation: Comparative analysis of bacterial and fungal lignin degraders. *Molecular BioSystems*, *6*(5), 815–821. https://doi.org/10.1039/b908966g.

Allison, S. D. (2005). Cheaters, diffusion and nutrients constrain decomposition by microbial enzymes in spatially structured environments. *Ecology Letters*, *8*(6), 626–635. https://doi.org/10.1111/j.1461-0248.2005.00756.x.

Anderson, K. L., & Salyers, A. A. (1989). Biochemical evidence that starch breakdown by *Bacteroides thetaiotaomicron* involves outer membrane starch-binding sites and periplasmic starch-degrading enzymes. *Journal of Bacteriology*, *171*(6), 3192. https://doi.org/10.1128/jb.171.6.3192-3198.1989.

Artzi, L., Bayer, E. A., & Moraïs, S. (2016). Cellulosomes: Bacterial nanomachines for dismantling plant polysaccharides. *Nature Reviews Microbiology*, *15*, 83. https://doi.org/10.1038/nrmicro.2016.164.

Bågenholm, V., Reddy, S. K., Bouraoui, H., Morrill, J., Kulcinskaja, E., Bahr, C. M., et al. (2017). Galactomannan catabolism conferred by a polysaccharide utilisation locus of *Bacteroides ovatus*: Enzyme synergy and crystal structure of a β-mannanase. *Journal of Biological Chemistry*, *292*, 229–243. https://doi.org/10.1074/jbc.M116.746438.

Bai, Y., Eijsink, V. G. H., Kielak, A. M., van Veen, J. A., & de Boer, W. (2016). Genomic comparison of chitinolytic enzyme systems from terrestrial and aquatic bacteria. *Environmental Microbiology*, *18*(1), 38–49. https://doi.org/10.1111/1462-2920.12545.

Baldrian, P., Vetrovsky, T., Cajthaml, T., Dobiasova, P., Petrankova, M., Snajdr, J., et al. (2013). Estimation of fungal biomass in forest litter and soil. *Fungal Ecology*, *6*(1), 1–11. https://doi.org/10.1016/j.funeco.2012.10.002.

Bardgett, R. D., & van der Putten, W. H. (2014). Belowground biodiversity and ecosystem functioning. *Nature*, *515*(7528), 505–511. https://doi.org/10.1038/nature13855.

Bates, S. T., Clemente, J. C., Flores, G. E., Walters, W. A., Parfrey, L. W., Knight, R., et al. (2013). Global biogeography of highly diverse protistan communities in soil. *The ISME Journal*, 7(3), 652–659. https://doi.org/10.1038/ismej.2012.147.

Bolam, D. N., & Sonnenburg, J. L. (2011). Mechanistic insight into polysaccharide use within the intestinal microbiota. *Gut Microbes*, *2*(2), 86–90. https://doi.org/10.4161/gmic.2.2.15232.

Brabcová, V., Nováková, M., Davidová, A., & Baldrian, P. (2016). Dead fungal mycelium in forest soil represents a decomposition hotspot and a habitat for a specific microbial community. *New Phytologist*, *210*(4), 1369–1381. https://doi.org/10.1111/nph.13849.

Breed, M. F., Harrison, P. A., Blyth, C., Byrne, M., Gaget, V., Gellie, N. J. C., et al. (2019). The potential of genomics for restoring ecosystems and biodiversity. *Nature Reviews Genetics*, *20*(10), 615–628. https://doi.org/10.1038/s41576-019-0152-0.

Burchard, R. P., & Bloodgood, R. A. (1990). Surface proteins of the gliding bacterium *Cytophaga* sp. strain U67 and its mutants defective in adhesion and motility. *Journal of Bacteriology*, *172*(6), 3379–3387. https://doi.org/10.1128/jb.172.6.3379-3387.1990.

Busse-Wicher, M., Li, A., Silveira, R. L., Pereira, C. S., Tryfona, T., Gomes, T. C., et al. (2016). Evolution of xylan substitution patterns in gymnosperms and angiosperms: Implications for xylan interaction with cellulose. *Plant Physiology*, *171*(4), 2418–2431. https://doi.org/10.1104/pp.16.00539.

Cameron, E. A., Kwiatkowski, K. J., Lee, B. H., Hamaker, B. R., Koropatkin, N. M., & Martens, E. C. (2014). Multifunctional nutrient-binding proteins adapt human symbiotic bacteria for glycan competition in the gut by separately promoting enhanced sensing and catalysis. *MBio*, *5*(5), e01441-14. https://doi.org/10.1128/mBio.01441-14.

Cameron, E. A., Maynard, M. A., Smith, C. J., Smith, T. J., Koropatkin, N. M., & Martens, E. C. (2012). Multidomain carbohydrate-binding proteins involved in *Bacteroides thetaiotaomicron* starch metabolism. *Journal of Biological Chemistry, 287*(41), 34614–34625. https://doi.org/10.1074/jbc.M112.397380.

Cao, Y., Rocha, E. R., & Smith, C. J. (2014). Efficient utilization of complex *N*-linked glycans is a selective advantage for *Bacteroides fragilis* in extraintestinal infections. *Proceedings of the National Academy of Sciences of the United States of America, 111*(35), 12901–12906. https://doi.org/10.1073/pnas.1407344111.

Carson, J. K., Campbell, L., Rooney, D., Clipson, N., & Gleeson, D. B. (2009). Minerals in soil select distinct bacterial communities in their microhabitats. *FEMS Microbiology Ecology, 67*(3), 381–388. https://doi.org/10.1111/j.1574-6941.2008.00645.x.

Carson, J. K., Rooney, D., Gleeson, D. B., & Clipson, N. (2007). Altering the mineral composition of soil causes a shift in microbial community structure. *FEMS Microbiology Ecology, 61*(3), 414–423. https://doi.org/10.1111/j.1574-6941.2007.00361.x.

Cho, K. H., & Salyers, A. A. (2001). Biochemical analysis of interactions between outer membrane proteins that contribute to starch utilization by *Bacteroides thetaiotaomicron. Journal of Bacteriology, 183*(24), 7224–7230. https://doi.org/10.1128/jb.183.24.7224-7230.2001.

Costa, T. R. D., Felisberto-Rodrigues, C., Meir, A., Prevost, M. S., Redzej, A., Trokter, M., et al. (2015). Secretion systems in Gram-negative bacteria: Structural and mechanistic insights. *Nature Reviews Microbiology, 13*, 343. https://doi.org/10.1038/nrmicro3456.

Costa, O. Y. A., Raaijmakers, J. M., & Kuramae, E. E. (2018). Microbial extracellular polymeric substances: Ecological function and impact on soil aggregation. *Frontiers in Microbiology, 9*, 1636. https://doi.org/10.3389/fmicb.2018.01636.

Couturier, M., Ladeveze, S., Sulzenbacher, G., Ciano, L., Fanuel, M., Moreau, C., et al. (2018). Lytic xylan oxidases from wood-decay fungi unlock biomass degradation. *Nature Chemical Biology, 14*(3), 306–310. https://doi.org/10.1038/nchembio.2558.

Cretoiu, M. S., Korthals, G. W., Visser, J. H. M., & van Elsas, J. D. (2013). Chitin amendment increases soil suppressiveness toward plant pathogens and modulates the Actinobacterial and Oxalobacteraceal communities in an experimental agricultural field. *Applied and Environmental Microbiology, 79*(17), 5291. https://doi.org/10.1128/AEM.01361-13.

Cuskin, F., Lowe, E. C., Temple, M. J., Zhu, Y., Cameron, E., Pudlo, N. A., et al. (2015). Human gut Bacteroidetes can utilize yeast mannan through a selfish mechanism. *Nature, 517*(7533), 165–169. https://doi.org/10.1038/nature13995.

de Diego, I., Ksiazek, M., Mizgalska, D., Koneru, L., Golik, P., Szmigielski, B., et al. (2016). The outer-membrane export signal of *Porphyromonas gingivalis* type IX secretion system (T9SS) is a conserved C-terminal β-sandwich domain. *Scientific Reports, 6*(1), 23123. https://doi.org/10.1038/srep23123.

Delgado-Baquerizo, M., Oliverio, A. M., Brewer, T. E., Benavent-González, A., Eldridge, D. J., Bardgett, R. D., et al. (2018). A global atlas of the dominant bacteria found in soil. *Science, 359*(6373), 320. https://doi.org/10.1126/science.aap9516.

Donaldson, G. P., Lee, S. M., & Mazmanian, S. K. (2015). Gut biogeography of the bacterial microbiota. *Nature Reviews Microbiology, 14*, 20. https://doi.org/10.1038/nrmicro3552.

D'Onofrio, A., Crawford, J. M., Stewart, E. J., Witt, K., Gavrish, E., Epstein, S., et al. (2010). Siderophores from neighboring organisms promote the growth of uncultured bacteria. *Chemical Biology, 17*(3), 254–264. https://doi.org/10.1016/j.chembiol.2010.02.010.

D'Souza, G., Shitut, S., Preussger, D., Yousif, G., Waschina, S., & Kost, C. (2018). Ecology and evolution of metabolic cross-feeding interactions in bacteria. *Natural Product Reports, 35*(5), 455–488. https://doi.org/10.1039/c8np00009c.

Fierer, N., Leff, J. W., Adams, B. J., Nielsen, U. N., Bates, S. T., Lauber, C. L., et al. (2012). Cross-biome metagenomic analyses of soil microbial communities and their functional attributes. *Proceedings of the National Academy of Sciences of the United States of America, 109*(52), 21390. https://doi.org/10.1073/pnas.1215210110.

Fierer, N., & Lennon, J. T. (2011). The generation and maintenance of diversity in microbial communities. *American Journal of Botany, 98*(3), 439–448. https://doi.org/10.3732/ajb.1000498.

Fierer, N., Strickland, M. S., Liptzin, D., Bradford, M. A., & Cleveland, C. C. (2009). Global patterns in belowground communities. *Ecology Letters, 12*(11), 1238–1249. https://doi.org/10.1111/j.1461-0248.2009.01360.x.

Foley, M. H., Cockburn, D. W., & Koropatkin, N. M. (2016). The Sus operon: A model system for starch uptake by the human gut Bacteroidetes. *Cellular and Molecular Life Sciences, 73*(14), 2603–2617. https://doi.org/10.1007/s00018-016-2242-x.

Freeman, C., Ostle, N., & Kang, H. (2001). An enzymic 'latch' on a global carbon store. *Nature, 409*(6817), 149. https://doi.org/10.1038/35051650.

Fulthorpe, R. R., Roesch, L. F., Riva, A., & Triplett, E. W. (2008). Distantly sampled soils carry few species in common. *The ISME Journal, 2*(9), 901–910. https://doi.org/10.1038/ismej.2008.55.

Giummarella, N., Balakshin, M., Koutaniemi, S., Kärkönen, A., & Lawoko, M. (2019). Nativity of lignin carbohydrate bonds substantiated by biomimetic synthesis. *Journal of Experimental Botany, 70*, 5591–5601. https://doi.org/10.1093/jxb/erz324.

Glavina Del Rio, T., Abt, B., Spring, S., Lapidus, A., Nolan, M., Tice, H., et al. (2010). Complete genome sequence of *Chitinophaga pinensis* type strain (UQM 2034). *Standards in Genomic Sciences, 2*(1), 87–95. https://doi.org/10.4056/sigs.661199.

Glenwright, A. J., Pothula, K. R., Bhamidimarri, S. P., Chorev, D. S., Baslé, A., Firbank, S. J., et al. (2017). Structural basis for nutrient acquisition by dominant members of the human gut microbiota. *Nature, 541*, 407. https://doi.org/10.1038/nature20828.

Glew, M. D., Veith, P. D., Peng, B., Chen, Y.-Y., Gorasia, D. G., Yang, Q., et al. (2012). PG0026 is the C-terminal signal peptidase of a novel secretion system of *Porphyromonas gingivalis*. *Journal of Biological Chemistry, 287*(29), 24605–24617.

Gonzalez, A., Clemente, J. C., Shade, A., Metcalf, J. L., Song, S., Prithiviraj, B., et al. (2011). Our microbial selves: What ecology can teach us. *EMBO Reports, 12*(8), 775–784. https://doi.org/10.1038/embor.2011.137.

Gow, N. A. R., Latge, J. P., & Munro, C. A. (2017). The fungal cell wall: Structure, biosynthesis, and function. *Microbiology Spectrum, 5*(3). https://doi.org/10.1128/microbiolspec.FUNK-0035-2016.

Green, E. R., & Mecsas, J. (2016). Bacterial secretion systems: An overview. *Microbiology Spectrum, 4*(1). https://doi.org/10.1128/microbiolspec.VMBF-0012-2015.

Grondin, J. M., Tamura, K., Déjean, G., Abbott, D. W., & Brumer, H. (2017). Polysaccharide utilization loci: Fueling microbial communities. *Journal of Bacteriology, 199*(15), e00860-16. https://doi.org/10.1128/JB.00860-16.

Haitjema, C. H., Gilmore, S. P., Henske, J. K., Solomon, K. V., de Groot, R., Kuo, A., et al. (2017). A parts list for fungal cellulosomes revealed by comparative genomics. *Nature Microbiology, 2*, 17087. https://doi.org/10.1038/nmicrobiol.2017.87.

Hehemann, J. H., Correc, G., Barbeyron, T., Helbert, W., Czjzek, M., & Michel, G. (2010). Transfer of carbohydrate-active enzymes from marine bacteria to Japanese gut microbiota. *Nature, 464*(7290), 908–912. https://doi.org/10.1038/nature08937.

Hehemann, J.-H., Kelly, A. G., Pudlo, N. A., Martens, E. C., & Boraston, A. B. (2012). Bacteria of the human gut microbiome catabolize red seaweed glycans with carbohydrate-active enzyme updates from extrinsic microbes. *Proceedings of the National Academy of Sciences of the United States of America, 109*(48), 19786. https://doi.org/10.1073/pnas.1211002109.

Horn, S. J., Vaaje-Kolstad, G., Westereng, B., & Eijsink, V. G. (2012). Novel enzymes for the degradation of cellulose. *Biotechnology for Biofuels*, *5*(1), 45. https://doi.org/10.1186/1754-6834-5-45.

Hüttner, S., Varnai, A., Petrovic, D. M., Bach, C. X., Kim Anh, D. T., Thanh, V. N., et al. (2019). Functional characterization of AA9 LPMOs in the thermophilic fungus *Malbranchea cinnamomea* reveals specific xylan activity. *Applied and Environmental Microbiology*. e01408-19. https://doi.org/10.1128/aem.01408-19.

Janssen, P. H. (2006). Identifying the dominant soil bacterial taxa in libraries of 16S rRNA and 16S rRNA genes. *Applied and Environmental Microbiology*, *72*(3), 1719. https://doi.org/10.1128/AEM.72.3.1719-1728.2006.

Janusz, G., Pawlik, A., Sulej, J., Swiderska-Burek, U., Jarosz-Wilkolazka, A., & Paszczynski, A. (2017). Lignin degradation: Microorganisms, enzymes involved, genomes analysis and evolution. *FEMS Microbiology Reviews*, *41*(6), 941–962. https://doi.org/10.1093/femsre/fux049.

Jin, Q., & Kirk, M. F. (2018). pH as a primary control in environmental microbiology: 1. Thermodynamic perspective. *Frontiers in Environmental Science*, *6*, 21. https://doi.org/10.3389/fenvs.2018.00021.

Johnston, J. J., Shrivastava, A., & McBride, M. J. (2018). Untangling *Flavobacterium johnsoniae* gliding motility and protein secretion. *Journal of Bacteriology*, *200*(2), e00362-17. https://doi.org/10.1128/jb.00362-17.

Kabisch, A., Otto, A., König, S., Becher, D., Albrecht, D., Schüler, M., et al. (2014). Functional characterization of polysaccharide utilization loci in the marine Bacteroidetes '*Gramella forsetii*' KT0803. *The ISME Journal*, *8*(7), 1492–1502. https://doi.org/10.1038/ismej.2014.4.

Kharade, S. S., & McBride, M. J. (2014). *Flavobacterium johnsoniae* chitinase ChiA is required for chitin utilization and is secreted by the type IX secretion system. *Journal of Bacteriology*, *196*(5), 961–970. https://doi.org/10.1128/jb.01170-13.

Kharade, S. S., & McBride, M. J. (2015). *Flavobacterium johnsoniae* PorV is required for secretion of a subset of proteins targeted to the type IX secretion system. *Journal of Bacteriology*, *197*(1), 147. https://doi.org/10.1128/JB.02085-14.

Koropatkin, N. M., Martens, E. C., Gordon, J. I., & Smith, T. J. (2008). Starch catabolism by a prominent human gut symbiont is directed by the recognition of amylose helices. *Structure*, *16*(7), 1105–1115. https://doi.org/10.1016/j.str.2008.03.017.

Kramer, S., Dibbern, D., Moll, J., Huenninghaus, M., Koller, R., Krueger, D., et al. (2016). Resource partitioning between bacteria, fungi, and protists in the detritusphere of an agricultural soil. *Frontiers in Microbiology*, 7, 1524. https://doi.org/10.3389/fmicb.2016.01524.

Kulkarni, S. S., Johnston, J. J., Zhu, Y., Hying, Z. T., & McBride, M. J. (2019). The carboxy-terminal region of *Flavobacterium johnsoniae* SprB facilitates its secretion by the Type IX secretion system and propulsion by the gliding motility machinery. *Journal of Bacteriology*, *201*(19), e00218-19. https://doi.org/10.1128/jb.00218-19.

Kulkarni, S. S., Zhu, Y., Brendel, C. J., & McBride, M. J. (2017). Diverse C-terminal sequences involved in *Flavobacterium johnsoniae* protein secretion. *Journal of Bacteriology*, *199*(12). e00884-16https://doi.org/10.1128/JB.00884-16.

Lapébie, P., Lombard, V., Drula, E., Terrapon, N., & Henrissat, B. (2019). Bacteroidetes use thousands of enzyme combinations to break down glycans. *Nature Communications*, *10*(1), 2043. https://doi.org/10.1038/s41467-019-10068-5.

Lapidus, I. R., & Berg, H. C. (1982). Gliding motility of *Cytophaga* sp. strain U67. *Journal of Bacteriology*, *151*(1), 384–398.

Larsbrink, J., Rogers, T. E., Hemsworth, G. R., McKee, L. S., Tauzin, A. S., Spadiut, O., et al. (2014). A discrete genetic locus confers xyloglucan metabolism in select human gut Bacteroidetes. *Nature*, *506*(7489), 498–502. https://doi.org/10.1038/nature12907.

Larsbrink, J., Tuveng, T. R., Pope, P. B., Bulone, V., Eijsink, V. G., Brumer, H., et al. (2017a). Proteomic insights into mannan degradation and protein secretion by the forest floor bacterium *Chitinophaga pinensis*. *Journal of Proteomics*, *156*, 63–74. https://doi.org/10.1016/j.jprot.2017.01.003.

Larsbrink, J., Tuveng, T. R., Pope, P. B., Bulone, V., Eijsink, V. G. H., Brumer, H., et al. (2017b). Proteomic data on enzyme secretion and activity in the bacterium *Chitinophaga pinensis*. *Data in Brief*, *11*, 484–490. https://doi.org/10.1016/j.dib.2017.02.032.

Larsbrink, J., Zhu, Y., Kharade, S. S., Kwiatkowski, K. J., Eijsink, V. G. H., Koropatkin, N. M., et al. (2016). A polysaccharide utilization locus from *Flavobacterium johnsoniae* enables conversion of recalcitrant chitin. *Biotechnology for Biofuels*, *9*(1), 260. https://doi.org/10.1186/s13068-016-0674-z.

Lasica, A. M., Goulas, T., Mizgalska, D., Zhou, X., de Diego, I., Ksiazek, M., et al. (2016). Structural and functional probing of PorZ, an essential bacterial surface component of the type-IX secretion system of human oral-microbiomic *Porphyromonas gingivalis*. *Scientific Reports*, *6*, 37708. https://doi.org/10.1038/srep37708.

Lasica, A. M., Ksiazek, M., Madej, M., & Potempa, J. (2017). The type IX secretion system (T9SS): Highlights and recent insights into its structure and function. *Frontiers in Cellular and Infection Microbiology*, 7, 215. https://doi.org/10.3389/fcimb.2017.00215.

Lauber, F., Deme, J. C., Lea, S. M., & Berks, B. C. (2018). Type 9 secretion system structures reveal a new protein transport mechanism. *Nature*, *564*(7734), 77–82. https://doi.org/10.1038/s41586-018-0693-y.

Lee, S., Kang, M., Bae, J. H., Sohn, J. H., & Sung, B. H. (2019). Bacterial valorization of lignin: Strains, enzymes, conversion pathways, biosensors, and perspectives. *Frontiers in Bioengineering and Biotechnology*, 7, 209. https://doi.org/10.3389/fbioe.2019.00209.

Liu, L., Gao, P., Chen, G., & Wang, L. (2014). Draft genome sequence of cellulose-digesting bacterium *Sporocytophaga myxococcoides* PG-01. *Genome Announcements*, *2*(6). https://doi.org/10.1128/genomeA.01154-14.

Lombard, V., Golaconda Ramulu, H., Drula, E., Coutinho, P. M., & Henrissat, B. (2014). The carbohydrate-active enzymes database (CAZy) in 2013. *Nucleic Acids Research*, *42*(Database issue), D490–D495. https://doi.org/10.1093/nar/gkt1178.

Lopez-Mondejar, R., Voriskova, J., Vetrovsky, T., & Baldrian, P. (2015). The bacterial community inhabiting temperate deciduous forests is vertically stratified and undergoes seasonal dynamics. *Soil Biology & Biochemistry*, *87*, 43–50. https://doi.org/10.1016/j.soilbio.2015.04.008.

Louca, S., Polz, M. F., Mazel, F., Albright, M. B. N., Huber, J. A., O'Connor, M. I., et al. (2018). Function and functional redundancy in microbial systems. *Nature Ecology & Evolution*, *2*(6), 936–943. https://doi.org/10.1038/s41559-018-0519-1.

Luis, A. S., & Martens, E. C. (2018). Interrogating gut bacterial genomes for discovery of novel carbohydrate degrading enzymes. *Current Opinion in Chemical Biology*, *47*, 126–133. https://doi.org/10.1016/j.cbpa.2018.09.012.

Martens, E. C., Kelly, A. G., Tauzin, A. S., & Brumer, H. (2014). The devil lies in the details: How variations in polysaccharide fine-structure impact the physiology and evolution of gut microbes. *Journal of Molecular Biology*, *426*(23), 3851–3865. https://doi.org/10.1016/j.jmb.2014.06.022.

Martens, E. C., Koropatkin, N. M., Smith, T. J., & Gordon, J. I. (2009). Complex glycan catabolism by the human gut microbiota: The Bacteroidetes Sus-like paradigm. *Journal of Biological Chemistry*, *284*(37), 24673–24677. https://doi.org/10.1074/jbc.R109.022848.

Martens, E. C., Lowe, E. C., Chiang, H., Pudlo, N. A., Wu, M., McNulty, N. P., et al. (2011). Recognition and degradation of plant cell wall polysaccharides by two human gut symbionts. *PLoS Biology*, *9*(12). e1001221. https://doi.org/10.1371/journal.pbio.1001221.

Martinez-Abad, A., Berglund, J., Toriz, G., Gatenholm, P., Henriksson, G., Lindstrom, M., et al. (2017). Regular motifs in xylan modulate molecular flexibility and interactions with cellulose surfaces. *Plant Physiology*, *175*(4), 1579–1592. https://doi.org/10.1104/pp.17.01184.

Martínez-Abad, A., Giummarella, N., Lawoko, M., & Vilaplana, F. (2018). Differences in extractability under subcritical water reveal interconnected hemicellulose and lignin recalcitrance in birch hardwoods. *Green Chemistry*, *20*(11), 2534–2546. https://doi.org/10.1039/C8GC00385H.

McBride, M. J. (2019). Bacteroidetes gliding motility and the type IX secretion system. *Microbiology Spectrum*, 7(1). https://doi.org/10.1128/microbiolspec.PSIB-0002-2018.

McBride, M. J., Braun, T. F., & Brust, J. L. (2003). *Flavobacterium johnsoniae* GldH is a lipoprotein that is required for gliding motility and chitin utilization. *Journal of Bacteriology*, *185*(22), 6648–6657. https://doi.org/10.1128/jb.185.22.6648-6657.2003.

McBride, M. J., & Nakane, D. (2015). *Flavobacterium* gliding motility and the type IX secretion system. *Current Opinion in Microbiology*, *28*, 72–77. https://doi.org/10.1016/j.mib.2015.07.016.

McBride, M. J., Xie, G., Martens, E. C., Lapidus, A., Henrissat, B., Rhodes, R. G., et al. (2009). Novel features of the polysaccharide-digesting gliding bacterium *Flavobacterium johnsoniae* as revealed by genome sequence analysis. *Applied and Environmental Microbiology*, *75*(21), 6864–6875. https://doi.org/10.1128/aem.01495-09.

McBride, M. J., & Zhu, Y. (2013). Gliding motility and Por secretion system genes are widespread among members of the phylum Bacteroidetes. *Journal of Bacteriology*, *195*(2), 270–278. https://doi.org/10.1128/jb.01962-12.

McKee, L. S., & Brumer, H. (2015). Growth of *Chitinophaga pinensis* on plant cell wall glycans and characterisation of a glycoside hydrolase family 27 beta-L-arabinopyranosidase implicated in arabinogalactan utilisation. *PLoS One*, *10*(10), e0139932. https://doi.org/10.1371/journal.pone.0139932.

McKee, L. S., Martinez-Abad, A., Ruthes, A. C., Vilaplana, F., & Brumer, H. (2019). Focused metabolism of beta-glucans by the soil Bacteroidetes species *Chitinophaga pinensis*. *Applied and Environmental Microbiology*, *85*(2), e02231-18. https://doi.org/10.1128/aem.02231-18.

Mélida, H., Sandoval-Sierra, J. V., Diéguez-Uribeondo, J., & Bulone, V. (2013). Analyses of extracellular carbohydrates in oomycetes unveil the existence of three different cell wall types. *Eukaryotic Cell*, *12*(2), 194. https://doi.org/10.1128/EC.00288-12.

Mohamad, N. R., Marzuki, N. H. C., Buang, N. A., Huyop, F., & Wahab, R. A. (2015). An overview of technologies for immobilization of enzymes and surface analysis techniques for immobilized enzymes. *Biotechnology and Biotechnological Equipment*, *29*(2), 205–220. https://doi.org/10.1080/13102818.2015.1008192.

Mohnen, D. (2008). Pectin structure and biosynthesis. *Current Opinion in Plant Biology*, *11*(3), 266–277. https://doi.org/10.1016/j.pbi.2008.03.006.

Mohr, K. I., Volz, C., Jansen, R., Wray, V., Hoffmann, J., Bernecker, S., et al. (2015). Pinensins: The first antifungal lantibiotics. *Angewandte Chemie (International ed. in English)*, *54*(38), 11254–11258. https://doi.org/10.1002/anie.201500927.

Naas, A. E., Mackenzie, A. K., Mravec, J., Schuckel, J., Willats, W. G., Eijsink, V. G., et al. (2014). Do rumen Bacteroidetes utilize an alternative mechanism for cellulose degradation? *MBio*, *5*(4). e01401-14https://doi.org/10.1128/mBio.01401-14.

Naas, A. E., Solden, L. M., Norbeck, A. D., Brewer, H., Hagen, L. H., Heggenes, I. M., et al. (2018). "*Candidatus Paraporphyromonas polyenzymogenes*" encodes multi-modular cellulases linked to the type IX secretion system. *Microbiome*, *6*(1), 44. https://doi.org/10.1186/s40168-018-0421-8.

Nan, B. (2017). Bacterial gliding motility: Rolling out a consensus model. *Current Biology*, *27*(4), R154–R156. https://doi.org/10.1016/j.cub.2016.12.035.

Nan, B., McBride, M. J., Chen, J., Zusman, D. R., & Oster, G. (2014). Bacteria that glide with helical tracks. *Current Biology*, *24*(4), R169–R173. https://doi.org/10.1016/j.cub.2013.12.034.

Ndeh, D., Rogowski, A., Cartmell, A., Luis, A. S., Basle, A., Gray, J., et al. (2017). Complex pectin metabolism by gut bacteria reveals novel catalytic functions. *Nature*, *544*(7648), 65–70. https://doi.org/10.1038/nature21725.

Nelson, S. S., Glocka, P. P., Agarwal, S., Grimm, D. P., & McBride, M. J. (2007). *Flavobacterium johnsoniae* SprA is a cell surface protein involved in gliding motility. *Journal of Bacteriology*, *189*(19), 7145–7150. https://doi.org/10.1128/jb.00892-07.

Nguyen, K. A., Travis, J., & Potempa, J. (2007). Does the importance of the C-terminal residues in the maturation of RgpB from *Porphyromonas gingivalis* reveal a novel mechanism for protein export in a subgroup of Gram-negative bacteria? *Journal of Bacteriology*, *189*(3), 833–843. https://doi.org/10.1128/jb.01530-06.

Nishimura, H., Kamiya, A., Nagata, T., Katahira, M., & Watanabe, T. (2018). Direct evidence for alpha ether linkage between lignin and carbohydrates in wood cell walls. *Scientific Reports*, *8*(1), 6538. https://doi.org/10.1038/s41598-018-24328-9.

Ochoa-Hueso, R. (2017). Global change and the soil microbiome: A human-health perspective. *Frontiers in Ecology and Evolution*, *5*, 71. https://doi.org/10.3389/fevo.2017.00071.

Payne, C. M., Knott, B. C., Mayes, H. B., Hansson, H., Himmel, M. E., Sandgren, M., et al. (2015). Fungal cellulases. *Chemical Reviews*, *115*(3), 1308–1448. https://doi.org/10.1021/cr500351c.

Qurashi, A. W., & Sabri, A. N. (2012). Bacterial exopolysaccharide and biofilm formation stimulate chickpea growth and soil aggregation under salt stress. *Brazilian Journal of Microbiology*, *43*(3), 1183–1191. https://doi.org/10.1590/S1517-838220120003000046.

Rakoff-Nahoum, S., Foster, K. R., & Comstock, L. E. (2016). The evolution of cooperation within the gut microbiota. *Nature*, *533*(7602), 255–259. https://doi.org/10.1038/nature17626.

Ralph, J., Lapierre, C., & Boerjan, W. (2019). Lignin structure and its engineering. *Current Opinion in Biotechnology*, *56*, 240–249. https://doi.org/10.1016/j.copbio.2019.02.019.

Ramakrishna, B., Vaikuntapu, P., Mallakuntla, M. K., Bhuvanachandra, B., Sivaramakrishna, D., Uikey, S., et al. (2018). Carboxy-terminal glycosyl hydrolase 18 domain of a carbohydrate active protein of *Chitinophaga pinensis* is a non-processive exochitinase. *International Journal of Biological Macromolecules*, *115*, 1225–1232. https://doi.org/10.1016/j.ijbiomac.2018.04.159.

Ramirez, K. S., Leff, J. W., Barberán, A., Bates, S. T., Betley, J., Crowther, T. W., et al. (2014). Biogeographic patterns in below-ground diversity in New York City's Central Park are similar to those observed globally. *Proceedings of the Royal Society B: Biological Sciences*, *281*(1795), 20141988. https://doi.org/10.1098/rspb.2014.1988.

Rashid, G. M., Taylor, C. R., Liu, Y., Zhang, X., Rea, D., Fulop, V., et al. (2015). Identification of manganese superoxide dismutase from *Sphingobacterium* sp. T2 as a novel bacterial enzyme for lignin oxidation. *ACS Chemical Biology*, *10*(10), 2286–2294. https://doi.org/10.1021/acschembio.5b00298.

Razeq, F. M., Jurak, E., Stogios, P. J., Yan, R., Tenkanen, M., Kabel, M. A., et al. (2018). A novel acetyl xylan esterase enabling complete deacetylation of substituted xylans. *Biotechnology for Biofuels*, *11*(1), 74. https://doi.org/10.1186/s13068-018-1074-3.

Reeves, A. R., Elia, J. N., Frias, J., & Salyers, A. A. (1996). A *Bacteroides thetaiotaomicron* outer membrane protein that is essential for utilization of maltooligosaccharides and starch. *Journal of Bacteriology*, *178*(3), 823. https://doi.org/10.1128/jb.178.3.823-830.1996.

Rhodes, R. G., Samarasam, M. N., Shrivastava, A., van Baaren, J. M., Pochiraju, S., Bollampalli, S., et al. (2010). *Flavobacterium johnsoniae* gldN and gldO are partially redundant genes required for gliding motility and surface localization of SprB. *Journal of Bacteriology*, *192*(5), 1201–1211. https://doi.org/10.1128/jb.01495-09.

Richnow, H. H., Seifert, R., Hefter, J., Link, M., Francke, W., Schaefer, G., et al. (1997). Organic pollutants associated with macromolecular soil organic matter: Mode of binding. *Organic Geochemistry*, *26*(11), 745–758. https://doi.org/10.1016/S0146-6380(97)00054-5.

Riviere, A., Gagnon, M., Weckx, S., Roy, D., & De Vuyst, L. (2015). Mutual cross-feeding interactions between *Bifidobacterium longum* subsp. *longum* NCC2705 and *Eubacterium rectale* ATCC 33656 explain the bifidogenic and butyrogenic effects of arabinoxylan oligosaccharides. *Applied and Environmental Microbiology*, *81*(22), 7767–7781. https://doi.org/10.1128/aem.02089-15.

Roesch, L. F. W., Fulthorpe, R. R., Riva, A., Casella, G., Hadwin, A. K. M., Kent, A. D., et al. (2007). Pyrosequencing enumerates and contrasts soil microbial diversity. *The ISME Journal*, *1*(4), 283–290. https://doi.org/10.1038/ismej.2007.53.

Rogowski, A., Briggs, J. A., Mortimer, J. C., Tryfona, T., Terrapon, N., Lowe, E. C., et al. (2015). Glycan complexity dictates microbial resource allocation in the large intestine. *Nature Communications*, *6*, 7481. https://doi.org/10.1038/ncomms8481.

Russell, A. B., Wexler, A. G., Harding, B. N., Whitney, J. C., Bohn, A. J., Goo, Y. A., et al. (2014). A type VI secretion-related pathway in Bacteroidetes mediates interbacterial antagonism. *Cell Host & Microbe*, *16*(2), 227–236. https://doi.org/10.1016/j.chom.2014.07.007.

Sangkhobol, V., & Skerman, V. B. D. (1981). *Chitinophaga*, a new genus of chitinolytic myxobacteria. *International Journal of Systematic and Evolutionary Microbiology*, *31*(3), 285–293. https://doi.org/10.1099/00207713-31-3-285.

Sato, K., Naito, M., Yukitake, H., Hirakawa, H., Shoji, M., McBride, M. J., et al. (2010). A protein secretion system linked to bacteroidete gliding motility and pathogenesis. *Proceedings of the National Academy of Sciences of the United States of America*, *107*(1), 276–281. https://doi.org/10.1073/pnas.0912010107.

Sato, K., Sakai, E., Veith, P. D., Shoji, M., Kikuchi, Y., Yukitake, H., et al. (2005). Identification of a new membrane-associated protein that influences transport/maturation of gingipains and adhesins of *Porphyromonas gingivalis*. *Journal of Biological Chemistry*, *280*(10), 8668–8677. https://doi.org/10.1074/jbc.M413544200.

Scheller, H. V., & Ulvskov, P. (2010). Hemicelluloses. *Annual Review of Plant Biology*, *61*(1), 263–289. https://doi.org/10.1146/annurev-arplant-042809-112315.

Schwarz, S., Hood, R. D., & Mougous, J. D. (2010). What is type VI secretion doing in all those bugs? *Trends in Microbiology*, *18*(12), 531–537. https://doi.org/10.1016/j.tim.2010.09.001.

Seers, C. A., Slakeski, N., Veith, P. D., Nikolof, T., Chen, Y. Y., Dashper, S. G., et al. (2006). The RgpB C-terminal domain has a role in attachment of RgpB to the outer membrane and belongs to a novel C-terminal-domain family found in *Porphyromonas gingivalis*. *Journal of Bacteriology*, *188*(17), 6376–6386. https://doi.org/10.1128/jb.00731-06.

Senesi, N. (1992). Binding mechanisms of pesticides to soil humic substances. *Science of The Total Environment*, *123–124*, 63–76. https://doi.org/10.1016/0048-9697(92)90133-D.

Sheth, U., Li, M., Jiang, W., Sims, P. A., Leong, K. W., & Wang, H. H. (2019). Spatial metagenomic characterization of microbial biogeography in the gut. *Nature Biotechnology*, *37*(8), 877–883. https://doi.org/10.1038/s41587-019-0183-2.

Shipman, J. A., Berleman, J. E., & Salyers, A. A. (2000). Characterization of four outer membrane proteins involved in binding starch to the cell surface of *Bacteroides thetaiotaomicron*. *Journal of Bacteriology*, *182*(19), 5365–5372. https://doi.org/10.1128/jb.182.19.5365-5372.2000.

Shipman, J. A., Cho, K. H., Siegel, H. A., & Salyers, A. A. (1999). Physiological characterization of SusG, an outer membrane protein essential for starch utilization by *Bacteroides thetaiotaomicron*. *Journal of Bacteriology*, *181*(23), 7206.

Shrivastava, A., & Berg, H. C. (2015). Towards a model for *Flavobacterium* gliding. *Current Opinion in Microbiology*, *28*, 93–97. https://doi.org/10.1016/j.mib.2015.07.018.

Shrivastava, A., Johnston, J. J., van Baaren, J. M., & McBride, M. J. (2013). *Flavobacterium johnsoniae* GldK, GldL, GldM, and SprA are required for secretion of the cell surface gliding motility adhesins SprB and RemA. *Journal of Bacteriology*, *195*(14), 3201–3212. https://doi.org/10.1128/jb.00333-13.

Shrivastava, A., Rhodes, R. G., Pochiraju, S., Nakane, D., & McBride, M. J. (2012). *Flavobacterium johnsoniae* RemA is a mobile cell surface lectin involved in gliding. *Journal of Bacteriology*, *194*(14), 3678–3688. https://doi.org/10.1128/jb.00588-12.

Sorensen, L. H. (1956). Decomposition of xylan by *Sporocytophaga myxococcoides*. *Nature*, *177*(4514), 845.

Soudzilovskaia, N. A., van der Heijden, M. G., Cornelissen, J. H., Makarov, M. I., Onipchenko, V. G., Maslov, M. N., et al. (2015). Quantitative assessment of the differential impacts of arbuscular and ectomycorrhiza on soil carbon cycling. *New Phytologist*, *208*(1), 280–293. https://doi.org/10.1111/nph.13447.

Stanier, R. Y. (1942). The *Cytophaga* group: A contribution to the biology of myxobacteria. *Bacteriological Reviews*, *6*(3), 143–196.

Taillefer, M., Arntzen, M. O., Henrissat, B., Pope, P. B., & Larsbrink, J. (2018). Proteomic dissection of the cellulolytic machineries used by soil-dwelling Bacteroidetes. *mSystems*, *3*(6), e00240-18. https://doi.org/10.1128/mSystems.00240-18.

Takahashi, N., & Koshijima, T. (1988). Ester linkages between lignin and glucuronoxylan in a lignin-carbohydrate complex from beech (*Fagus crenata*) wood. *Wood Science and Technology*, *22*(3), 231–241. https://doi.org/10.1007/BF00386018.

Tauzin, A. S., Kwiatkowski, K. J., Orlovsky, N. I., Smith, C. J., Creagh, A. L., Haynes, C. A., et al. (2016). Molecular dissection of xyloglucan recognition in a prominent human gut symbiont. *MBio*, 7(2). e02134-15https://doi.org/10.1128/mBio.02134-15.

Taylor, C. R., Hardiman, E. M., Ahmad, M., Sainsbury, P. D., Norris, P. R., & Bugg, T. D. (2012). Isolation of bacterial strains able to metabolize lignin from screening of environmental samples. *Journal of Applied Microbiology*, *113*(3), 521–530. https://doi.org/10.1111/j.1365-2672.2012.05352.x.

Tedersoo, L., Bahram, M., Toots, M., Diedhiou, A. G., Henkel, T. W., Kjoller, R., et al. (2012). Towards global patterns in the diversity and community structure of ectomycorrhizal fungi. *Molecular Ecology*, *21*(17), 4160–4170. https://doi.org/10.1111/j.1365-294X.2012.05602.x.

Terrapon, N., Lombard, V., Drula, E., Lapebie, P., Al-Masaudi, S., Gilbert, H. J., et al. (2018). PULDB: The expanded database of Polysaccharide Utilization Loci. *Nucleic Acids Research*, *46*(D1), D677–D683. https://doi.org/10.1093/nar/gkx1022.

Terrapon, N., Lombard, V., Gilbert, H. J., & Henrissat, B. (2015). Automatic prediction of polysaccharide utilization loci in Bacteroidetes species. *Bioinformatics*, *31*(5), 647–655. https://doi.org/10.1093/bioinformatics/btu716.

Thomas, F., Hehemann, J.-H., Rebuffet, E., Czjzek, M., & Michel, G. (2011). Environmental and gut bacteroidetes: The food connection. *Frontiers in Microbiology*, *2*, 93. https://doi.org/10.3389/fmicb.2011.00093.

Tiedje, J. M., Asuming-Brempong, S., Nusslein, K., Marsh, T. L., & Flynn, S. J. (1999). Opening the black box of soil microbial diversity. *Applied Soil Ecology*, *13*(2), 109–122. https://doi.org/10.1016/s0929-1393(99)00026-8.

Tlaskal, V., Voriskova, J., & Baldrian, P. (2016). Bacterial succession on decomposing leaf litter exhibits a specific occurrence pattern of cellulolytic taxa and potential decomposers of fungal mycelia. *FEMS Microbiology Ecology*, *92*(11), fiw177. https://doi.org/10.1093/femsec/fiw177.

Totsche, K. U., Rennert, T., Gerzabek, M. H., Kögel-Knabner, I., Smalla, K., Spiteller, M., et al. (2010). Biogeochemical interfaces in soil: The interdisciplinary challenge for soil science. *Journal of Plant Nutrition and Soil Science*, *173*(1), 88–99. https://doi.org/10.1002/jpln.200900105.

Tsirigotaki, A., De Geyter, J., Šoštaric´, N., Economou, A., & Karamanou, S. (2016). Protein export through the bacterial Sec pathway. *Nature Reviews Microbiology*, *15*, 21. https://doi.org/10.1038/nrmicro.2016.161.

Tuson, H. H., Foley, M. H., Koropatkin, N. M., & Biteen, J. S. (2018). The starch utilization system assembles around stationary starch-binding proteins. *Biophysical Journal*, *115*(2), 242–250. https://doi.org/10.1016/j.bpj.2017.12.015.

Veith, P. D., Talbo, G. H., Slakeski, N., Dashper, S. G., Moore, C., Paolini, R. A., et al. (2002). Major outer membrane proteins and proteolytic processing of RgpA and Kgp of *Porphyromonas gingivalis* W50. *Biochemical Journal*, *363*(Pt. 1), 105–115. https://doi.org/10.1042/0264-6021:3630105.

Velicer, G. J. (2003). Social strife in the microbial world. *Trends in Microbiology*, *11*(7), 330–337. https://doi.org/10.1016/s0966-842x(03)00152-5.

Vincent, M. S., Canestrari, M. J., Leone, P., Stathopulos, J., Ize, B., Zoued, A., et al. (2017). Characterization of the *Porphyromonas gingivalis* type IX secretion trans-envelope PorKLMNP core complex. *Journal of Biological Chemistry*, *292*(8), 3252–3261.

Voriskova, J., & Baldrian, P. (2013). Fungal community on decomposing leaf litter undergoes rapid successional changes. *The ISME Journal*, 7(3), 477–486. https://doi.org/10.1038/ismej.2012.116.

Voriskova, J., Brabcova, V., Cajthaml, T., & Baldrian, P. (2014). Seasonal dynamics of fungal communities in a temperate oak forest soil. *New Phytologist*, *201*(1), 269–278. https://doi.org/10.1111/nph.12481.

Vu, V. V., Beeson, W. T., Span, E. A., Farquhar, E. R., & Marletta, M. A. (2014). A family of starch-active polysaccharide monooxygenases. *Proceedings of the National Academy of Sciences of the United States of America*, *111*(38), 13822–13827. https://doi.org/10.1073/pnas.1408090111.

Wagg, C., Schlaeppi, K., Banerjee, S., Kuramae, E. E., & van der Heijden, M. G. A. (2019). Fungal-bacterial diversity and microbiome complexity predict ecosystem functioning. *Nature Communications*, *10*(1), 4841. https://doi.org/10.1038/s41467-019-12798-y.

Wang, X., Han, Q., Chen, G., Zhang, W., & Liu, W. (2017). A putative type II secretion system is involved in cellulose utilization in *Cytophaga hutchisonii*. *Frontiers in Microbiology*, *8*, 1482. https://doi.org/10.3389/fmicb.2017.01482.

Wang, D., Lin, Y., Du, W., Liang, J., & Ning, Y. (2013). Optimization and characterization of lignosulfonate biodegradation process by a bacterial strain, *Sphingobacterium* sp. HY-H. *International Biodeterioration & Biodegradation*, *85*, 365–371. https://doi.org/10.1016/j.ibiod.2013.06.032.

Wang, X., Zhang, W., Zhou, H., Chen, G., & Liu, W. (2019). An extracytoplasmic function sigma factor controls cellulose utilization by regulating the expression of an outer membrane protein in *Cytophaga hutchinsonii*. *Applied and Environmental Microbiology*, *85*(5), e02606-18. https://doi.org/10.1128/aem.02606-18.

Wexler, A. G., & Goodman, A. L. (2017). An insider's perspective: *Bacteroides* as a window into the microbiome. *Nature Microbiology*, *2*, 17026. https://doi.org/10.1038/nmicrobiol.2017.26.

Wilson, M. M., Anderson, D. E., & Bernstein, H. D. (2015). Analysis of the outer membrane proteome and secretome of *Bacteroides fragilis* reveals a multiplicity of secretion mechanisms. *PLoS One*, *10*(2). e0117732https://doi.org/10.1371/journal.pone.0117732.

Xu, J., Bjursell, M. K., Himrod, J., Deng, S., Carmichael, L. K., Chiang, H. C., et al. (2003). A genomic view of the human-*Bacteroides thetaiotaomicron* symbiosis. *Science*, *299*(5615), 2074–2076. https://doi.org/10.1126/science.1080029.

Yuan, T.-Q., Sun, S.-N., Xu, F., & Sun, R.-C. (2011). Characterization of lignin structures and lignin–carbohydrate complex (LCC) linkages by quantitative 13C and 2D HSQC NMR spectroscopy. *Journal of Agricultural and Food Chemistry, 59*(19), 10604–10614. https://doi.org/10.1021/jf2031549.

Zegeye, E. K., Brislawn, C. J., Farris, Y., Fansler, S. J., Hofmockel, K. S., Jansson, J. K., et al. (2019). Selection, succession, and stabilization of soil microbial consortia. *mSystems*, 4(4), e00055-19. https://doi.org/10.1128/mSystems.00055-19.

Zhu, Y., Kwiatkowski, K. J., Yang, T., Kharade, S. S., Bahr, C. M., Koropatkin, N. M., et al. (2015). Outer membrane proteins related to SusC and SusD are not required for *Cytophaga hutchinsonii* cellulose utilization. *Applied Microbiology and Biotechnology, 99*(15), 6339–6350. https://doi.org/10.1007/s00253-015-6555-8.

Zhu, Y., & McBride, M. J. (2014). Deletion of the *Cytophaga hutchinsonii* type IX secretion system gene sprP results in defects in gliding motility and cellulose utilization. *Applied Microbiology and Biotechnology, 98*(2), 763–775. https://doi.org/10.1007/s00253-013-5355-2.

Zhu, Y., & McBride, M. J. (2017). The unusual cellulose utilization system of the aerobic soil bacterium *Cytophaga hutchinsonii*. *Applied Microbiology and Biotechnology, 101*(19), 7113–7127. https://doi.org/10.1007/s00253-017-8467-2.

CHAPTER THREE

Anaerobic and hydrogenogenic carbon monoxide-oxidizing prokaryotes: Versatile microbial conversion of a toxic gas into an available energy

Yuto Fukuyama, Masao Inoue, Kimiho Omae, Takashi Yoshida, Yoshihiko Sako*

Laboratory of Marine Microbiology, Graduate School of Agriculture, Kyoto University, Kyoto, Japan

*Corresponding author: e-mail address: sako@kais.kyoto-u.ac.jp

Contents

Advances in Applied Microbiology, Volume 110
ISSN 0065-2164
https://doi.org/10.1016/bs.aambs.2019.12.001

Abstract

Carbon monoxide (CO) is a gas that is toxic to various organisms including humans and even microbes; however, it has low redox potential, which can fuel certain microbes, namely, CO oxidizers. Hydrogenogenic CO oxidizers utilize an energy conservation system via a CO dehydrogenase/energy-converting hydrogenase complex to produce hydrogen gas, a zero emission fuel, by CO oxidation coupled with proton reduction. Biochemical and molecular biological studies using a few model organisms have revealed their enzymatic reactions and transcriptional response mechanisms using CO. Biotechnological studies for CO-dependent hydrogen production have also been carried out with these model organisms. In this chapter, we review recent advances in the studies of these microbes, which reveal their unique and versatile metabolic profiles and provides future perspectives on ecological roles and biotechnological applications. Over the past decade, the number of isolates has doubled (37 isolates in 5 phyla, 20 genera, and 32 species). Some of the recently isolated ones show broad specificity to electron acceptors. Moreover, accumulating genomic information predicts their unique physiologies and reveals their phylogenomic relationships with novel potential hydrogenogenic CO oxidizers. Combined with genomic database surveys, a molecular ecological study has unveiled the wide distribution and low abundance of these microbes. Finally, recent biotechnological applications of hydrogenogenic CO oxidizers have been achieved via diverse approaches (e.g., metabolic engineering and co-cultivation), and the identification of thermophilic facultative anaerobic CO oxidizers will promote industrial applications as oxygen-tolerant biocatalysts for efficient hydrogen production by genomic engineering.

1. Introduction

Carbon monoxide (CO) is a trace gas in the atmosphere which occurs at a mixing ratio of 0.06–0.15 ppm (Mörsdorf, Frunzke, Gadkari, & Meyer, 1992). The total budget of atmospheric CO is 2000–3000 Tg year^{-1} (Khalil & Rasmussen, 1990; Mörsdorf et al., 1992; Schade & Crutzen, 1999), and about half of CO emissions are from anthropogenic sources. The remaining half of the CO emissions are from natural chemical processes (e.g., photochemical and thermochemical degradation of organic matter) and biological processes (e.g., production by microbes, leaves, roots, and animals) in volcanic, fresh water, marine, and terrestrial environments

(Conrad, 1996; Conte, Szopa, Séférian, & Bopp, 2019; Khalil & Rasmussen, 1990; King & Weber, 2007; Mörsdorf et al., 1992).

CO binds with high affinity to many ferrous heme-containing proteins, including hemoglobin, myoglobin, and cytochrome oxidase, which results in the inhibition of these proteins (Wu & Wang, 2005). Many people are killed every year from poisoning by this colorless, tasteless, and odorless gas called a "silent killer" (Rose et al., 2017). CO is also toxic for many microbes (Wareham, Southam, & Poole, 2018). CO-releasing molecules show antimicrobial action against pathogens like *Escherichia coli* and *Staphylococcus aureus* (Nobre, Seixas, Romão, & Saraiva, 2007), by inhibiting the intracellular respiratory heme enzymes (Davidge et al., 2009) and the tricarboxylic acid cycle iron-sulfur enzymes (Carvalho, Marques, Romão, & Saraiva, 2019), or by generation of reactive oxygen species (Tavares et al., 2011). Growth inhibition by CO gas has also been observed in sulfate reducers, methanogens, and acetogens (Davidova, Tarasova, Mukhitova, & Karpilova, 1994; Parshina, Kijlstra, et al., 2005).

Meanwhile, CO has one of the lowest-redox potentials ($E^{0\prime} = -520$ mV, $CO + H_2O \leftrightarrow CO_2 + 2H^+ + 2e^-$) in inorganic compounds available in the biosphere (Thauer, Jungermann, & Decker, 1977) (Fig. 1). Therefore, CO oxidation couples with the reduction of various electron acceptors, including low potential electron carriers like ferredoxin. Some microbes (so-called carboxydotrophs) can oxidize CO and obtain reducing power, which drives various metabolic pathways including energy conservation and carbon

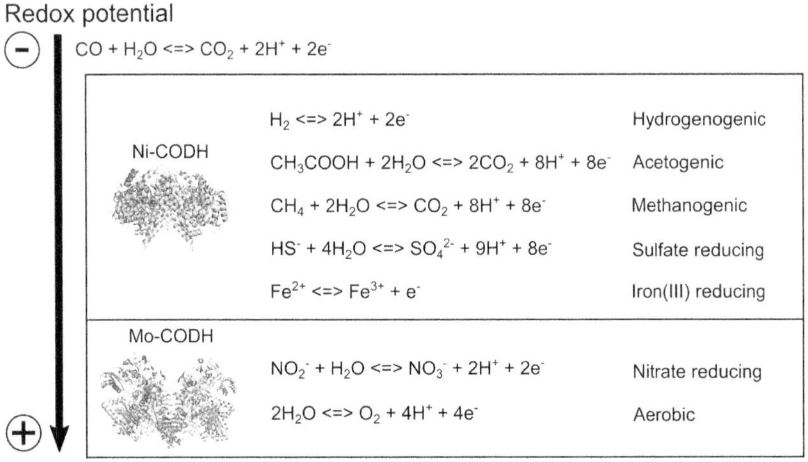

Fig. 1 Carbon monoxide oxidation by two types of carbon monoxide dehydrogenases and their coupled reactions for energy conservation.

fixation (Diender, Stams, & Sousa, 2015; Oelgeschläger & Rother, 2008; Sokolova et al., 2009; Techtmann, Colman, & Robb, 2009). Note that the term "carboxydotrophs" was originally coined for microbes with aerobic, respiratory, and chemolithoautotrophic utilization of CO as a sole source of carbon and energy (Meyer & Schlegel, 1983). However, to date, anaerobic or heterotrophic members which use CO only as an energy source are also reported and called "carboxydotrophs" (see Section 2). In this review, microbes that can grow with CO as an energy source will be called "CO oxidizers" and distinguished from "carboxydotrophs," which use CO as a metabolic building block.

To the best of our knowledge, all enzymes involving CO are related to CO dehydrogenases (CODHs) (Xavier, Preiner, & Martin, 2018). CODHs are divided into two phylogenetically and structurally distinct groups: an aerobic molybdenum- and copper-containing CODH (Mo-CODH) belonging to the xanthine oxidase family (Hille, Dingwall, & Wilcoxen, 2015; King & Weber, 2007), and an anaerobic nickel-containing CODH (Ni-CODH) belonging to the hybrid-cluster protein family (Aragão, Mitchell, Frazão, Carrondo, & Lindley, 2008; Inoue, Nakamoto, et al., 2019) (Fig. 1). Mo-CODHs can reduce quinones and utilize O_2 and nitrate as terminal electron acceptors in aerobic CO oxidizers (King, 2006; Wilcoxen, Zhang, & Hille, 2011), whereas Ni-CODHs can directly reduce ferredoxin and therefore utilize various types of terminal electron acceptors, such as protons, CO_2, sulfur compounds, and ferric iron (Fe(III)) in anaerobic CO oxidizers (Diender et al., 2015; Oelgeschläger & Rother, 2008; Sokolova et al., 2009). Ni-CODH is also a component of the CO-methylating acetyl-CoA synthase complex involved in CO_2 fixation via the carbonyl branch of the Wood-Ljungdahl pathway, with the CO intermediate finally being combined with coenzyme-A and a methyl group from a corrinoid-iron sulfur protein to synthesize acetyl-CoA (Can, Armstrong, & Ragsdale, 2014; Doukov, Iverson, Seravalli, Ragsdale, & Drennan, 2002; Gong et al., 2008; Nitschke & Russell, 2013; Schuchmann & Müller, 2014). In this pathway, CO can also be a direct substrate of the complex (Tan, Loke, Fitch, & Lindahl, 2005). Ni-CODHs are found in phylogenetically and physiologically diverse anaerobes which can grow by autotrophy, heterotrophy, fermentation, or anaerobic respiration (Inoue, Nakamoto, et al., 2019); hydrogenogens, acetogens, methanogens, sulfurcompounds reducers, and Fe(III) reducers have been described as anaerobic CO oxidizers (Diender et al., 2015; Oelgeschläger & Rother, 2008; Sokolova et al., 2009).

Among them, hydrogenogenic CO oxidizers couple CO oxidation with proton reduction to produce hydrogen (H_2), in which the proton or sodium motive force is generated with residual energy via a Ni-CODH and energy-converting hydrogenase (ECH) complex (Fox, Kerby, Roberts, & Ludden, 1996; Schoelmerich & Müller, 2019; Soboh, Linder, & Hedderich, 2002). This machinery is very simple because only water is required as an electron acceptor, and only a hydrogenase and an ATP synthase are required for the respiration system, which provides an implication for ancient energy conservation (Schoelmerich & Müller, 2019). Moreover, CO-dependent H_2 production by hydrogenogenic CO oxidizers is considered a "safety valve" to reduce toxic CO and supply H_2, which is an energy source for H_2-utilizing microbial communities (Techtmann et al., 2009). H_2 fuel has now expanded to human society as a zero emission fuel and CO is contained in syngas or industrial waste gases (Dürre & Eikmanns, 2015). Thus, hydrogenogenic carboxydotrophy has potential for biotechnological applications to produce H_2 from CO in syngas or industrial waste gases.

In the past decade, novel hydrogenogenic CO oxidizers have been isolated and several important milestones have been achieved in the biochemistry, physiology, ecology, and biotechnology of the organisms. Moreover, current progress in next-generation sequencing techniques has expanded the available information associated with microbial genomes including hydrogenogenic CO oxidizers, which has enabled comprehensive and comparative analyses of their genomic information and physiology. In this chapter, the physiology, genomic features, evolution, ecology, and biotechnological applications of hydrogenogenic CO oxidizers have been reviewed. Further, we provide the state-of-the-art on the study of hydrogenogenic CO oxidizers and future perspectives.

2. Isolates of hydrogenogenic CO oxidizing prokaryotes

In the 1970s, microbial hydrogenogenic CO oxidizing activity was reported for the first time in *Rubrivivax gelatinosus* 1 (formerly *Rhodopseudomonas gelatinosa* 1) by Uffen and coworkers (Dashekvicz & Uffen, 1979; Uffen, 1976). They also reported on *Rhodospirillum rubrum* ATCC 11170 (Dashekvicz & Uffen, 1979), a model of hydrogenogenic CO oxidizer whose Ni-CODH/ECH enzyme complex was purified and biochemically characterized (Ensign & Ludden, 1991; Fox, Kerby, et al., 1996). After these discoveries in mesophilic phototrophs isolated from fresh water, *Carboxydothermus hydrogenoformans* Z-2901 was the first thermophilic, hydrogenogenic CO oxidizer isolated from a hot spring (Svetlichny et al., 1991), which also possessed the biochemically characterized Ni-CODH/ECH enzyme complex (*Ch*CODH-I/ECH) (Soboh et al., 2002). The hydrogenogenic CO oxidizers have also been isolated from deep-sea hydrothermal vents; *Thermococcus* sp. AM4 was the first hyperthermophilic archaeon discovered which can grow by hydrogenogenic CO oxidation (Sokolova, Jeanthon, et al., 2004). *Thermococcus onnurineus* NA1 is another representative hydrogenogenic CO oxidizing archaeon (Bae et al., 2006), whose hydrogen production has been remarkably enhanced by biotechnological studies (see Section 6).

In addition to these representative species, phylogenetically diverse prokaryotes which can grow by hydrogenogenic CO oxidation have also been reported, mainly due to the isolation efforts by Russian research groups in hydrothermal fields (Novikov, Sokolova, Lebedinsky, Kolganova, & Bonch-Osmolovskaya, 2011; Slepova, Sokolova, Kolganova, Tourova, & Bonch-Osmolovskaya, 2009; Slepova et al., 2006; Sokolova, González, et al., 2004; Sokolova et al., 2001, 2007, 2009, 2005, 2002; Zavarzina et al., 2007). Hydrogenogenic CO oxidizers were reviewed in detail in 2009 (Sokolova et al., 2009; Techtmann et al., 2009); however, 15 isolates have been reported as novel hydrogenogenic CO oxidizers since that time (Table 1). The purpose of this section is to organize and update information on hydrogenogenic CO oxidizing isolates to understand their diversity.

2.1 Overview of isolates of hydrogenogenic CO oxidizing prokaryotes

Today, 37 hydrogenogenic CO oxidizing isolates in 5 phyla, 20 genera, and 32 species have been reported (Table 1). Thermophilic bacteria of the phylum Firmicutes (11 genera; 21 species; 23 isolates), mesophilic bacteria of the phylum Proteobacteria (6 genera; 6 species; 7 isolates), hyperthermophilic archaea of the phyla Euryarchaeota (1 genus; 3 species; 5 isolates) and Crenarchaeota (1 genus; 1 species; 1 isolate), and hyperthermophilic bacteria of the phylum Dictyoglomi (1 genus; 1 species; 1 isolate) have been described so far. Hydrogenogenic CO oxidizers have been isolated from various settings such as marine environments (deep-sea hydrothermal vent and marine sediment), terrestrial aquatic environments (hot spring and fresh water), soils, and bioreactors (Table 1).

Firmicutes include the most hydrogenogenic CO oxidizers in prokaryotic phyla. As described below, some members of Firmicutes show versatile CO metabolism or obligate carboxydotrophy, which implies that they are

Table 1 Isolates of hydrogenogenic CO oxidizing prokaryotes.

Organism	Optimum growth conditions[a]	Feature[b]	Electron acceptor[c]	Isolation source	Reference	RefSeq Genome accession
Archaea; Crenarchaeota; Thermoprotei; Thermoproteales; Thermofilaceae						
Thermofilum carboxyditrophus 1505	90/6.5/n.r.	L	H^{+}*	Hot spring	Sokolova et al. (2009)	GCF_000813245.1
Archaea; Euryarchaeota; Thermococci; Thermococcales; Thermococcaceae						
Thermococcus barophilus CH5	(80)/(6.5)/(18–35)	O/L	H^{+}*, S^{0}*	Deep-sea hydrothermal fields	Kim et al. (2010) and Kozhevnikova et al. (2016)	GCF_001433455.1
Thermococcus barophilus CH1	(80)/(6.5)/(18–35)	L	H^{+}*	Deep-sea hydrothermal fields	Kim et al. (2010) and Kozhevnikova et al. (2016)	
Thermococcus barophilus MP	85/7.0/20–30	O/L	H^{+}*, S^{0}*	Deep-sea hydrothermal fields	Kozhevnikova et al. (2016) and Marteinsson et al. (1999)	GCF_000151105.2
Thermococcus onnurineus NA1	80/8.5/35	O/L	H^{+}*, S^{0}	Deep-sea hydrothermal fields	Bae et al. (2006)	GCF_000018365.1
Thermococcus sp. AM4	87.5/7.0/20	O/L	H^{+}*, S^{0}*	Active chimney	Sokolova, Jeanthon, et al. (2004)	GCF_000151205.2

Continued

Table 1 Isolates of hydrogenogenic CO oxidizing prokaryotes.—cont'd

Organism	Optimum growth conditions[a]	Feature[b]	Electron acceptor[c]	Isolation source	Reference	RefSeq Genome accession
Bacteria; Dictyoglomi; Dictyoglomia; Dictyoglomales; Dictyoglomaceae						
Dictyoglomus carboxydivorans	80/6.5/(0)	L	H^{+}*	Hot spring	Sokolova et al. (2009)	
Bacteria; Firmicutes; Bacilli; Bacillales; Bacillaceae						
Parageobacillus thermoglucosidasius DSM 2542[a]	61–63/6.5–8.5/(5)	O/L/F	H^{+}*, O_2	Soil	Mohr, Aliyu, Küchlin, Polliack, et al. (2018) and Suzuki et al. (1983)	GCF_001295365.1
Parageobacillus thermoglucosidasius TG4	(65)/(N)/(18)	O/L/F	H^{+}*, O_2	Marine sediment	Inoue, Tanimura, et al. (2019)	GCF_003865195.1
Bacteria; Firmicutes; Clostridia; Clostridiales; Family XVI						
Carboxydocella sp. JDF658	(T)/(N)/n.r.	L	H^{+}*	Open-air stream from a hot spring well	Fukuyama, Oguro, et al. (2017)	GCF_002049395.1
Carboxydocella sp. ULO1	(T)/(N)/n.r.	L	H^{+}*	Sediment of a maar lake	Fukuyama, Oguro, et al. (2017)	GCF_002049255.1
Carboxydocella sporoproducens DSM 16521	60/6.8/(0)	L/F	H^{+}*	Hot spring	Slepova et al. (2006)	GCF_900167165.1
Carboxydocella thermautotrophica 019	58/7.0/(0)	O/L/F	H^{+}*, Fe(III)*	Thermal field	Toshchakov et al. (2018)	GCF_003047205.1
Carboxydocella thermautotrophica 41	58/7.0/(0)	L	H^{+}*	Terrestrial hot vent	Sokolova et al. (2002)	GCF_003054495.1

Bacteria; Firmicutes; Clostridia; Clostridiales; Peptococcaceae						
Desulfotomaculum nigrificans CO-1-SRB	55/6.8–7.2/8	O/L/F	H^+*, SO_4^{2-}*	Anaerobic bioreactor sludge	Parshina, Kijlstra, et al. (2005) and Parshina, Sipma, et al. (2005)	GCF_000214435.1
Thermincola carboxydiphila 2204	55/8.0/(0)	L	H^+*	Hot spring	Sokolova et al. (2005)	
Thermincola ferriacetica Z-0001	57–60/7.0–7.2/0	O/L	H^+*, Fe(III), magnetite, AQDS, MnO_2	Ferric deposits of a terrestrial hydrothermal spring	Zavarzina et al. (2007)	GCF_001263415.1
Thermincola potens JR	(56)/(6.8)/n.r.	O/L	H^+*, AQDS, Fe(III)	Thermophilic microbial fuel cell	Byrne-Bailey et al. (2010) and Wrighton et al. (2008)	GCF_000092945.1
Bacteria; Firmicutes; Clostridia; Thermoanaerobacterales; Thermoanaerobacteraceae						
Caldanaerobacter subterraneus subsp. *pacificus* DSM 12653	70/6.8–7.2/20.5–25.5	L/F	H^+*	Deep-sea hydrothermal fields	Sokolova et al. (2001)	GCF_000156275.2
Calderihabitans maritimus KKC1	65/7.0–7.5/24.6	O/L	H^+*, Fe(III)*, Fe_2O_3*, AQDS*, SO_3^{2-}*, $S_2O_3^{2-}$*, fumarate*	Submerged marine caldera	Yoneda, Yoshida, Yasuda, et al. (2013)	GCF_002207765.1
Carboxydothermus hydrogenoformans Z-2901	70–72/6.8–7.0/n.r.	O/L	H^+*, AQDS*, fumarate*, $S_2O_3^{2-}$, SO_3^{2-}, S^0, NO_3^-	Hot swamp	Henstra and Stams (2004) and Svetlichny et al. (1991)	GCF_000012865.1

Continued

Table 1 Isolates of hydrogenogenic CO oxidizing prokaryotes.—cont'd

Organism	Optimum growth conditions[a]	Feature[b]	Electron acceptor[c]	Isolation source	Reference	RefSeq Genome accession
Carboxydothermus islandicus SET	65/5.5–6.0/(0)	L/F	H^{+}*	Hot spring	Novikov et al. (2011)	GCF_001950325.1
Carboxydothermus pertinax Ug1	65/6.0–6.5/(0.1)	O/L	H^{+}*, Fe(III)*, AQDS*, $S_2O_3^{2-}$*, S^0*	Acidic hot spring	Yoneda et al. (2012)	GCF_001950255.1
Carboxydothermus siderophilus 1315	65/6.5–7.2/(0)	O/L	H^{+}*, Fe(III)*, AQDS*	Hot spring	Slepova et al. (2009)	
Moorella stamsii DSM 26271	65/7.5/(0.3)	O/L/F	H^{+}*, AQDS, NO_3^-, ClO_4^-	Anaerobic sludge	Alves, van Gelder, Alves, Sousa, and Plugge (2013)	GCF_002995805.1
Moorella thermoacetica DSM 21394	60–65/6.9/0–8.7	O/L/F	H^{+}*, $S_2O_3^{2-}$	Anaerobic bioreactors	Jiang et al. (2009)	GCF_001875325.1
Thermoanaerobacter kivui DSM 2030	66/6.4/(0.45)	O/L	H^{+}*, CO_2	Lake sediment	Leigh et al. (1981) and Weghoff and Müller (2016)	GCF_000763575.1
Thermoanaerobacter thermohydrosulfuricus subsp. *carboxydovorans* TLO	70/6.3–6.8/(0.45)	O/L/F	H^{+}*, $S_2O_3^{2-}$, S^0, SO_3^{2-}, Fe(III), MnO_2, AQDS, AsO_4^{3-}	Hot spring	Balk et al. (2009)	

Bacteria; Firmicutes; Negativicutes; Selenomonadales; Sporomusaceae						
Thermosinus carboxydivorans Nor1	60/6.8–7.0/(0)	O/L/F	H^{+}*, Fe(III)*, $S_2O_3^{2-}$	Hot spring	Sokolova, González, et al. (2004)	GCF_000169155.1
Bacteria; Firmicutes; Thermolithobacteria; Thermolithobacterales; Thermolithobacteraceae						
Thermolithobacter carboxydivorans R1	73/7.1–7.3/(0)	L	H^{+}*	Hot spring	Sokolova et al. (2007)	
Bacteria; Proteobacteria; Alphaproteobacteria; Rhizobiales; Bradyrhizobiaceae						
Rhodopseudomonas palustris P4	(30)/(N)/n.r.	O/L/P	H^{+}*, O_2	Anaerobic wastewater sludge	Jung, Jung, Kim, Ahn, and Park (1999)	
Bacteria; Proteobacteria; Alphaproteobacteria; Rhizobiales; Methylocystaceae						
Pleomorphomonas carboxyditropha SVCO-16	30/6.5–7.3/(0)	O/L/F	H^{+}*, NO_3^-	Anaerobic sludge	Esquivel-Elizondo, Maldonado, and Krajmalnik-Brown (2018)	GCF_002770725.1
Bacteria; Proteobacteria; Alphaproteobacteria; Rhodospirillales; Rhodospirillaceae						
Rhodospirillum rubrum ATCC 11170	(30)/(6.7–6.9)/n.r.	O/L/P/F	H^{+}*, O_2	Fresh water	Dashekvicz and Uffen (1979) and Munk et al. (2011)	GCF_000013085.1
Bacteria; Proteobacteria; Betaproteobacteria; Burkholderiales; unclassified Burkholderiales; Burkholderiales Genera incertae sedis						
Rubrivivax gelatinosus 1	(34)/(7.2–7.4)/n.r.	O/L/P/F	H^{+}*, O_2	Fresh water	Dashekvicz and Uffen (1979) and Uffen (1976)	
Rubrivivax gelatinosus CBS	(M)/(N)/n.r.	L/P	H^{+}*	Soil	Maness and Weaver (1994, 2002)	GCF_000257145.1

Continued

Table 1 Isolates of hydrogenogenic CO oxidizing prokaryotes.—cont'd

Organism	Optimum growth conditions[a]	Feature[b]	Electron acceptor[c]	Isolation source	Reference	RefSeq Genome accession
Bacteria; Proteobacteria; Gammaproteobacteria; Enterobacterales; Enterobacteriaceae						
Citrobacter amalonaticus Y19	36/6.0–7.0/n.r.	O/L	H^{+}*, O_2	Anaerobic wastewater sludge digester	Jung, Kim, Jung, Park, and Park (1999) and Oh, Seol, Kim, and Park (2003)	GCF_000981805.1
Bacteria; Proteobacteria; delta/epsilon subdivisions; Epsilonproteobacteria; Campylobacterales; Campylobacteraceae						
Sulfurospirillum carboxydovorans MV	30.2/7.3/20	O/L/F	H^{+}*, $S_2O_3^{2-}$*, S^0*, DMSO*, NO_3^-, NO_2^-, C_2Cl_4, AsO_4^{3-}, SeO_4^{2-}, O_2	Marine sedimen	Jensen and Finster (2005)	

"n.r." indicates "not reported."

[a]Optimum growth conditions are shown by optimum temperature (°C)/pH/salinity (g/L NaCl). When the optimum growth conditions could not be identified, the growth conditions where the strains are cultivated are shown in parenthesis, or indicated by T (thermophilic), M (mesophilic), or N (neutrophilic). Growth conditions which are not optimal are shown in parenthesis.

[b]Metabolic features are shown by O (chemoorganotrophy), L (chemolithotrophy), P (phototrophy), or F (fermentation).

[c]An electron acceptor whose reduction is coupled with CO oxidation is indicated by asterisk.

"expert CO utilizers." On the other hand, hydrogenogenic CO oxidizers in the other phyla (e.g., Proteobacteria) show sparse phylogenetic distribution unlike Firmicutes. All of them can grow by chemoorganotrophy and phototrophy, which indicates that hydrogenogenic CO oxidation is an accessory energy conservation pathway for these prokaryotes. Notably, the ability of hydrogenogenic CO oxidation is not usually tested in most prokaryotic isolates, and the diversity of hydrogenogenic CO oxidizers might be underestimated (see Section 5 for a comprehensive genomic survey of hydrogenogenic CO oxidizers). Further isolation efforts are needed to understand the diversity and evolution of hydrogenogenic CO oxidizers.

2.2 Firmicutes

More than half of the hydrogenogenic CO oxidizing isolates have been found in the phylum Firmicutes (Table 1). These members are composed of thermophilic, neutrophilic bacteria, whose optimal growth temperature and pH are 55–73 and 6.5–7.5 °C, respectively, except for slightly alkaliphilic *Thermincola carboxydiphila* 2204 (Sokolova et al., 2005) and slightly acidophilic *Carboxydothermus islandicus* SET (Novikov et al., 2011), *Carboxydothermus pertinax* Ug1 (Yoneda et al., 2012), and *Thermoanaerobacter kivui* DSM 2030 (Leigh, Mayer, & Wolfe, 1981). Most of the Firmicutes hydrogenogenic CO oxidizers are included in strictly anaerobic groups of the class Clostridia (8 genera; 18 species; 19 isolates), Negativicutes (1 genus; 1 species; 1 isolate), and Thermolithobacteria (1 genus; 1 species; 1 isolate). This might be because Ni-CODH is an anaerobic enzyme and sensitive to O_2 (Can et al., 2014). However, facultative anaerobic hydrogenogenic CO oxidizers of Bacilli (1 genus; 1 species; 2 isolates) have also been recently reported (Inoue, Tanimura, et al., 2019; Mohr, Aliyu, Küchlin, Polliack, et al., 2018). These thermophilic, O_2-tolerant, hydrogenogenic CO oxidizers have a potential for biotechnological application (see Section 6).

Strictly anaerobic, hydrogenogenic CO oxidizers of Firmicutes include chemoorganotrophic and chemolithotrophic bacteria, some of which can also grow by fermentation (Table 1). Many members can utilize sulfur compounds and iron as terminal electron acceptors; several can couple CO oxidation not only with proton reduction (H_2 production), but also with sulfate reduction (e.g., *Desulfotomaculum nigrificans* CO-1-SRB) and Fe(III) reduction (e.g., *Carboxydocella thermautotrophica* 019 and *Thermosinus carboxydivorans* Nor1) (Table 1). In particular, the recently isolated clostridia strain *C. pertinax* Ug1 (Yoneda et al., 2012) and *Calderihabitans maritimus*

KKC1 (Yoneda, Yoshida, Yasuda, Imada, & Sako, 2013) can utilize various substrates such as Fe(III), thiosulfate, and AQDS as electron acceptors during CO oxidation. On the other hand, the Clostridia class also includes obligate chemolithoautotrophs like *C. thermautotrophica* 41 and *T. carboxydiphila* 2204, which grow exclusively by hydrogenogenic CO oxidation (Sokolova & Lebedinsky, 2013).

In Bacilli, which are facultative anaerobes, only two strains of *Parageobacillus thermoglucosidasius* are reported to be hydrogenogenic CO oxidizers (Table 1). *P. thermoglucosidasius* DSM 2542 can grow by aerobic chemoorganotrophy and fermentation. Under 50% CO and 50% air conditions, it first consumes residual oxygen and then shifts to anaerobic hydrogenogenic CO oxidation (Mohr, Aliyu, Küchlin, Polliack, et al., 2018). *P. thermoglucosidasius* TG4 can grow hydrogenogenically under 100% CO (Inoue, Tanimura, et al., 2019).

The thermophilic, hydrogenogenic CO oxidizers in Firmicutes described before 2010 had been isolated from hot environments such as deep-sea hydrothermal vent, hot spring, and anaerobic bioreactors. However, despite being thermophilic, some Firmicutes members reported as hydrogenogenic CO oxidizer after 2010 were from mesophilic environments such as marine sediment (e.g., *P. thermoglucosidasius* TG4 and *C. maritimus* KKC1), lake sediment (e.g., *Carboxydocella* sp. ULO1 and *T. kivui* DSM 2030), and soil (e.g., *P. thermoglucosidasius* DSM 2542) (Table 1). Whether these bacteria are active or not in the mesophilic environments remains unclear; nevertheless, the distribution of thermophilic, hydrogenogenic CO oxidizers might be wider than previously thought. It is possible that endospore formation might contribute to the wide distribution of the Firmicutes thermophilic members, as previous studies have indicated (Hubert et al., 2009; Müller et al., 2014).

2.3 Proteobacteria

Proteobacteria is the second largest phylum of hydrogenogenic CO oxidizers, which are composed of mesophilic and neutrophilic bacteria (Table 1). The members of this phylum are found in diverse classes (alpha-, beta-, gamma-, and epsilonproteobacteria). Hydrogenogenic CO oxidizers in Proteobacteria include phototrophs such as *R. rubrum* ATCC 11170 and *R. gelatinosus* CBS (Table 1). Other members are anaerobes (e.g., *Pleomorphomonas carboxyditropha* SVCO-16) or facultative anaerobes (e.g., *Citrobacter amalonaticus* Y19 and *Sulfurospirillum carboxydovorans* MV) which can grow by chemoorganotrophy or chemolithotrophy (Table 1). Among the members of Proteobacteria, only *S. carboxydovorans* MV can couple CO oxidation with non-hydrogenogenic reactions such as the reduction of thiosulfate, elemental sulfur, and DMSO (Table 1). Proteobacteria members have been isolated from mesophilic environments including fresh water, marine sediment, soil, and anaerobic sludge (Table 1).

2.4 Euryarchaeota

Five members of the genus *Thermococcus* isolated from deep-sea hydrothermal vent are included in hydrogenogenic CO oxidizers of Euryarchaeota (Table 1). *Thermococcus* archaea are hyperthermophiles whose optimal growth temperature exceeds 80 °C, and are neutrophiles except for *T. onnurineus* NA1 (Table 1). They are obligate anaerobic chemoorganotrophs and utilize elemental sulfur as an electron acceptor. *Thermococcus barophilus* CH5, *T. barophilus* MP, and *Thermococcus* sp. AM4 can couple CO oxidation with the reduction of elemental sulfur (Table 1). It may be assumed that just CO of abiotic origin present in deep-sea hydrothermal environments is the substrate for *Thermococcus* representatives (Kozhevnikova, Taranov, Lebedinsky, Bonch-Osmolovskaya, & Sokolova, 2016).

2.5 Crenarchaeota and dictyoglomi

Hydrogenogenic CO oxidizers are also reported in two other phyla, Crenarchaeota and Dictyoglomi (Table 1), although the isolates have never been validly described. *Thermofilum carboxyditrophus* 1505 is a hyperthermophilic, neutrophilic archaeon of the Crenarchaeota (the order Thermoproteales) isolated from the terrestrial hot spring of Uzon Caldera (Kamchatka) (Sokolova et al., 2009). *Dictyoglomus carboxydivorans* is a hyperthermophilic, neutrophilic bacteria of the Dictyoglomi isolated from the terrestrial hot spring of Uzon Caldera (Sokolova et al., 2009).

3. Energy conservation mechanisms of hydrogenogenic CO oxidizing prokaryotes

Species that undergo hydrogenogenic carboxydotrophy have unique molecular mechanisms. Owing to the very low redox potential of CO ($E^{0\prime} = -520$ mV), hydrogenogenic CO oxidizers gain energy by coupling CO oxidation with reducing protons ($E^{0\prime} = -414$ mV) via an energy conservation system mediated by ferredoxin ($E^{0\prime} = -414$ to -500 mV)

(Buckel & Thauer, 2013, 2018; Diender et al., 2015; Müller, Chowdhury, & Basen, 2018). The Ni-CODH/ECH complex is considered the key enzyme of hydrogenogenic CO metabolism, as described in Section 1. In this section, we introduce an energy conservation mechanism and the versatility of hydrogenogenic CO metabolism with the increasing number of hydrogenogenic CO oxidizing isolates and their genomic information.

3.1 Variation of ferredoxin-mediated energy conservation systems

Ferredoxin with low redox potentials can mediate several energy conservation systems. To reduce ferredoxin, most of microbes use reducing power from organic compounds. For instance, energy conservation via formate- and pyruvate-dependent H_2 production is conducted by a formate-hydrogenlyase complex (Kim et al., 2010; Sinha, Roy, & Das, 2015) and a pyruvate-ferredoxin oxidoreductase/ECH pathway (Hallenbeck, 2009; Kruse, Goris, Westermann, Adrian, & Diekert, 2018), respectively. On the other hand, Ni-CODH can directly reduce ferredoxin using very low redox potential of CO ($E^{0\prime} = -520$ mV) for energy conservation in CO oxidizers (Buckel & Thauer, 2013, 2018; Diender et al., 2015; Müller et al., 2018). As another option of ferredoxin-mediated energy conservation, flavin-based electron bifurcation (FBEB) was recently discovered as a fundamental mechanism in which low potential electrons for ferredoxin reduction are generated from compounds with higher redox potentials than ferredoxin, such as H_2, NAD(P)H, and $F_{420}H_2$ (Buckel & Thauer, 2018). The FBEB enzyme complexes split hydride electron pairs into one electron with a more positive reduction potential (exergonic reaction) and one with a more negative reduction potential (endergonic reaction) using flavoproteins. For instance, *T. kivui* reduces ferredoxin from H_2 by FBEB hydrogenase (HydABC complex) for energy conservation with acetate production (Hess, Poehlein, Weghoff, Daniel, & Müller, 2014).

The reduced ferredoxin is consumed in generation of proton or sodium motive force by energy conservation machineries, such as ECH and Rnf (*Rhodobacter* nitrogen fixation). The ECH machinery is a multimeric transmembrane protein complex and is evolutionarily related to the NADH-oxidizing quinone-reducing respiratory complex I (Yu et al., 2018). The catalytic subunits of ECH contain a H_2-evolving [NiFe] center, electron-transferring iron-sulfur clusters, and are phylogenetically classified into group 4 [NiFe] hydrogenases (Greening et al., 2015; Vignais & Billoud, 2007; Yu et al., 2018). The coupling of Ni-CODH with ECH, which form

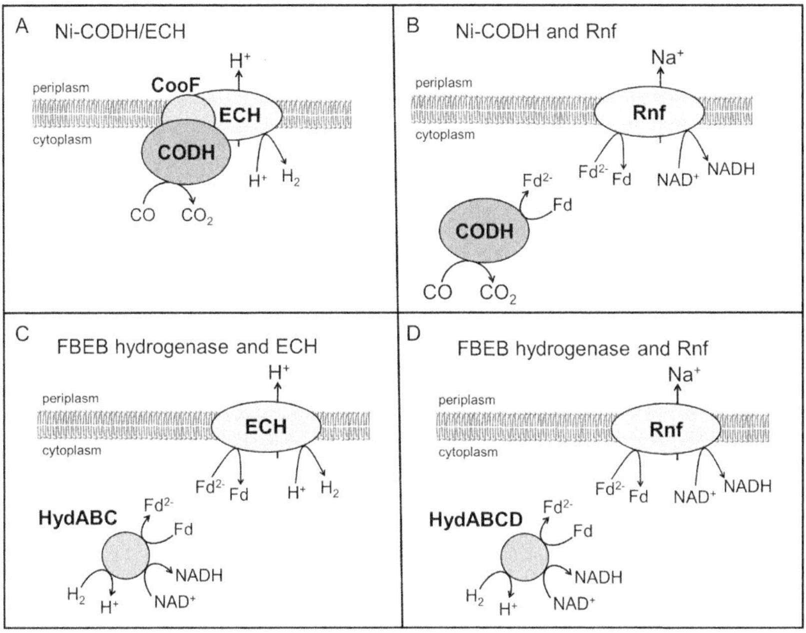

Fig. 2 Schemes of ferredoxin-mediated energy conservations. (A) Ni-CODH/ECH complex in *Carboxydothermus hydrogenoformans*; (B) Ni-CODH and Rnf in *Clostridium ljungdahlii*; (C) FBEB hydrogenase and ECH in *Moorella thermoacetica*; (D) FBEB hydrogenase and Rnf in *Acetobacterium woodii*. Ni-CODH, Nickel-containing carbon monoxide dehydrogenase; ECH, energy-converting hydrogenase; Rnf, energy-converting ferredoxin: NAD^+ reductase; FBEB, flavin-based electron bifurcation.

complex on membrane, is responsible for energy conservation in hydrogenogenic CO metabolism (Fig. 2A). As another option, ferredoxin-mediated energy conservation is conducted by an Rnf complex (Saeki & Kumagai, 1998). The Rnf complex functions as an energy-converting ferredoxin: NAD^+ reductase couples ferredoxin oxidation with NAD^+ reduction to generate a proton or sodium ion gradient (Biegel, Schmidt, González, & Müller, 2011; Imkamp, Biegel, Jayamani, Buckel, & Müller, 2007; Müller, Imkamp, Biegel, Schmidt, & Dilling, 2008). The coupling of Ni-CODH with the Rnf complex is considered a key pathway for CO-dependent acetogenesis in various acetogens, such as *Clostridium ljungdahlii* (Köpke et al., 2010; Younesi, Najafpour, & Mohamed, 2005), *Acetobacterium woodii* (Bertsch & Müller, 2015), and *Eubacterium limosum* (Jeong et al., 2015) (Fig. 2B). Moreover, in *Methanosarcina acetivorans*, a Rnf-like complex is proposed to generate reduced F_{420} via the methanopterin cycle from ferredoxin in CO-dependent methanogenesis (Lessner et al., 2006; Rother & Metcalf, 2004). The reduced ferredoxin by FBEB enzymes is also consumed by ECH or Rnf machineries (Schuchmann & Müller, 2014) (Fig. 2C and D). Collectively, energy conservation via the Ni-CODH/ECH complex is quite simple compared with other ferredoxin-mediated energy conservation mechanisms, since the Ni-CODH/ECH complex can conserve energy without cofactors as electron acceptors and respiratory chains. Next, we introduce details of energy conservation mechanism of Ni-CODH/ECH.

3.2 How do hydrogenogenic CO oxidizers conserve energy?

Sequential reactions from CO oxidation to proton motive force generation are responsible for energy conservation in the Ni-CODH/ECH complex (Fig. 2A). The initial step of hydrogenogenic CO metabolism, CO oxidation is catalyzed by Ni-CODH (Bonam & Ludden, 1987; Soboh et al., 2002). Structural analyses of Ni-CODHs, such as *C. hydrogenoformans* Ni-CODH-II and *R. rubrum* Ni-CODH have shown that the homodimeric Ni-CODH contains five metal clusters; each subunit contains a catalytic C-cluster comprising of Ni, Fe, and S along with a [4Fe-4S] cubane-type cluster (B-cluster), and a subunit interface of the dimer that accommodates an additional [4Fe-4S] cluster (D-cluster) (Dobbek, Svetlitchnyi, Gremer, Huber, & Meyer, 2001; Doukov et al., 2002; Drennan, Heo, Sintchak, Schreiter, & Ludden, 2001). The acid-base catalysts His and Lys are located around the C-cluster (Fesseler, Jeoung, & Dobbek, 2015; Jeoung & Dobbek, 2007). Intramolecular electron transfer is conducted among these three metal clusters, whereas intermolecular electron transfer occurs between the solvent-accessible D-cluster and ferredoxin-like proteins, such as CooF (Dobbek et al., 2001; Ensign & Ludden, 1991; Kerby et al., 1992; Singer, Hirst, & Ludden, 2006). In next step, proton reduction with ferredoxin oxidation is catalyzed by the ECH machinery. The transmembrane proton/sodium translocation modules conduct energy conservation coupled with the redox reaction by the catalytic subunits (Yu et al., 2018). The generated electrochemical proton gradient then fuels ATP synthesis (Schoelmerich & Müller, 2019).

3.3 Ni-CODH–ECH gene clusters for hydrogenogenic CO metabolism

The Ni-CODH associated genes and ECH-associated genes forms Ni-CODH–ECH gene cluster (Sokolova et al., 2009). The functions of other Ni-CODHs also have been predicted from the genomic context of the gene clusters, including the Ni-CODH catalytic gene (*cooS*) (Inoue, Nakamoto, et al., 2019; Techtmann, Lebedinsky, Colman, Sokolova, Woyke, Goodwin, & Robb, 2012). In Ni-CODH gene clusters, the minimum components are genes encoding the catalytic subunit of Ni-CODH (*cooS*), ferredoxin-like protein (*cooF*), and Ni insertion protein (*cooC*) (Jeon, Cheng, & Ludden, 2001). However, the Ni-CODH gene cluster associated-ECH gene cluster are divided into three types: Coo type (group 4c), Hyc/Hyf type (group 4a), and Mbh-Mrp (group 4b) based on the phylogenetic subgroup of catalytic subunit gene in group 4 ECH gene cluster (Greening et al., 2015; Vignais & Billoud, 2007; Yu et al., 2018).

First, the Coo-type Ni-CODH–ECH gene cluster is represented in *R. rubrum* (Kerby et al., 1992) and *C. hydrogenoformans* (Wu et al., 2005) and includes genes for transcriptional regulation (*cooA*) (Youn, Kerby, Conrad, & Roberts, 2004). In the *C. hydrogenoformans* genome, the Coo-type ECH gene cluster is a component of genes for two membrane integral subunits (*cooMK*), four hydrophilic subunits (*cooLXUH*), and a Ni insertion protein (*hypA*) (Watanabe et al., 2009; Wu et al., 2005). The Coo-type ECH gene cluster inserts into the Ni-CODH gene cluster (*cooAC–cooMKLXUHhypA–cooFS*) (Wu et al., 2005). Second, the Hyc/Hyf-type Ni-CODH–ECH gene cluster is represented by *P. thermoglucosidasius* (Inoue, Tanimura, et al., 2019; Mohr, Aliyu, Küchlin, Polliack, et al., 2018) and *Moorella* species (Fukuyama, Tanimura, et al., 2019; Poehlein, Böer, Steensen, & Daniel, 2018). Genes for the transcriptional regulation of the Hyc/Hyf-type Ni-CODH—ECH gene cluster are unclear. *Caldanaerobacter subterraneus* acquired the Hyc/Hyf-type Ni-CODH–ECH gene cluster via horizontal gene transfer from *P. thermoglucosidasius*, which enabled it to grow with CO-dependent H_2 production (Sant'Anna, Lebedinsky, Sokolova, Robb, & Gonzalez, 2015). In the *P. thermoglucosidasius* genome, the Hyc/Hyf-type ECH gene cluster is a component of genes for the large and small catalytic subunits (*hycEG*), a membrane subunit (*hyfC*), a hydrogenase component (*hyfBEFDhycH*), and maturation proteins (*hycIhypAB*) (Sant'Anna, Lebedinsky, Sokolova, Robb, & Gonzalez, 2015). The Hyc/Hyf-type ECH gene cluster flanks the Ni-CODH gene cluster (*cooCSF–hyfBCEFFGHhycGHIhypAB*) (Sant'Anna, Lebedinsky, Sokolova, Robb, & Gonzalez, 2015). Finally, the Mbh-Mrp type Ni-CODH–ECH gene cluster is represented by *T. onnurineus* (Lee et al., 2008; Lim, Kang, Lebedinsky, Lee, & Lee, 2010). In the *T. onnurineus* genome, the Mbh-Mrp type ECH gene cluster is largely divided into two components: genes for [NiFe] hydrogenase subunits including the large and small catalytic subunit, and Na^+/H^+ antiporter subunits (Lim et al., 2010). The Mbh-Mrp type ECH gene cluster flanks the Ni-CODH gene cluster with genes for transcriptional regulation (CorQR) (Kim et al., 2015) and a hypothetical protein (Lee et al., 2008). During hydrogenogenic CO metabolism, Mbh-Mrp type Ni-CODH/ECH complexes promote CO-dependent H_2 production with a proton ion gradient, and then the Na^+/H^+ antiporter catalyzes sodium ion extrusion in exchange for proton ions to conserve energy via sodium ion driven ATP synthase.

3.4 Hydrogenogenic CO metabolism independent of the Ni-CODH–ECH gene cluster

Recently, hydrogenogenic CO metabolism independent of the Ni-CODH–ECH gene cluster has been reported (Fukuyama, Omae, Yoneda, Yoshida, & Sako, 2017; Weghoff & Müller, 2016). Originally, *T. kivui* was isolated as thermophilic and H_2 oxidizing acetogen (Leigh et al., 1981) using FBEB enzyme for energy conservation (see Section 3.1). Further, this species possess two gene clusters for Ni-CODH and two gene clusters for the ECH complex. However, these gene clusters do not form a Ni-CODH–ECH gene cluster (Hess et al., 2014). Interestingly, *T. kivui* acquired the ability of hydrogenogenic CO oxidation by cultivation with increasing CO concentrations (Weghoff & Müller, 2016). Under 100% CO, *T. kivui* consumes CO with H_2 production with Ni-CODH and ECH enzymatic activity (Weghoff & Müller, 2016), suggesting that its hydrogenogenic CO metabolism is independent of the Ni-CODH – ECH gene cluster.

The genus *Carboxydothermus* has been one of the most studied models of thermophilic carboxydotrophy. Comparative genome analysis in the genus *Carboxydothermus* revealed that *C. pertinax* lacks the Ni-CODH-I and respective transcriptional regulator-encoding gene (*cooA*) in its Ni-CODH–ECH gene cluster, despite undergoing hydrogenogenic carboxydotrophy (Fukuyama, Omae, et al., 2017). Gene expression analysis of *C. pertinax* during hydrogenogenic CO metabolism has shown that the transcripts of other Ni-CODH and ECH catalytic genes are remarkably upregulated. These results imply that they can alternatively couple other Ni-CODH to the distal ECH (Fukuyama, Omae, Yoneda, Yoshida, & Sako, 2018).

3.5 Diverse electron flows from CO in hydrogenogenic CO oxidizers

In the past decade, more than 10 genomes have been reported as anaerobic, thermophilic, hydrogenogenic CO oxidizers (Omae et al., 2019), which has extended our knowledge on hydrogenogenic CO metabolism. The genomic information combined with growth studies of hydrogenogenic CO oxidizers implies variations in energy conservation via additional coupling between CO oxidation and reduction of another electron acceptor. *C. thermautotrophica* 019 can grow by coupling hydrogenogenic CO oxidation with the dissimilatory reduction of Fe(III), unlike type 41^T (Toshchakov et al., 2018). The both strains harbor six Ni-CODH genes, which is the highest number of Ni-CODH genes among the known genomes of all prokaryotes (Toshchakov et al., 2018). The gene organization of their Ni-CODH–ECH gene cluster is unique, because two Ni-CODH are simultaneously encoded within the same gene cluster: *cooASC* (Ni-CODH gene cluster)–*cooMKLXUHhypA* (ECH gene cluster)–*cooFSC* (Ni-CODH gene cluster). Of the two *cooS*, the latter is phylogenetically similar to the *cooS* in the Ni-CODH–ECH gene cluster from *C. hydrogenoformans* (*Ch*CODH-I gene) (Wu et al., 2005), suggesting that this *cooS* is responsible for hydrogenogenic CO metabolism. Meanwhile, the former is phylogenetically distant from other ECH-associated Ni-CODHs and its function is currently unknown (Toshchakov et al., 2018). Further, only strain 019 harbors extra genes of a 17-heme cytochrome for dissimilatory iron reduction when compared with strain 41^T (Toshchakov et al., 2018). This difference correlates with the ability of Fe(III) reduction during hydrogenogenic CO metabolism in strain 019 (Toshchakov et al., 2018).

C. maritimus, isolated from marine sediment, can couple CO oxidation with the reduction of sulfite, thiosulfate, fumarate, Fe(III), 9,10-anthraquinone 2,6-disulfonate, or possibly MnO_2 as electron acceptors during hydrogenogenic CO metabolism (Yoneda, Yoshida, Yasuda, et al., 2013). Genome analysis of *C. maritimus* shows that it also harbors six Ni-CODH genes (Omae, Yoneda, Fukuyama, Yoshida, & Sako, 2017): three Ni-CODH genes, including ECH-associated Ni-CODH genes with well-known genomic contexts, and three others found in novel genomic contexts with unclear functions (Omae et al., 2017). Thus, some hydrogenogenic CO oxidizers, especially those possessing multiple Ni-CODH genes, appear to be highly dependent on CO as a very low potential electron donor and utilize it flexibly by simultaneously reducing various electron acceptors other than protons.

C. pertinax performs hydrogenogenic CO metabolism with iron reduction, in which H_2 production corresponds to approximately 69% of CO uptake (Yoneda et al., 2012), whereas CO utilization results in nearly equimolar H_2 release in other hydrogenogenic members. Further, *C. pertinax* can utilize various substrates as electron acceptors, including thiosulfate during hydrogenogenic CO metabolism (Yoneda et al., 2012) (see Section 2). A cultivation study of *C. pertinax* under 100% CO with or without thiosulfate showed that they efficiently grow with thiosulfate. Therefore, *C. pertinax* can simultaneously couple CO oxidation with H_2 and hydrogen sulfide production. In this case, *C. pertinax* conserves energy by shifting the electron flow from CO oxidation to the reduction of thiosulfate, probably due to the obscure coupling of Ni-CODH with distal ECH (Fukuyama et al., 2018) (see Section 3.3).

In addition to these organisms with diverse electron flows from CO, a recent biochemical study on *T. kivui* shows that H_2 from hydrogenogenic CO metabolism is recycled by FBEB hydrogenase for their energy conservation. In this process, an additional ferredoxin is reduced for proton motive force generation and NADH for acetogenesis (Schoelmerich & Müller, 2019). In this efficient energy conservation system, *T. kivui* acquires the reducing power from CO in multiple phases by using ferredoxin, H_2, and NADH as intermediators.

Overall, hydrogenogenic CO metabolism via the Ni-CODH/ECH complex is featured by simple energy conservation producing H_2; however, the electron flows from CO and the associated energy conservation systems are quite diverse. The reducing power from CO oxidation is simultaneously coupled not only to H_2 production, but also to other metabolisms inherent in each prokaryote. Moreover, the high energy gas product H_2 may be used for H_2 cycling (Eckert et al., 2019; Heidelberg et al., 2004) or other H_2 oxidizing microbes (Techtmann et al., 2009). Coupling of hydrogenogenic CO metabolism and other novel energy conservation systems may be found in other microbes or even in microbial interactions.

4. Response mechanisms to CO in hydrogenogenic CO metabolism

So far, the known CO-responsive transcriptional activator during hydrogenogenic CO metabolism is CooA (Aono, Nakajima, Saito, & Okada, 1996; Shelver, Kerby, He, & Roberts, 1995), RcoM (Kerby, Youn, & Roberts, 2008), and CorQR (Kim et al., 2015). All of the enzymes are located within or beside the Ni-CODH–ECH gene cluster. In this section, we introduce the response mechanism to CO in hydrogenogenic CO metabolism.

4.1 Known CO-responsive transcriptional activators

CooA is widespread CO-responsive transcriptional activator in most of the hydrogenogenic CO oxidizers (Youn et al., 2004) and characterization of CooA has been advanced in the hydrogenogenic CO oxidizer, *R. rubrum* (Aono et al., 1996; Roberts, Youn, & Kerby, 2004). CooA is a heme-containing homodimeric protein in the cyclic AMP receptor protein family (Shelver et al., 1995). Cys^{75} or His^{77} in CooA is coordinated to the ferric and ferrous hemes, respectively, to respond the effector, CO (Aono et al., 2000). In the absence of CO, CooA is inactive (Shelver, Kerby, He, & Roberts, 1997). When CooA senses CO, replacement of one heme ligand makes CooA active for transcription with conformational change (Leduc, Thorsteinsson, Gaal, & Roberts, 2001; Shelver et al., 1997). In the genome of *R. rubrum*, the Ni-CODH gene cluster and ECH gene cluster are closely arranged with the CooA-binding site (5′-TGTCA-N_6-CGACA-3′) (He, Shelver, Kerby, & Roberts, 1996) upstream of each gene cluster (Fox, He, Shelver, Roberts, & Ludden, 1996; Rajeev et al., 2012). Phylogenetically, CooA are divided into two distinct groups (CooA-1 and CooA-2). CooA-1 is found in the most of the hydrogenogenic CO oxidizers, whereas CooA-2 is limited in few CO oxidizers who harbor multiple Ni-CODH gene clusters (Techtmann et al., 2011). Further, promoter binding analyses of two CooA groups in *C. hydrogenoformans* shows that both CooA groups are active in high CO concentrations, whereas only CooA-2 keeps its active form even in low CO concentrations (Techtmann et al., 2011). The different activation thresholds by CO in the two CooA groups suggest that a part of hydrogenogenic CO oxidizers can regulate transcription of genes in multiple Ni-CODH gene clusters across a wide range of CO concentrations (Techtmann et al., 2011). In addition, RcoM and CorQR have been reported as anaerobic CO-responsive transcriptional factor. RcoM possesses a potential CO-sensing heme-bearing domain and regulates aerobic Mo-CODH and in some case anaerobic Ni-CODH (Kerby et al., 2008). Similar to CooA genes, the gene encoding RcoM is also flanked with Ni-CODH–ECH gene clusters in the hydrogenogenic CO oxidizer, *Rubrivivax gelatinosus* (Wawrousek et al., 2014). The CorQR pair is composed of CorQ and CorR. CorQ possesses a DNA-binding domain of LysR-type transcriptional regulator family. CorR possesses 4-vinyl reductase domains instead of heme for CO sensing (Kim et al., 2015). Genes encoding a CorQR pair are also adjacent to Ni-CODH–ECH gene clusters in the hydrogenogenic CO oxidizing archaeron, *Thermococcus onnurineus* (Kim et al., 2015).

4.2 Transcriptional changes during hydrogenogenic CO metabolism

Recently, whole transcriptome analysis from *C. pertinax* has been reported (Fukuyama, Omae, Yoshida, & Sako, 2019). *C. pertinax* possesses one CooA homolog (CooA-2 group), whereas a model of hydrogenogenic CO oxidizer, *C. hydrogenoformans* possesses two CooA homologs (Wu et al., 2005). Thus, transcriptional regulation in *C. pertinax* in response to CO is expected to be simpler than *C. hydrogenoformans*. Therefore, this organism is helpful to understand the regulatory mechanism of hydrogenogenic CO metabolism. Whole transcriptome analysis suggested that *C. pertinax* performs hydrogenogenic CO metabolism under 100% CO instead of anaerobic respiration, via significant expression changes of genes in a relatively low number of gene clusters (Fukuyama, Omae, et al., 2019). When *C. pertinax* switches to CO metabolism, nine gene clusters including Ni-CODH-II and ECH gene clusters (the genomic features of hydrogenogenic CO metabolism are described in Section 3) are significantly upregulated. Of these, only the ECH gene cluster is regulated by the active CooA. The other gene clusters are separately activated by unknown CO-responsive transcriptional factors or in the same transcriptional cascade as the ECH gene cluster. Some hydrogenogenic CO oxidizers possess no putative CO-responsive transcriptional activator described above in their Ni-CODH–ECH gene cluster (e.g., *P. thermoglucosidasius* strain TG4) (Inoue, Tanimura, et al., 2019). Further, comprehensive analyses of Ni-CODHs including their genomic context showed transcriptional factors as novel Ni-CODH-associated proteins (Inoue, Nakamoto, et al., 2019). These findings suggest that unknown CO-responsive transcription exist.

5. Phylogenetic diversity of Ni-CODH/ECH and distribution of its owner

As described in Section 3, hydrogenogenic carboxydotrophy is conducted by a Ni-CODH/ECH complex. Thus far, bioinformatics-based studies on the key enzymes, Ni-CODHs and ECHs, have been performed to investigate their phylogenetic and functional diversity (Greening et al., 2015; Sant'Anna, Lebedinsky, Sokolova, Robb, & Gonzalez, 2015; Techtmann et al., 2012). Conversely, the distribution and ecology of hydrogenogenic CO oxidizers are very limited. In this section, we reviewed the state-of-the-art on bioinformatics-based and molecular ecological studies to demonstrate the phylogeny, diversity, distribution, and ecology of hydrogenogenic CO oxidizers.

5.1 Phylogenetic diversity of Ni-CODH/ECH revealed by bioinformatics

Recently, comprehensive phylogenomic analyses of Ni-CODHs show the distribution and diversity of the Ni-CODH–ECH gene clusters in a current massive genome database (Inoue, Nakamoto, et al., 2019; Omae et al., 2019). Classification of ~2000 Ni-CODHs from ~130,000 prokaryotic genomes have revealed their structural and phylogenetic diversity with variety of genomic contexts including the Ni-CODH–ECH gene cluster (Inoue, Nakamoto, et al., 2019); whereas, taxonomic information of microbes encoding the Ni-CODH–ECH gene clusters from ~140,000 prokaryotic genomes have applied to microbial community analysis using 16S rRNA amplicon sequencing (Omae et al., 2019).

Phylogenetically, the Ni-CODHs are divided into seven clades (A to G) (Inoue, Nakamoto, et al., 2019). Among them, the ECH-associated Ni-CODHs are limited to five subclades in clades E and F, and possess complete residues for metal cluster formation and catalysis (Fig. 3) (Inoue, Nakamoto, et al., 2019; Omae et al., 2019). Meanwhile, Ni-CODH-related ECHs are divided into three types, Coo (Group 4c), Hyc/Hyf (Group 4a), and Mbh-Mrp (Group 4b) by their phylogeny, components, and functions (Greening et al., 2015; Mohr, Aliyu, Küchlin, Polliack, et al., 2018; Sant'Anna, Lebedinsky, Sokolova, Robb, & Gonzalez, 2015; Schut, Lipscomb, Nguyen, Kelly, & Adams, 2016). Genes for the ECH-associated Ni-CODHs in clade F are found in gene clusters with those for the Coo- and Hyc/Hyf-type ECHs. Among them, the Coo-related Ni-CODHs are distributed in the Firmicutes (Clostridia) and the Proteobacteria, whereas the Hyc/Hyf-related Ni-CODHs are distributed in the Firmicutes (Clostridia and Bacilli) (Inoue, Nakamoto, et al., 2019; Omae et al., 2019). On the other hand, genes for the ECH-associated Ni-CODHs in clade E are found in gene clusters with those for the Mbh-Mrp-type ECHs and are mainly distributed in the Euryarchaeota and the Crenarchaeota (Inoue, Nakamoto, et al., 2019; Omae et al., 2019; Sant'Anna, Lebedinsky, Sokolova, Robb, & Gonzalez, 2015). There are two exceptions: clade E Ni-CODH genes in *C. subterraneus* subsp. *pacificus* and *Thermoanaerobacter* sp. YS13 form gene clusters with genes for the Hyc/Hyf-type ECHs in place of clade

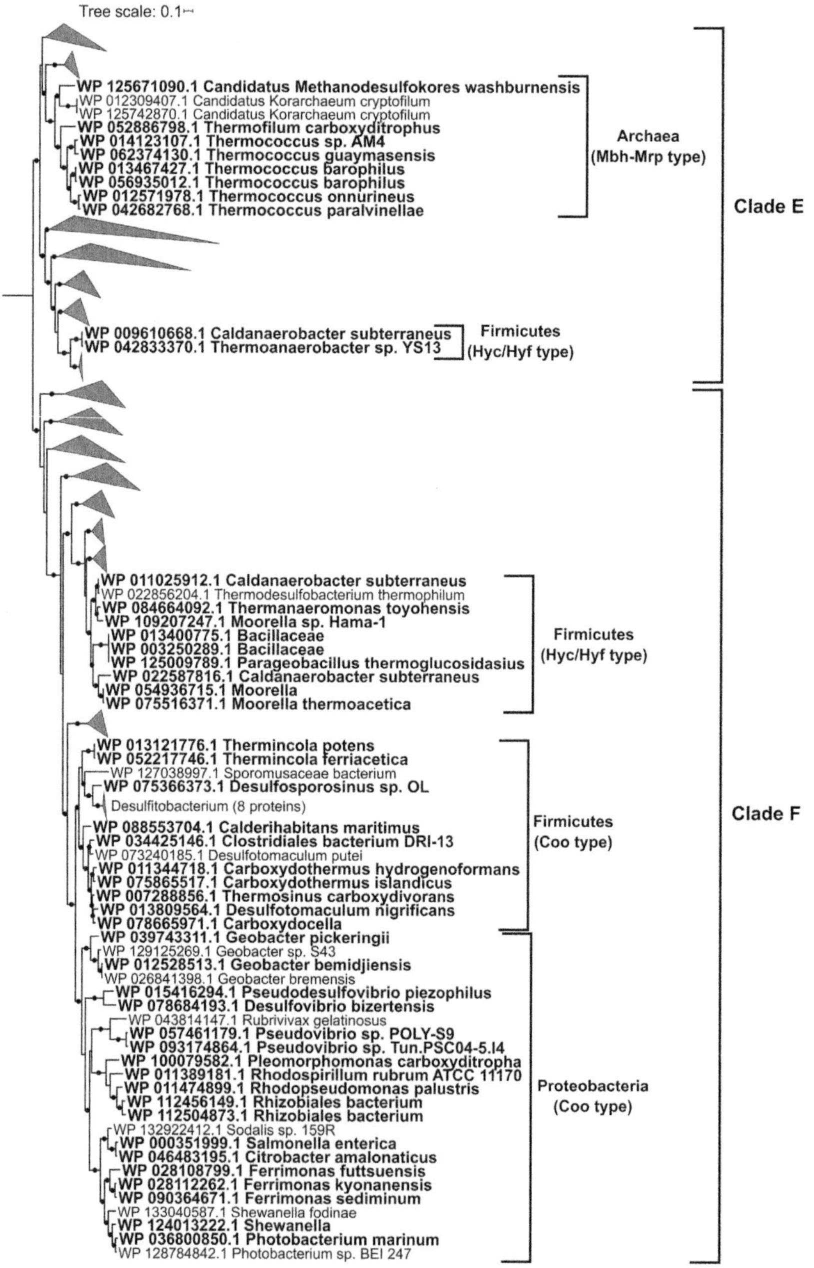

Fig. 3 Phylogenetic relationships of Ni-CODHs in clades E and F. An unrooted phylogenetic tree was constructed by FastML (Price, Dehal, & Arkin, 2010) using an alignment of 1642 Ni-CODH amino acid sequences, and only clades E and F are shown. Ni-CODH genes in Ni-CODH–ECH gene clusters are shown in bold. Nodes supported by a bootstrap value >80% are indicated by black circles.

F Ni-CODH genes in phylogenetically closely related *C. subterraneus* subsp. *tengcongensis* and *C. subterraneus* subsp. *yonseiensis*, which suggests horizontal gene transfer (Sant'Anna, Lebedinsky, Sokolova, Robb, & Gonzalez, 2015).

5.2 Potential hydrogenogenic CO oxidizers predicted by bioinformatics

The Ni-CODH–ECH gene clusters are identified in 71 genomes, including 46 genomes whose carriers have never been reported as hydrogenogenic CO oxidizers and are candidates for novel hydrogenogenic CO oxidizers, although their growth with CO-dependent H_2 production should be tested (Table 2) (Omae et al., 2019). These include 2, 7, and 15 species of Euryarchaeota, Firmicutes, and Proteobacteria, respectively, indicating an underestimation of the diversity of hydrogenogenic CO oxidizers by

Table 2 Potential hydrogenogenic CO oxidizing prokaryotes found in RefSeq genome database.

Organism	Isolation source	Reference[a]	RefSeq Genome accession
Archaea; Euryarchaeota; Thermococci; Thermococcales; Thermococcaceae			
Thermococcus guaymasensis DSM 11113	Deep-sea hydrothermal vent	Canganella, Jones, Gambacorta, and Antranikian (1998)	GCF_000816105.1
Thermococcus paralvinellae ES1	Polychaete worms at deep-sea hydrothermal vent chimneys	Hensley, Jung, Park, and Holden (2014)	GCF_000517445.1
Bacteria; Firmicutes; Bacilli; Bacillales; Bacillaceae			
Parageobacillus thermoglucosidasius B4168	Food	Berendsen et al. (2016)	GCF_001587555.1
Parageobacillus thermoglucosidasius C56-YS93	Hot spring	Brumm, Land, and Mead (2015)	GCF_000178395.2
Parageobacillus thermoglucosidasius DSM 2542	Soil	SAMN03447989[a]	GCF_000966225.1
Parageobacillus thermoglucosidasius GT23	Casein pipeline	SAMN04532072[a]	GCF_001651535.1

Continued

Table 2 Potential hydrogenogenic CO oxidizing prokaryotes found in RefSeq genome database.—cont'd

Organism	Isolation source	Reference[a]	RefSeq Genome accession
Parageobacillus thermoglucosidasius NBRC 107763	n.r.		GCF_000648295.1
Parageobacillus thermoglucosidasius NCIMB 11955	n.r.		GCF_001700985.1
Parageobacillus thermoglucosidasius TM242	n.r.		GCF_001902495.1
Parageobacillus thermoglucosidasius TNO-09.020	Dairy factory	Zhao, Caspers, Abee, Siezen, and Kort (2012)	GCF_000258725.1
Parageobacillus thermoglucosidasius Y4.1MC1	Hot spring	Brumm, Land, Hauser, et al. (2015)	GCF_000166075.1
Bacteria; Firmicutes; Clostridia; Clostridiales; Peptococcaceae			
Desulfosporosinus sp. OL	n.r.		GCF_001936615.1
Bacteria; Firmicutes; Clostridia; Thermoanaerobacterales; Thermoanaerobacteraceae			
Caldanaerobacter subterraneus subsp. *tengcongensis* MB4	Hot spring	Bao et al. (2002)	GCF_000007085.1
Caldanaerobacter subterraneus subsp. *yonseiensis* KB-1	Geothermal hot stream	Kim, Grote, Lee, Antranikian, and Pyun (2001)	GCF_000473865.1
Moorella glycerini NMP	Underground gas storage	Liebensteiner et al. (2015)	GCF_001373375.1
Moorella sp. Hama-1	Thermophilic digestion reactor	Harada et al. (2018)	GCF_003116935.1
Thermanaeromonas toyohensis ToBE	Geothermal aquifer at deep subsurface	Mori, Hanada, Maruyama, and Marumo (2002)	GCF_900176005.1
Thermoanaerobacter sp. YS13	Hot spring	Peng, Pan, Christopher, Sparling, and Levin (2015)	GCF_000806225.2

Table 2 Potential hydrogenogenic CO oxidizing prokaryotes found in RefSeq genome database.—cont'd

Organism	Isolation source	Reference[a]	RefSeq Genome accession
Bacteria; Proteobacteria; Alphaproteobacteria; Rhizobiales; Bradyrhizobiaceae			
Rhodopseudomonas palustris BisB18	n.r.		GCF_000013745.1
Bacteria; Proteobacteria; Alphaproteobacteria; Rhodobacterales; Rhodobacteraceae			
Pseudovibrio sp. POLY-S9	Marine sponge symbiont	Alex and Antunes (2015)	GCF_001431305.1
Pseudovibrio sp. Tun. PSC04–5.I4	Tunicate symbiont (marine)	SAMN04515695[a]	GCF_900104145.1
Bacteria; Proteobacteria; Alphaproteobacteria; Rhodospirillales; Rhodospirillaceae			
Rhodospirillum rubrum F11	n.r.	Lonjers et al. (2012)	GCF_000225955.1
Bacteria; Proteobacteria; Deltaproteobacteria; Desulfovibrionales; Desulfovibrionaceae			
Desulfovibrio bizertensis DSM 18034	Marine sediment	Haouari et al. (2006)	GCF_900167065.1
Pseudodesulfovibrio piezophilus C1TLV30	Wood falls at deep sea	Khelaifia et al. (2011)	GCF_000341895.1
Bacteria; Proteobacteria; Deltaproteobacteria; Desulfuromonadales; Geobacteraceae			
Geobacter bemidjiensis Bem	Subsurface sediments	Nevin et al. (2005)	GCF_000020725.1
Geobacter pickeringii G13	Subsurface sediments	Shelobolina et al. (2007)	GCF_000817955.1
Bacteria; Proteobacteria; Gammaproteobacteria; Alteromonadales; Ferrimonadaceae			
Ferrimonas futtsuensis DSM 18154	Mudflat sediment at beach	Nakagawa, Iino, Suzuki, and Harayama (2006)	GCF_000422645.1
Ferrimonas kyonanensis DSM 18153	Mudflat sediment at beach	Nakagawa et al. (2006)	GCF_000425405.1
Ferrimonas sediminum DSM 23317	Coastal sediment	Ji et al. (2013)	GCF_900100175.1
Bacteria; Proteobacteria; Gammaproteobacteria; Alteromonadales; Shewanellaceae			
Shewanella sp. M2	Antarctic deep-sea sediments	SAMN10397594[a]	GCF_003855155.1
Shewanella sp. R106	Antarctic deep-sea sediments	SAMN10397511[a]	GCF_003797165.1

Continued

Table 2 Potential hydrogenogenic CO oxidizing prokaryotes found in RefSeq genome database.—cont'd

Organism	Isolation source	Reference[a]	RefSeq Genome accession
Bacteria; Proteobacteria; Gammaproteobacteria; Enterobacterales; Enterobacteriaceae			
Salmonella enterica subsp. *enterica* serovar Montevideo 50,262	n.r.		GCF_001276695.1
Salmonella enterica subsp. *enterica* serovar Montevideo 50,270	n.r.		GCF_001276905.1
Salmonella enterica subsp. *enterica* serovar Senftenberg 50,263	n.r.		GCF_001276745.1
Salmonella enterica subsp. *enterica* serovar Senftenberg 50,264	n.r.		GCF_001276765.1
Salmonella enterica subsp. *enterica* serovar Senftenberg 50,265	n.r.		GCF_001276775.1
Salmonella enterica subsp. *enterica* serovar Senftenberg 50,271	n.r.		GCF_001276825.1
Salmonella enterica subsp. *enterica* serovar Senftenberg 50,272	n.r.		GCF_001276925.1
Salmonella enterica subsp. *enterica* serovar Senftenberg SS209	Hatchery of broiler chickens	Grépinet et al. (2012)	GCF_000263295.1
Bacteria; Proteobacteria; Gammaproteobacteria; Vibrionales; Vibrionaceae			
Photobacterium marinum AK15	Marine sediment	Srinivas, Vijaya Bhaskar, Bhumika, and Anil Kumar (2013)	GCF_000331515.1
[Uncultured] Archaea; TACK group; Candidatus Korarchaeota;			
Candidatus Methanodesulfokores washburnensis	Hot spring metagenome	McKay et al. (2019)	GCF_003947435.1
[Uncultured] Unclassified			
Clostridiales bacterium DRI-13	n.r.	n.r.	GCF_000746025.1

Table 2 Potential hydrogenogenic CO oxidizing prokaryotes found in RefSeq genome database.—cont'd

Organism	Isolation source	Reference[a]	RefSeq Genome accession
Rhizobiales bacterium AFS016371	n.r.	n.r.	GCF_900466875.1
Rhizobiales bacterium AFS041951	n.r.	n.r.	GCF_900468955.1
Rhizobiales bacterium AFS049984	n.r.	n.r.	GCF_900469445.1
Rhizobiales bacterium AFS089140	n.r.	n.r.	GCF_900472805.1

[a]When there is no appropriate reference work, NCBI BioSample accessions are shown if available.
"n.r." indicates "not reported."

cultivation study bias, especially in Proteobacteria. In the phylum Firmicutes, *Desulfosporosinus* sp. OL and *Thermanaeromonas toyohensis* ToBE, respectively, isolated from a soil surface and a geothermal aquifer in deep mines, are predicted to be phylogenetically novel, thermophilic, hydrogenogenic CO oxidizers. In the phylum Proteobacteria, diverse mesophiles are predicted to be hydrogenogenic CO oxidizers which were isolated from various environments: marine anaerobes or facultative anaerobes isolated from coastal environments (e.g., *Ferrimonas futtsuensis* DSM 18154 and *Photobacterium marinum* AK15), deep-sea environments (e.g., *Pseudodesulfovibrio piezophilus* C1TLV30 and *Shewanella* sp. M2) and marine sponge symbiont environments (e.g., *Pseudovibrio* sp. POLY-S9); iron reducers isolated from subsurface environments (e.g., *Geobacter bemidjiensis* Bem and *Geobacter pickeringii* G13); and potential pathogens isolated from a hatchery of broiler chickens (e.g., *Salmonella enterica* subsp. *enterica* serovar Senftenberg SS209). Furthermore, this analysis identified the Ni-CODH–ECH gene clusters in six metagenome-assembled genomes (MAGs) of uncultivated microbes. In particular, *Candidatus* Methanodesulfokores washburnensis was a recently characterized Korarchaeon with the potential of anaerobic methane oxidation, whose MAG was constructed from Yellowstone National Park hot spring (McKay et al., 2019). Meanwhile, genomic information for 28 out of 37 isolates of the hydrogenogenic CO oxidizers (as described in Section 2) revealed that all of the hydrogenogenic CO oxidizers with genomic information harbor Ni-CODH – ECH gene clusters except for *C. pertinax* and *T. kivui*. All isolates with the Ni-CODH–ECH gene clusters show hydrogenogenic carboxydotrophy except for isolates without experimental data for CO utilization and H_2 production. Thus, existence of the Ni-CODH–ECH gene cluster would be a marker-gene set for hydrogenogenic carboxydotrophy.

5.3 Distribution, diversity, and ecology of hydrogenogenic CO oxidizers

To demonstrate the distribution, diversity, and ecology of hydrogenogenic CO oxidizers and to identify novel types of hydrogenogenic CO oxidizers, culture-independent techniques are considered effective by replacing isolation-based approaches. However, only a few ECH-associated Ni-CODHs have been detected in MAGs as described above, possibly because of low abundance of the hydrogenogenic CO oxidizers. Indeed, molecular ecological research revealed a wide and relatively low-abundance (<0.1%) distribution of potential thermophilic, hydrogenogenic CO oxidizers of Firmicutes in terrestrial hot springs by quantifying putative phylotypes harboring Ni-CODH–ECH gene clusters using 16S rRNA amplicon sequencing (Omae et al., 2019). Other studies based on ^{13}CO DNA stable isotope probing (Brady, Sharp, Grasby, & Dunfield, 2015) and quantification of Ni-CODH genes of *Carboxydothermus* (Yoneda et al., 2015) also showed low abundance of thermophilic, hydrogenogenic CO oxidizers of Firmicutes in hydrothermal environments. Conversely, they can also be dominant (~10% of bacterial population) in certain hydrothermal environments (Yoneda, Yoshida, Daifuku, et al., 2013). The features of thermophilic, hydrogenogenic CO oxidizers of Firmicutes correspond to those of a "rare biosphere," which is a collection of dormant or active but zero-growth microorganisms (Hausmann, Pelikan, Rattei, Loy, & Pester, 2019; Lynch & Neufeld, 2015). Note that, to date, little is known about the ecology of hydrogenogenic CO oxidizing Proteobacteria or Archaea, despite of their unexpected diversity as described above. We need to expand our ecological studies into various environments, except for terrestrial hydrothermal fields, to understand the true importance of the metabolic process by hydrogenogenic CO oxidation in microbial communities.

The low abundance in environments makes it difficult to understand the distribution, diversity, and ecology of Firmicutes hydrogenogenic CO oxidizers by metagenomics. In addition, sparse distribution of the Ni-CODH–ECH gene clusters especially in Proteobacteria and Archaea cause difficulty in the exploration of hydrogenogenic CO oxidizers using 16S rRNA amplicon sequencing. One solution for these issues would be environmental DNA metabarcoding technology using universal primers to amplify target regions of the Ni-CODH–ECH gene clusters, or other marker-gene sets, for hydrogenogenic carboxydotrophy. This method is also applicable to reverse-transcribed DNAs to reveal in situ activity. However, these approaches might be challenging because of sequence diversity and sporadic distribution of the ECH-associated Ni-CODHs or the Ni-CODH-related ECHs. The other clues would be metagenomics/transcriptomics or single-cell genomics/transcriptomics. To apply metagenomics/transcriptomics technology to hydrogenogenic CO oxidizers, an enrichment step would be required for deep sequencing to construct MAGs.

6. Biotechnological application of hydrogenogenic CO oxidizers

Hydrogenogenic carboxydotrophy has the potential for biotechnological application to produce H_2 from CO in syngas or industrial waste gases (Alfano & Cavazza, 2018; Diender et al., 2015). Bioconversion of CO in syngas to H_2 has been first investigated in *R. rubrum*, a mesophilic, photosynthetic bacterium (Najafpour & Younesi, 2007). Meanwhile, assumptions of temperature effects on hydrogenogenic carboxydotrophy have made by a recent critical review (Diender et al., 2015). Temperature increase causes decreased gas solubility and increased gas diffusion rates. Because of the high affinity to CO in Ni-CODHs, the gas diffusion rates can be major determinants for efficient catalysis in the application of hydrogenogenic carboxydotrophy. At increased temperatures, increased gas diffusion rates supply CO more rapidly and avoid H_2 accumulation. Thus, thermophilic organisms would be better than mesophilic organisms as biocatalysts in CO-dependent H_2 production.

6.1 Biotechnological application of thermophilic, hydrogenogenic CO oxidizers in CO-dependent H_2 production

Thus far, biotechnological application of thermophilic, hydrogenogenic CO oxidizers in H_2 production with CO in syngas have been demonstrated in *C. hydrogenoformans* (Haddad, Cimpoia, & Guiot, 2014; Zhao, Haddad, Cimpoia, Liu, & Guiot, 2013) and *T. onnurineus* (Lee et al., 2016), which are obligate anaerobes. In particular, *T. onnurineus* acts as a model organism for the biotechnological application of CO-dependent H_2 production (Kim et al., 2013, 2015; Lee et al., 2014, 2015, 2016). Genetic engineering of *T. onnurineus* to introduce strong promoters to its Ni-CODH–ECH gene cluster or by upregulating the expression of a transcription factor (CorQR) enables fast growth, increased cell density, and high CO consumption and H_2 production rates (Kim et al., 2013, 2015; Lee et al., 2014, 2015). Moreover, an adaptive evolution strategy with CO over 150 serial transfers of cell cultures (over 2000 generations) led *T. onnurineus* to possess higher CO consumption and H_2 production rates, which was associated with increased cell density (Lee et al., 2016).

Meanwhile, heterologous expression of the *T. onnurineus* Ni-CODH/ECH complex in *Pyrococcus furiosus*, which has no CO-dependent H_2 production activity, converts it to a hydrogenogenic CO oxidizer, which indicates the Ni-CODH/ECH complex is fully active across genera (Basen et al., 2014; Schut et al., 2016). Moreover, the engineered *P. furiosus* converts organic acids to the corresponding alcohols, which suggests that the Ni-CODH/ECH complex can be utilized to provide low potential electrons from CO to reduce organic acids as well as to evolve H_2 (Basen et al., 2014; Schut et al., 2016). These studies provide hints for efficient (re)utilization of carbon sources in non-carbon fixing organisms by Ni-CODHs.

6.2 Application of thermophilic, facultative anaerobic, hydrogenogenic CO oxidizers

As described above, biotechnological application of thermophilic, hydrogenogenic CO oxidizers has only been investigated in strict anaerobes. Recently, a thermophilic, facultative anaerobic bacterium, *P. thermoglucosidasius*, has been identified with the ability to catalyze CO-dependent H_2 production as described in Section 2 (Inoue, Tanimura, et al., 2019; Mohr, Aliyu, Küchlin, Polliack, et al., 2018; Mohr, Aliyu, Küchlin, Zwick, et al., 2018). Surprisingly, *P. thermoglucosidasius* can grow under 50% CO and 50% air, first by consuming O_2 by aerobic respiration and then by utilizing CO by CO-dependent H_2 production (Mohr, Aliyu, Küchlin, Polliack, et al., 2018; Mohr, Aliyu, Küchlin, Zwick, et al., 2018). *P. thermoglucosidasius* can also grow with high concentrations of CO (100% CO) to produce H_2 (Inoue, Tanimura, et al., 2019). These data indicate that *P. thermoglucosidasius* has tolerance and utilizes both CO and O_2. Over 40 years after its initial isolation, various biotechnological applications for metabolic engineering using *P. thermoglucosidasius* are now available (Bacon, Hamley-Bennett, Danson, & Leak, 2017; Cripps et al., 2009; Gilman et al., 2019; Marcano-Velazquez, Lo, Nag, Maness, & Chou, 2019; Reeve, Martinez-Klimova, de Jonghe, Leak, & Ellis, 2016; Taylor, Esteban, & Leak, 2008). Thus, engineered *P. thermoglucosidasius* with efficient catalysis for CO-dependent H_2 production will come out in the immediate future.

6.3 Co-cultivation of hydrogenogenic CO oxidizers with hydrogen utilizers

The combination of hydrogenogenic CO oxidizers with H_2 utilizers also has new possibilities in the bioconversion of CO. A recent study has shown that co-cultivation of *Methanothermobacter thermoautotrophicus* with *C. hydrogenoformans* accelerates methanogenesis from CO-rich gases compared to the monoculture of *M. thermoautotrophicus*. This is due to the removal of CO by *C. hydrogenoformans*, which lifts the toxic effects on the methanogen while simultaneously providing substrate in the form of H_2 and CO_2 (Diender, Uhl, Bitter, Stams, & Sousa, 2018). Another study has also shown that co-cultivation of *Citrobacter amalonaticus* and *Sporomusa ovata* results in higher acetate production from CO compared to the monoculture counterparts, where CO_2 and H_2 evolved by *C. amalonaticus* are further utilized by *S. ovata* (Lee et al., 2018).

6.4 Future perspectives on biotechnological applications

For better application, fundamental studies to understand the molecular mechanisms of Ni-CODH/ECH complex (i.e., reaction coupling, electron transfer, and proton translocation) are essential. Recent biochemical study of *T. kivui* first investigates energy conservation with proton/sodium translocation by a Ni-CODH/ECH complex (Schoelmerich & Müller, 2019). This study also shows feedback inhibition of CO-dependent H_2 production by membrane potential and a weak interaction between Ni-CODH and ECH, which provides a clue for the efficient catalysis of H_2 production. Moreover, another structural and biochemical study revealed that a Ni-CODH paralog from *C. hydrogenoformans* (*Ch*CODH-IV) has a very high affinity to CO, it oxidizes CO at a diffusion-limited rate over a wide range of temperatures, and is more tolerant to O_2 than *Ch*CODH-II; this finding highlights that Ni-CODHs are more diverse in terms of structure and reactivity than expected (Domnik et al., 2017), although *Ch*CODH-IV might not be able to complex with ECH (Soboh et al., 2002). The structural diversity of Ni-CODHs is also indicated by recent structural and bioinformatics-based analyses (Inoue, Nakamoto, et al., 2019; Wittenborn et al., 2018). Unfortunately, the information for the quaternary structure of the Ni-CODH/ECH complex are unavailable; however, the quaternary structure of the membrane-bound hydrogenase (MBH) complex from *P. furiosus* has recently been solved by cryo-electron microscopy technology, which unveiled possible routes and mechanistic insights for electron transfer and proton/sodium translocation (Yu et al., 2018). Thus, the quaternary structure of the Ni-CODH/ECH complex might be solved in the near future, which would enable structure-based applications to improve enzyme catalysis and to produce "superbiocatalysts" for CO-dependent H_2 production.

7. Concluding remarks

From the 1970s to 2000s, hydrogenogenic CO metabolism had been understood by a few models of hydrogenogenic CO oxidizers isolated from fresh water, hot springs, or deep-sea hydrothermal vents. In the past decade, however, physiologically and phylogenetically novel hydrogenogenic CO oxidizers isolated from an unexpectedly wide range of environments like marine and lake sediments have been reported. Their genomic information and physiology provide novel insights into hydrogenogenic CO metabolism. Simply put, hydrogenogenic CO oxidation is performed by the Ni-CODH/ECH complex, a simple mechanism encoded in a single gene cluster (Ni-CODH–ECH) whose transcription is regulated by a CO-responsive transcriptional activator, and featured by a primitive respiratory system which directly reduces protons without any extra electron carriers. However, diverse electron flows from CO to associated energy conservation systems during hydrogenogenic growth is revealed in the recent hydrogenogenic CO oxidizing isolates, which prompts us to assume the versatility of their CO metabolism owing to the very low redox potential of CO. In addition, recent findings of hydrogenogenic CO oxidation independent of the continuous Ni-CODH–ECH gene cluster and the known CO-responsive transcriptional activators suggest that unknown CO-responsive transcription mechanisms exist. Further, state-of-the-art bioinformatics-based studies identified additional potential hydrogenogenic CO oxidizers harboring Ni-CODH–ECH gene clusters from diverse environments including marine and subsurface, which suggests the importance of the metabolic process by hydrogenogenic CO oxidation in microbial communities. In fact, molecular ecological studies that targeted these potential hydrogenogenic CO oxidizers revealed a wide and relatively low-abundance distribution of Firmicutes members which constituted a "rare biosphere." In addition, recent progress in the biotechnological application of thermophilic, hydrogenogenic CO oxidizers has evolved diverse approaches (e.g., metabolic engineering and co-cultivation) for H_2, methane, and acetate production from syngas. In particular, recently reported thermophilic hydrogenogenic CO oxidizers of the genus *Parageobacillus* are facultative anaerobes with useful techniques of gene manipulation and have the potential for the application of O_2-tolerant CO-dependent H_2 production.

Acknowledgments

This work was supported by Grants-in-Aid for Scientific Research (S) 16H06381 from The Ministry of Education, Culture, Sports, Science and Technology (MEXT). Part of the computational analysis was completed at the Supercomputer System at the Institute for Chemical Research, Kyoto University.

References

Alex, A., & Antunes, A. (2015). Whole genome sequencing of the symbiont *Pseudovibrio* sp. from the intertidal marine sponge *Polymastia penicillus* revealed a gene repertoire for host-switching permissive lifestyle. *Genome Biology and Evolution*, 7(11), 3022–3032. https://doi.org/10.1093/gbe/evv199.

Alfano, M., & Cavazza, C. (2018). The biologically mediated water–gas shift reaction: Structure, function and biosynthesis of monofunctional [NiFe]-carbon monoxide dehydrogenases. *Sustainable Energy & Fuels*, *2*(8), 1653–1670. https://doi.org/10.1039/C8SE00085A.

Alves, J. I., van Gelder, A. H., Alves, M. M., Sousa, D. Z., & Plugge, C. M. (2013). *Moorella stamsii* sp. nov., a new anaerobic thermophilic hydrogenogenic carboxydotroph isolated from digester sludge. *International Journal of Systematic and Evolutionary Microbiology*, *63*(11), 4072–4076. https://doi.org/10.1099/ijs.0.050369-0.

Aono, S., Honma, Y., Ohkubo, K., Tawara, T., Kamiya, T., & Nakajima, H. (2000). CO sensing and regulation of gene expression by the transcriptional activator CooA. *Journal of Inorganic Biochemistry*, *82*(1–4), 51–56. https://doi.org/10.1016/S0162-0134(00)00139-2.

Aono, S., Nakajima, H., Saito, K., & Okada, M. (1996). A novel heme protein that acts as a carbon monoxide-dependent transcriptional activator in *Rhodospirillum rubrum*. *Biochemical and Biophysical Research Communications*, *228*(3), 752–756. https://doi.org/10.1006/bbrc.1996.1727.

Aragão, D., Mitchell, E. P., Frazão, C. F., Carrondo, M. A., & Lindley, P. F. (2008). Structural and functional relationships in the hybrid cluster protein family: Structure of the anaerobically purified hybrid cluster protein from *Desulfovibrio vulgaris* at 1.35 Å resolution. *Acta Crystallographica Section D*, *64*(6), 665–674. https://doi.org/10.1107/S0907444908009165.

Bacon, L. F., Hamley-Bennett, C., Danson, M. J., & Leak, D. J. (2017). Development of an efficient technique for gene deletion and allelic exchange in *Geobacillus* spp. *Microbial Cell Factories*, *16*, 58. https://doi.org/10.1186/s12934-017-0670-4.

Bae, S. S., Kim, Y. J., Yang, S. H., Lim, J. K., Jeon, J. H. O., Lee, H. S., et al. (2006). *Thermococcus onnurineus* sp. nov., a hyperthermophilic archaeon isolated from a deep-sea hydrothermal vent area at the PACMANUS field. *Journal of Microbiology and Biotechnology*, *16*(11), 1826–1831.

Balk, M., Heilig, H. G. H. J., van Eekert, M. H. A., Stams, A. J. M., Rijpstra, I. C., Sinninghe-Damsté, J. S., et al. (2009). Isolation and characterization of a new CO-utilizing strain, *Thermoanaerobacter thermohydrosulfuricus* subsp. carboxydovorans, isolated from a geothermal spring in Turkey. *Extremophiles*, *13*(6), 885–894. https://doi.org/10.1007/s00792-009-0276-9.

Bao, Q., Tian, Y., Li, W., Xu, Z., Xuan, Z., Hu, S., et al. (2002). A complete sequence of the *T. tengcongensis* genome. *Genome Research, 12*(5), 689–700. https://doi.org/10.1101/gr.219302.

Basen, M., Schut, G. J., Nguyen, D. M., Lipscomb, G. L., Benn, R. A., Prybol, C. J., et al. (2014). Single gene insertion drives bioalcohol production by a thermophilic archaeon. *Proceedings of the National Academy of Sciences of the United States of America, 111*(49), 17618–17623. https://doi.org/10.1073/pnas.1413789111.

Berendsen, E. M., Wells-Bennik, M. H. J., Krawczyk, A. O., de Jong, A., van Heel, A., Holsappel, S., et al. (2016). Draft genome sequences of seven thermophilic spore-forming bacteria isolated from foods that produce highly heat-resistant spores, comprising *Geobacillus* spp., *Caldibacillus debilis*, and *Anoxybacillus flavithermus*. *Genome Announcements, 4*(3), e00105-16. https://doi.org/10.1128/genomeA.00105-16.

Bertsch, J., & Müller, V. (2015). CO metabolism in the acetogen *Acetobacterium woodii*. *Applied and Environmental Microbiology, 81*(17), 5949–5956. https://doi.org/10.1128/AEM.01772-15.

Biegel, E., Schmidt, S., González, J. M., & Müller, V. (2011). Biochemistry, evolution and physiological function of the Rnf complex, a novel ion-motive electron transport complex in prokaryotes. *Cellular and Molecular Life Sciences, 68*(4), 613–634. https://doi.org/10.1007/s00018-010-0555-8.

Bonam, D., & Ludden, P. W. (1987). Purification and characterization of carbon monoxide dehydrogenase, a nickel, zinc, iron-sulfur protein, from *Rhodospirillum rubrum*. *Journal of Biological Chemistry, 262*(7), 2980–2987.

Brady, A. L., Sharp, C. E., Grasby, S. E., & Dunfield, P. F. (2015). Anaerobic carboxydotrophic bacteria in geothermal springs identified using stable isotope probing. *Frontiers in Microbiology, 6*, 897. https://doi.org/10.3389/fmicb.2015.00897.

Brumm, P., Land, M. L., Hauser, L. J., Jeffries, C. D., Chang, Y.-J., & Mead, D. A. (2015). Complete genome sequence of *Geobacillus* strain Y4.1MC1, a novel CO-utilizing *Geobacillus thermoglucosidasius* strain isolated from bath hot spring in Yellowstone National Park. *Bioenergy Research, 8*(3), 1039–1045. https://doi.org/10.1007/s12155-015-9585-2.

Brumm, P. J., Land, M. L., & Mead, D. A. (2015). Complete genome sequence of *Geobacillus thermoglucosidasius* C56-YS93, a novel biomass degrader isolated from obsidian hot spring in Yellowstone National Park. *Standards in Genomic Sciences, 10*, 73. https://doi.org/10.1186/s40793-015-0031-z.

Buckel, W., & Thauer, R. K. (2013). Energy conservation via electron bifurcating ferredoxin reduction and proton/Na^+ translocating ferredoxin oxidation. *Biochimica et Biophysica Acta (BBA)—Bioenergetics, 1827*(2), 94–113. https://doi.org/10.1016/j.bbabio.2012.07.002.

Buckel, W., & Thauer, R. K. (2018). Flavin-based electron bifurcation, ferredoxin, flavodoxin, and anaerobic respiration with protons (Ech) or NAD^+ (Rnf) as electron acceptors: A historical review. *Frontiers in Microbiology, 9*, 401. https://doi.org/10.3389/fmicb.2018.00401.

Byrne-Bailey, K. G., Wrighton, K. C., Melnyk, R. A., Agbo, P., Hazen, T. C., & Coates, J. D. (2010). Complete genome sequence of the electricity-producing "*Thermincola potens*" strain JR. *Journal of Bacteriology, 192*(15), 4078–4079. https://doi.org/10.1128/JB.00044-10.

Can, M., Armstrong, F. A., & Ragsdale, S. W. (2014). Structure, function, and mechanism of the nickel metalloenzymes, CO dehydrogenase, and acetyl-CoA synthase. *Chemical Reviews, 114*(8), 4149–4174. https://doi.org/10.1021/cr400461p.

Canganella, F., Jones, W. J., Gambacorta, A., & Antranikian, G. (1998). *Thermococcus guaymasensis* sp. nov. and *Thermococcus aggregans* sp. nov., two novel thermophilic archaea isolated from the Guaymas Basin hydrothermal vent site. *International Journal of Systematic and Evolutionary Microbiology, 48*(4), 1181–1185. https://doi.org/10.1099/00207713-48-4-1181.

Carvalho, S. M., Marques, J., Romão, C. C., & Saraiva, L. M. (2019). Metabolomics of *Escherichia coli* treated with the antimicrobial carbon monoxide-releasing molecule CORM-3 reveals tricarboxylic acid cycle as major target. *Antimicrobial Agents and Chemotherapy*, *63*(10), e00643-19. https://doi.org/10.1128/AAC.00643-19.

Conrad, R. (1996). Soil microorganisms as controllers of atmospheric trace gases (H_2, CO, CH_4, OCS, N_2O, and NO). *Microbiology and Molecular Biology Reviews*, *60*(4), 609–640.

Conte, L., Szopa, S., Séférian, R., & Bopp, L. (2019). The oceanic cycle of carbon monoxide and its emissions to the atmosphere. *Biogeosciences*, *16*(4), 881–902. https://doi.org/10.5194/bg-16-881-2019.

Cripps, R. E., Eley, K., Leak, D. J., Rudd, B., Taylor, M., Todd, M., et al. (2009). Metabolic engineering of *Geobacillus thermoglucosidasius* for high yield ethanol production. *Metabolic Engineering*, *11*(6), 398–408. https://doi.org/10.1016/j.ymben.2009.08.005.

Dashekvicz, M. P., & Uffen, R. L. (1979). Identification of a carbon monoxide-metabolizing bacterium as a strain of *Rhodopseudomonas gelatinosa* (Molisch) van niel. *International Journal of Systematic Bacteriology*, *29*(2), 145–148. https://doi.org/10.1099/00207713-29-2-145.

Davidge, K. S., Sanguinetti, G., Yee, C. H., Cox, A. G., McLeod, C. W., Monk, C. E., et al. (2009). Carbon monoxide-releasing antibacterial molecules target respiration and global transcriptional regulators. *The Journal of Biological Chemistry*, *284*(7), 4516–4524. https://doi.org/10.1074/jbc.M808210200.

Davidova, M. N., Tarasova, N. B., Mukhitova, F. K., & Karpilova, I. U. (1994). Carbon monoxide in metabolism of anaerobic bacteria. *Canadian Journal of Microbiology*, *40*(6), 417–425. https://doi.org/10.1139/m94-069.

Diender, M., Stams, A. J. M., & Sousa, D. Z. (2015). Pathways and bioenergetics of anaerobic carbon monoxide fermentation. *Frontiers in Microbiology*, *6*, 1275. https://doi.org/10.3389/fmicb.2015.01275.

Diender, M., Uhl, P. S., Bitter, J. H., Stams, A. J. M., & Sousa, D. Z. (2018). High rate biomethanation of carbon monoxide-rich gases via a thermophilic synthetic coculture. *ACS Sustainable Chemistry & Engineering*, *6*(2), 2169–2176. https://doi.org/10.1021/acssuschemeng.7b03601.

Dobbek, H., Svetlitchnyi, V., Gremer, L., Huber, R., & Meyer, O. (2001). Crystal structure of a carbon monoxide dehydrogenase reveals a [Ni-4Fe-5S] cluster. *Science*, *293*(5533), 1281–1285. https://doi.org/10.1126/science.1061500.

Domnik, L., Merrouch, M., Goetzl, S., Jeoung, J.-H., Léger, C., Dementin, S., et al. (2017). CODH-IV: A high-efficiency CO-scavenging CO dehydrogenase with resistance to O_2. *Angewandte Chemie International Edition*, *56*(48), 15466–15469. https://doi.org/10.1002/anie.201709261.

Doukov, T. I., Iverson, T. M., Seravalli, J., Ragsdale, S. W., & Drennan, C. L. (2002). A Ni-Fe-Cu center in a bifunctional carbon monoxide dehydrogenase/acetyl-CoA synthase. *Science*, *298*(5593), 567–572. https://doi.org/10.1126/science.1075843.

Drennan, C. L., Heo, J., Sintchak, M. D., Schreiter, E., & Ludden, P. W. (2001). Life on carbon monoxide: X-ray structure of *Rhodospirillum rubrum* Ni-Fe-S carbon monoxide dehydrogenase. *Proceedings of the National Academy of Sciences of the United States of America*, *98*(21), 11973–11978. https://doi.org/10.1073/pnas.211429998.

Dürre, P., & Eikmanns, B. J. (2015). C1-carbon sources for chemical and fuel production by microbial gas fermentation. *Current Opinion in Biotechnology*, *35*, 63–72. https://doi.org/10.1016/j.copbio.2015.03.008.

Eckert, C. A., Freed, E., Wawrousek, K., Smolinski, S., Yu, J., & Maness, P.-C. (2019). Inactivation of the uptake hydrogenase in the purple non-sulfur photosynthetic bacterium *Rubrivivax gelatinosus* CBS enables a biological water–gas shift platform for H_2 production. *Journal of Industrial Microbiology & Biotechnology*, *46*(7), 993–1002. https://doi.org/10.1007/s10295-019-02173-7.

Ensign, S. A., & Ludden, P. W. (1991). Characterization of the CO oxidation/H_2 evolution system of *Rhodospirillum rubrum*. Role of a 22-kDa iron-sulfur protein in mediating electron transfer between carbon monoxide dehydrogenase and hydrogenase. *Journal of Biological Chemistry*, *266*(27), 18395–18403.

Esquivel-Elizondo, S., Maldonado, J., & Krajmalnik-Brown, R. (2018). Anaerobic carbon monoxide metabolism by *Pleomorphomonas carboxyditropha* sp. nov., a new mesophilic hydrogenogenic carboxydotroph. *FEMS Microbiology Ecology*, *94*(6). fiy056, https://doi.org/10.1093/femsec/fiy056.

Fesseler, J., Jeoung, J.-H., & Dobbek, H. (2015). How the [$NiFe_4S_4$] cluster of CO dehydrogenase activates CO_2 and NCO^-. *Angewandte Chemie International Edition*, *54*(29), 8560–8564. https://doi.org/10.1002/anie.201501778.

Fox, J. D., He, Y., Shelver, D., Roberts, G. P., & Ludden, P. W. (1996). Characterization of the region encoding the CO-induced hydrogenase of *Rhodospirillum rubrum*. *Journal of Bacteriology*, *178*(21), 6200–6208. https://doi.org/10.1128/jb.178.21.6200-6208.1996.

Fox, J. D., Kerby, R. L., Roberts, G. P., & Ludden, P. W. (1996). Characterization of the CO-induced, CO-tolerant hydrogenase from *Rhodospirillum rubrum* and the gene encoding the large subunit of the enzyme. *Journal of Bacteriology*, *178*(6), 1515–1524. https://doi.org/10.1128/jb.178.6.1515-1524.1996.

Fukuyama, Y., Oguro, T., Omae, K., Yoneda, Y., Yoshida, T., & Sako, Y. (2017). Draft genome sequences of two hydrogenogenic carboxydotrophic bacteria, *Carboxydocella* sp. strains JDF658 and ULO1, isolated from two distinct volcanic fronts in Japan. *Genome Announcements*, *5*(16), e00242-17. https://doi.org/10.1128/genomeA.00242-17.

Fukuyama, Y., Omae, K., Yoneda, Y., Yoshida, T., & Sako, Y. (2017). Draft genome sequences of *Carboxydothermus pertinax* and *C. islandicus*, hydrogenogenic carboxydotrophic bacteria. *Genome Announcements*, *5*(8), e01648-16. https://doi.org/10.1128/genomeA.01648-16.

Fukuyama, Y., Omae, K., Yoneda, Y., Yoshida, T., & Sako, Y. (2018). Insight into energy conservation via alternative carbon monoxide metabolism in *Carboxydothermus pertinax* revealed by comparative genome analysis. *Applied and Environmental Microbiology*, *84*(14), e00458-18. https://doi.org/10.1128/AEM.00458-18.

Fukuyama, Y., Omae, K., Yoshida, T., & Sako, Y. (2019). Transcriptome analysis of a thermophilic and hydrogenogenic carboxydotroph *Carboxydothermus pertinax*. *Extremophiles*, *23*(4), 389–398. https://doi.org/10.1007/s00792-019-01091-x.

Fukuyama, Y., Tanimura, A., Inoue, M., Omae, K., Yoshida, T., & Sako, Y. (2019). Draft genome sequences of two thermophilic *Moorella* sp. strains, isolated from an acidic hot spring in Japan. *Microbiology Resource Announcements*, *8*(31), e00663-19. https://doi.org/10.1128/MRA.00663-19.

Gilman, J., Singleton, C., Tennant, R. K., James, P., Howard, T. P., Lux, T., et al. (2019). Rapid, heuristic discovery and design of promoter collections in non-model microbes for industrial applications. *ACS Synthetic Biology*, *8*(5), 1175–1186. https://doi.org/10.1021/acssynbio.9b00061.

Gong, W., Hao, B., Wei, Z., Ferguson, D. J., Tallant, T., Krzycki, J. A., et al. (2008). Structure of the $\alpha_2\varepsilon_2$ Ni-dependent CO dehydrogenase component of the *Methanosarcina barkeri* acetyl-CoA decarbonylase/synthase complex. *Proceedings of the National Academy of Sciences of the United States of America*, *105*(28), 9558–9563. https://doi.org/10.1073/pnas.0800415105.

Greening, C., Biswas, A., Carere, C. R., Jackson, C. J., Taylor, M. C., Stott, M. B., et al. (2015). Genomic and metagenomic surveys of hydrogenase distribution indicate H_2 is a widely utilised energy source for microbial growth and survival. *The ISME Journal*, *10*, 761–777. https://doi.org/10.1038/ismej.2015.153.

Grépinet, O., Boumart, Z., Virlogeux-Payant, I., Loux, V., Chiapello, H., Gendrault, A., et al. (2012). Genome sequence of the persistent *Salmonella enterica* subsp. *enterica* serotype senftenberg strain SS209. *Journal of Bacteriology*, *194*(9), 2385–2386. https://doi.org/10.1128/JB.00255-12.

Haddad, M., Cimpoia, R., & Guiot, S. R. (2014). Performance of *Carboxydothermus hydrogenoformans* in a gas-lift reactor for syngas upgrading into hydrogen. *International Journal of Hydrogen Energy*, *39*(6), 2543–2548. https://doi.org/10.1016/j.ijhydene.2013.12.022.

Hallenbeck, P. C. (2009). Fermentative hydrogen production: Principles, progress, and prognosis. *International Journal of Hydrogen Energy*, *34*(17), 7379–7389. https://doi.org/10.1016/j.ijhydene.2008.12.080.

Haouari, O., Fardeau, M.-L., Casalot, L., Tholozan, J.-L., Hamdi, M., & Ollivier, B. (2006). Isolation of sulfate-reducing bacteria from Tunisian marine sediments and description of *Desulfovibrio bizertensis* sp. nov. *International Journal of Systematic and Evolutionary Microbiology*, *56*(12), 2909–2913. https://doi.org/10.1099/ijs.0.64530-0.

Harada, J., Yamada, T., Giri, S., Hamada, M., Nobu, M. K., Narihiro, T., et al. (2018). Draft genome sequence of *Moorella* sp. strain hama-1, a novel acetogenic bacterium isolated from a thermophilic digestion reactor. *Genome Announcements*, *6*(24), e00517-18. https://doi.org/10.1128/genomeA.00517-18.

Hausmann, B., Pelikan, C., Rattei, T., Loy, A., & Pester, M. (2019). Long-term transcriptional activity at zero growth of a cosmopolitan rare biosphere member. *MBio*, *10*(1), e02189-18. https://doi.org/10.1128/mBio.02189-18.

He, Y., Shelver, D., Kerby, R. L., & Roberts, G. P. (1996). Characterization of a CO-responsive transcriptional activator from *Rhodospirillum rubrum*. *Journal of Biological Chemistry*, *271*(1), 120–123. https://doi.org/10.1074/jbc.271.1.120.

Heidelberg, J. F., Seshadri, R., Haveman, S. A., Hemme, C. L., Paulsen, I. T., Kolonay, J. F., et al. (2004). The genome sequence of the anaerobic, sulfate-reducing bacterium *Desulfovibrio vulgaris* Hildenborough. *Nature Biotechnology*, *22*(5), 554–559. https://doi.org/10.1038/nbt959.

Hensley, S. A., Jung, J.-H., Park, C.-S., & Holden, J. F. (2014). *Thermococcus paralvinellae* sp. nov. and *Thermococcus cleftensis* sp. nov. of hyperthermophilic heterotrophs from deep-sea hydrothermal vents. *International Journal of Systematic and Evolutionary Microbiology*, *64*(11), 3655–3659. https://doi.org/10.1099/ijs.0.066100-0.

Henstra, A. M., & Stams, A. J. M. (2004). Novel physiological features of *Carboxydothermus hydrogenoformans* and *Thermoterrabacterium ferrireducens*. *Applied and Environmental Microbiology*, *70*(12), 7236–7240. https://doi.org/10.1128/AEM.70.12.7236-7240.2004.

Hess, V., Poehlein, A., Weghoff, M. C., Daniel, R., & Müller, V. (2014). A genome-guided analysis of energy conservation in the thermophilic, cytochrome-free acetogenic bacterium *Thermoanaerobacter kivui*. *BMC Genomics*, *15*, 1139. https://doi.org/10.1186/1471-2164-15-1139.

Hille, R., Dingwall, S., & Wilcoxen, J. (2015). The aerobic CO dehydrogenase from *Oligotropha carboxidovorans*. *JBIC, Journal of Biological Inorganic Chemistry*, *20*(2), 243–251. https://doi.org/10.1007/s00775-014-1188-4.

Hubert, C., Loy, A., Nickel, M., Arnosti, C., Baranyi, C., Brüchert, V., et al. (2009). A constant flux of diverse thermophilic bacteria into the cold Arctic seabed. *Science*, *325*(5947), 1541–1544. https://doi.org/10.1126/science.1174012.

Imkamp, F., Biegel, E., Jayamani, E., Buckel, W., & Müller, V. (2007). Dissection of the caffeate respiratory chain in the acetogen *Acetobacterium woodii*: Identification of an Rnf-type NADH dehydrogenase as a potential coupling site. *Journal of Bacteriology*, *189*(22), 8145–8153. https://doi.org/10.1128/JB.01017-07.

Inoue, M., Nakamoto, I., Omae, K., Oguro, T., Ogata, H., Yoshida, T., et al. (2019). Structural and phylogenetic diversity of anaerobic carbon-monoxide dehydrogenases. *Frontiers in Microbiology*, *9*, 3353. https://doi.org/10.3389/fmicb.2018.03353.

Inoue, M., Tanimura, A., Ogami, Y., Hino, T., Okunishi, S., Maeda, H., et al. (2019). Draft genome sequence of *Parageobacillus thermoglucosidasius* strain TG4, a hydrogenogenic carboxydotrophic bacterium isolated from a marine sediment. *Microbiology Resource Announcements*, *8*(5), e01666-18. https://doi.org/10.1128/MRA.01666-18.

Jensen, A., & Finster, K. (2005). Isolation and characterization of *Sulfurospirillum carboxydovorans* sp. nov., a new microaerophilic carbon monoxide oxidizing epsilon proteobacterium. *Antonie Van Leeuwenhoek, 87*(4), 339–353. https://doi.org/10.1007/s10482-004-6839-y.

Jeon, W. B., Cheng, J., & Ludden, P. W. (2001). Purification and characterization of membrane-associated CooC protein and its functional role in the insertion of nickel into carbon monoxide dehydrogenase from *Rhodospirillum rubrum*. *Journal of Biological Chemistry, 276*(42), 38602–38609. https://doi.org/10.1074/jbc.M104945200.

Jeong, J., Bertsch, J., Hess, V., Choi, S., Choi, I.-G., Chang, I. S., et al. (2015). Energy conservation model based on genomic and experimental analyses of a carbon monoxide-utilizing, butyrate-forming acetogen, *Eubacterium limosum* KIST612. *Applied and Environmental Microbiology, 81*(14), 4782–4790. https://doi.org/10.1128/AEM.00675-15.

Jeoung, J.-H., & Dobbek, H. (2007). Carbon dioxide activation at the Ni,Fe-cluster of anaerobic carbon monoxide dehydrogenase. *Science, 318*(5855), 1461–1464. https://doi.org/10.1126/science.1148481.

Ji, S., Zhao, R., Li, Z., Li, B., Shi, X., & Zhang, X.-H. (2013). *Ferrimonas sediminum* sp. nov., isolated from coastal sediment of an amphioxus breeding zone. *International Journal of Systematic and Evolutionary Microbiology, 63*(3), 977–981. https://doi.org/10.1099/ijs.0.042408-0.

Jiang, B., Henstra, A.-M., Paulo, P. L., Balk, M., van Doesburg, W., & Stams, A. J. M. (2009). Atypical one-carbon metabolism of an acetogenic and hydrogenogenic *Moorella thermoacetica* strain. *Archives of Microbiology, 191*(2), 123–131. https://doi.org/10.1007/s00203-008-0435-x.

Jung, G. Y., Jung, H. O., Kim, J. R., Ahn, Y., & Park, S. (1999). Isolation and characterization of *Rhodopseudomonas palustris* P4 which utilizes CO with the production of H_2. *Biotechnology Letters, 21*(6), 525–529. https://doi.org/10.1023/A:1005560630351.

Jung, G. Y., Kim, J. R., Jung, H. O., Park, J.-Y., & Park, S. (1999). A new chemoheterotrophic bacterium catalyzing water-gas shift reaction. *Biotechnology Letters, 21*(10), 869–873. https://doi.org/10.1023/A:1005599600510.

Kerby, R. L., Hong, S. S., Ensign, S. A., Coppoc, L. J., Ludden, P. W., & Roberts, G. P. (1992). Genetic and physiological characterization of the *Rhodospirillum rubrum* carbon monoxide dehydrogenase system. *Journal of Bacteriology, 174*(16), 5284–5294. https://doi.org/10.1128/jb.174.16.5284-5294.1992.

Kerby, R. L., Youn, H., & Roberts, G. P. (2008). RcoM: A new single-component transcriptional regulator of CO metabolism in bacteria. *Journal of Bacteriology, 190*(9), 3336–3343. https://doi.org/10.1128/JB.00033-08.

Khalil, M. A. K., & Rasmussen, R. A. (1990). The global cycle of carbon monoxide: Trends and mass balance. *Chemosphere, 20*(1–2), 227–242. https://doi.org/10.1016/0045-6535(90)90098-E.

Khelaifia, S., Fardeau, M.-L., Pradel, N., Aussignargues, C., Garel, M., Tamburini, C., et al. (2011). *Desulfovibrio piezophilus* sp. nov., a piezophilic, sulfate-reducing bacterium isolated from wood falls in the Mediterranean Sea. *International Journal of Systematic and Evolutionary Microbiology, 61*(11), 2706–2711. https://doi.org/10.1099/ijs.0.028670-0.

Kim, M.-S., Bae, S. S., Kim, Y. J., Kim, T. W., Lim, J. K., Lee, S. H., et al. (2013). CO-dependent H_2 production by genetically engineered *Thermococcus onnurineus* NA1. *Applied and Environmental Microbiology, 79*(6), 2048–2053. https://doi.org/10.1128/AEM.03298-12.

Kim, M.-S., Choi, A. R., Lee, S. H., Jung, H.-C., Bae, S. S., Yang, T.-J., et al. (2015). A novel CO-responsive transcriptional regulator and enhanced H_2 production by an engineered *Thermococcus onnurineus* NA1 strain. *Applied and Environmental Microbiology, 81*(5), 1708–1714. https://doi.org/10.1128/AEM.03019-14.

Kim, B. C., Grote, R., Lee, D. W., Antranikian, G., & Pyun, Y. R. (2001). *Thermoanaerobacter yonseiensis* sp. nov., a novel extremely thermophilic, xylose-utilizing bacterium that grows at up to 85 degrees C. *International Journal of Systematic and Evolutionary Microbiology*, *51*(4), 1539–1548. https://doi.org/10.1099/00207713-51-4-1539.

Kim, Y. J., Lee, H. S., Kim, E. S., Bae, S. S., Lim, J. K., Matsumi, R., et al. (2010). Formate-driven growth coupled with H_2 production. *Nature*, *467*(7313), 352–355. https://doi.org/10.1038/nature09375.

King, G. M. (2006). Nitrate-dependent anaerobic carbon monoxide oxidation by aerobic CO-oxidizing bacteria. *FEMS Microbiology Ecology*, *56*(1), 1–7. https://doi.org/10.1111/j.1574-6941.2006.00065.x.

King, G. M., & Weber, C. F. (2007). Distribution, diversity and ecology of aerobic CO-oxidizing bacteria. *Nature Reviews. Microbiology*, *5*(2), 107–118. https://doi.org/10.1038/nrmicro1595.

Köpke, M., Held, C., Hujer, S., Liesegang, H., Wiezer, A., Wollherr, A., et al. (2010). *Clostridium ljungdahlii* represents a microbial production platform based on syngas. *Proceedings of the National Academy of Sciences of the United States of America*, *107*(29), 13087–13092. https://doi.org/10.1073/pnas.1004716107.

Kozhevnikova, D. A., Taranov, E. A., Lebedinsky, A. V., Bonch-Osmolovskaya, E. A., & Sokolova, T. G. (2016). Hydrogenogenic and sulfidogenic growth of *Thermococcus* archaea on carbon monoxide and formate. *Microbiology*, *85*(4), 400–410. https://doi.org/10.1134/S0026261716040135.

Kruse, S., Goris, T., Westermann, M., Adrian, L., & Diekert, G. (2018). Hydrogen production by *Sulfurospirillum* species enables syntrophic interactions of epsilonproteobacteria. *Nature Communications*, *9*(1), 4872. https://doi.org/10.1038/s41467-018-07342-3.

Leduc, J., Thorsteinsson, M. V., Gaal, T., & Roberts, G. P. (2001). Mapping CooA·RNA polymerase interactions: Identification of activating regions 2 and 3 in CooA, the CO-sensing transcriptional activator. *Journal of Biological Chemistry*, *276*(43), 39968–39973. https://doi.org/10.1074/jbc.M105758200.

Lee, H. S., Kang, S. G., Bae, S. S., Lim, J. K., Cho, Y., Kim, Y. J., et al. (2008). The complete genome sequence of *Thermococcus onnurineus* NA1 reveals a mixed heterotrophic and carboxydotrophic metabolism. *Journal of Bacteriology*, *190*(22), 7491–7499. https://doi.org/10.1128/JB.00746-08.

Lee, S. H., Kim, M.-S., Bae, S. S., Choi, A. R., Lee, J.-W., Kim, T. W., et al. (2014). Comparison of CO-dependent H_2 production with strong promoters in *Thermococcus onnurineus* NA1. *Applied Microbiology and Biotechnology*, *98*(2), 979–986. https://doi.org/10.1007/s00253-013-5448-y.

Lee, S. H., Kim, M.-S., Jung, H. C., Lee, J., Lee, J.-H., Lee, H. S., et al. (2015). Screening of a novel strong promoter by RNA sequencing and its application to H_2 production in a hyperthermophilic archaeon. *Applied Microbiology and Biotechnology*, *99*(9), 4085–4092. https://doi.org/10.1007/s00253-015-6444-1.

Lee, S. H., Kim, M.-S., Lee, J.-H., Kim, T. W., Bae, S. S., Lee, S.-M., et al. (2016). Adaptive engineering of a hyperthermophilic archaeon on CO and discovering the underlying mechanism by multi-omics analysis. *Scientific Reports*, *6*(1), 22896. https://doi.org/10.1038/srep22896.

Lee, C. R., Kim, C., Song, Y. E., Im, H., Oh, Y.-K., Park, S., et al. (2018). Co-culture-based biological carbon monoxide conversion by *Citrobacter amalonaticus* Y19 and *Sporomusa ovata* via a reducing-equivalent transfer mediator. *Bioresource Technology*, *259*, 128–135. https://doi.org/10.1016/j.biortech.2018.02.129.

Leigh, J. A., Mayer, F., & Wolfe, R. S. (1981). *Acetogenium kivui*, a new thermophilic hydrogen-oxidizing acetogenic bacterium. *Archives of Microbiology*, *129*(4), 275–280. https://doi.org/10.1007/BF00414697.

Lessner, D. J., Li, L., Li, Q., Rejtar, T., Andreev, V. P., Reichlen, M., et al. (2006). An unconventional pathway for reduction of CO_2 to methane in CO-grown *Methanosarcina acetivorans* revealed by proteomics. *Proceedings of the National Academy of Sciences of the United States of America, 103*(47), 17921–17926. https://doi.org/10.1073/pnas.0608833103.

Liebensteiner, M. G., Pinkse, M. W. H., Nijsse, B., Verhaert, P. D. E. M., Tsesmetzis, N., Stams, A. J. M., et al. (2015). Perchlorate and chlorate reduction by the crenarchaeon *Aeropyrum pernix* and two thermophilic firmicutes. *Environmental Microbiology Reports*, 7(6), 936–945. https://doi.org/10.1111/1758-2229.12335.

Lim, J. K., Kang, S. G., Lebedinsky, A. V., Lee, J.-H., & Lee, H. S. (2010). Identification of a novel class of membrane-bound [NiFe]-hydrogenases in *Thermococcus onnurineus* NA1 by in silico analysis. *Applied and Environmental Microbiology, 76*(18), 6286–6289. https://doi.org/10.1128/AEM.00123-10.

Lonjers, Z. T., Dickson, E. L., Chu, T.-P. T., Kreutz, J. E., Neacsu, F. A., Anders, K. R., et al. (2012). Identification of a new gene required for the biosynthesis of rhodoquinone in *Rhodospirillum rubrum*. *Journal of Bacteriology, 194*(5), 965–971. https://doi.org/10.1128/JB.06319-11.

Lynch, M. D. J., & Neufeld, J. D. (2015). Ecology and exploration of the rare biosphere. *Nature Reviews. Microbiology, 13*(4), 217–229. https://doi.org/10.1038/nrmicro3400.

Maness, P.-C., & Weaver, P. F. (1994). Production of poly-3-hydroxyalkanoates from CO and H_2 by a novel photosynthetic bacterium. *Applied Biochemistry and Biotechnology, 45*(1), 395–406. https://doi.org/10.1007/BF02941814.

Maness, P.-C., & Weaver, P. F. (2002). Hydrogen production from a carbon-monoxide oxidation pathway in *Rubrivivax gelatinosus*. *International Journal of Hydrogen Energy, 27*(11–12), 1407–1411. https://doi.org/10.1016/S0360-3199(02)00107-6.

Marcano-Velazquez, J. G., Lo, J., Nag, A., Maness, P.-C., & Chou, K. J. (2019). Developing riboswitch-mediated gene regulatory controls in Thermophilic bacteria. *ACS Synthetic Biology, 8*(4), 633–640. https://doi.org/10.1021/acssynbio.8b00487.

Marteinsson, V. T., Birrien, J.-L., Reysenbach, A.-L., Vernet, M., Marie, D., Gambacorta, A., et al. (1999). *Thermococcus barophilus* sp. nov., a new barophilic and hyperthermophilic archaeon isolated under high hydrostatic pressure from a deep-sea hydrothermal vent. *International Journal of Systematic Bacteriology, 49*(2), 351–359. https://doi.org/10.1099/00207713-49-2-351.

McKay, L. J., Dlakić, M., Fields, M. W., Delmont, T. O., Eren, A. M., Jay, Z. J., et al. (2019). Co-occurring genomic capacity for anaerobic methane and dissimilatory sulfur metabolisms discovered in the Korarchaeota. *Nature Microbiology, 4*(4), 614–622. https://doi.org/10.1038/s41564-019-0362-4.

Meyer, O., & Schlegel, H. G. (1983). Biology of aerobic carbon monoxide-oxidizing bacteria. *Annual Reviews in Microbiology, 37*, 277–310.

Mohr, T., Aliyu, H., Küchlin, R., Polliack, S., Zwick, M., Neumann, A., et al. (2018). CO-dependent hydrogen production by the facultative anaerobe *Parageobacillus thermoglucosidasius*. *Microbial Cell Factories, 17*, 108. https://doi.org/10.1186/s12934-018-0954-3.

Mohr, T., Aliyu, H., Küchlin, R., Zwick, M., Cowan, D., Neumann, A., et al. (2018). Comparative genomic analysis of *Parageobacillus thermoglucosidasius* strains with distinct hydrogenogenic capacities. *BMC Genomics, 19*, 880. https://doi.org/10.1186/s12864-018-5302-9.

Mori, K., Hanada, S., Maruyama, A., & Marumo, K. (2002). *Thermanaeromonas toyohensis* gen. nov., sp. nov., a novel thermophilic anaerobe isolated from a subterranean vein in the Toyoha mines. *International Journal of Systematic and Evolutionary Microbiology, 52*(5), 1675–1680. https://doi.org/10.1099/00207713-52-5-1675.

Mörsdorf, G., Frunzke, K., Gadkari, D., & Meyer, O. (1992). Microbial growth on carbon monoxide. *Biodegradation*, *3*(1), 61–82. https://doi.org/10.1007/BF00189635.

Müller, V., Chowdhury, N. P., & Basen, M. (2018). Electron bifurcation: A long-hidden energy-coupling mechanism. *Annual Review of Microbiology*, *72*, 331–353. https://doi.org/10.1146/annurev-micro-090816-093440.

Müller, A. L., de Rezende, J. R., Hubert, C. R. J., Kjeldsen, K. U., Lagkouvardos, I., Berry, D., et al. (2014). Endospores of thermophilic bacteria as tracers of microbial dispersal by ocean currents. *The ISME Journal*, *8*(6), 1153–1165. https://doi.org/10.1038/ismej.2013.225.

Müller, V., Imkamp, F., Biegel, E., Schmidt, S., & Dilling, S. (2008). Discovery of a ferredoxin:NAD^+-oxidoreductase (Rnf) in *Acetobacterium woodii*. *Annals of the New York Academy of Sciences*, *1125*(1), 137–146. https://doi.org/10.1196/annals.1419.011.

Munk, A. C., Copeland, A., Lucas, S., Lapidus, A., Del Rio, T. G., Barry, K., et al. (2011). Complete genome sequence of *Rhodospirillum rubrum* type strain ($S1^T$). *Standards in Genomic Sciences*, *4*(3), 293–302. https://doi.org/10.4056/sigs.1804360.

Najafpour, G. D., & Younesi, H. (2007). Bioconversion of synthesis gas to hydrogen using a light-dependent photosynthetic bacterium, *Rhodospirillum rubrum*. *World Journal of Microbiology and Biotechnology*, *23*(2), 275–284. https://doi.org/10.1007/s11274-006-9225-2.

Nakagawa, T., Iino, T., Suzuki, K., & Harayama, S. (2006). *Ferrimonas futtsuensis* sp. nov. and *Ferrimonas kyonanensis* sp. nov., selenate-reducing bacteria belonging to the gammaproteobacteria isolated from Tokyo Bay. *International Journal of Systematic and Evolutionary Microbiology*, *56*(11), 2639–2645. https://doi.org/10.1099/ijs.0.64399-0.

Nevin, K. P., Holmes, D. E., Woodard, T. L., Hinlein, E. S., Ostendorf, D. W., & Lovley, D. R. (2005). *Geobacter bemidjiensis* sp. nov. and *Geobacter psychrophilus* sp. nov., two novel Fe(III)-reducing subsurface isolates. *International Journal of Systematic and Evolutionary Microbiology*, *55*(4), 1667–1674. https://doi.org/10.1099/ijs.0.63417-0.

Nitschke, W., & Russell, M. J. (2013). Beating the acetyl coenzyme A-pathway to the origin of life. *Philosophical Transactions of the Royal Society, B: Biological Sciences*, *368*(1622), 20120258. https://doi.org/10.1098/rstb.2012.0258.

Nobre, L. S., Seixas, J. D., Romão, C. C., & Saraiva, L. M. (2007). Antimicrobial action of carbon monoxide-releasing compounds. *Antimicrobial Agents and Chemotherapy*, *51*(12), 4303–4307. https://doi.org/10.1128/AAC.00802-07.

Novikov, A. A., Sokolova, T. G., Lebedinsky, A. V., Kolganova, T. V., & Bonch-Osmolovskaya, E. A. (2011). *Carboxydothermus islandicus* sp. nov., a thermophilic, hydrogenogenic, carboxydotrophic bacterium isolated from a hot spring. *International Journal of Systematic and Evolutionary Microbiology*, *61*(10), 2532–2537. https://doi.org/10.1099/ijs.0.030288-0.

Oelgeschläger, E., & Rother, M. (2008). Carbon monoxide-dependent energy metabolism in anaerobic bacteria and archaea. *Archives of Microbiology*, *190*(3), 257–269. https://doi.org/10.1007/s00203-008-0382-6.

Oh, Y.-K., Seol, E.-H., Kim, J. R., & Park, S. (2003). Fermentative biohydrogen production by a new chemoheterotrophic bacterium *Citrobacter* sp. Y19. *International Journal of Hydrogen Energy*, *28*(12), 1353–1359. https://doi.org/10.1016/S0360-3199(03)00024-7.

Omae, K., Fukuyama, Y., Yasuda, H., Mise, K., Yoshida, T., & Sako, Y. (2019). Diversity and distribution of thermophilic hydrogenogenic carboxydotrophs revealed by microbial community analysis in sediments from multiple hydrothermal environments in Japan. *Archives of Microbiology*, *201*(7), 969–982. https://doi.org/10.1007/s00203-019-01661-9.

Omae, K., Yoneda, Y., Fukuyama, Y., Yoshida, T., & Sako, Y. (2017). Genomic analysis of *Calderihabitans maritimus* KKC1, a thermophilic, hydrogenogenic, carboxydotrophic bacterium isolated from marine sediment. *Applied and Environmental Microbiology*, *83*(15), e00832-17. https://doi.org/10.1128/AEM.00832-17.

Parshina, S. N., Kijlstra, S., Henstra, A. M., Sipma, J., Plugge, C. M., & Stams, A. J. M. (2005). Carbon monoxide conversion by thermophilic sulfate-reducing bacteria in pure culture and in co-culture with *Carboxydothermus hydrogenoformans*. *Applied Microbiology and Biotechnology*, *68*(3), 390–396. https://doi.org/10.1007/s00253-004-1878-x.

Parshina, S. N., Sipma, J., Nakashimada, Y., Henstra, A. M., Smidt, H., Lysenko, A. M., et al. (2005). *Desulfotomaculum carboxydivorans* sp. nov., a novel sulfate-reducing bacterium capable of growth at 100% CO. *International Journal of Systematic and Evolutionary Microbiology*, *55*(5), 2159–2165. https://doi.org/10.1099/ijs.0.63780-0.

Peng, T., Pan, S., Christopher, L., Sparling, R., & Levin, D. B. (2015). Draft genome sequence of *Thermoanaerobacter* sp. strain YS13, a novel thermophilic bacterium. *Genome Announcements*, *3*(3), e00584-15. https://doi.org/10.1128/genomeA.00584-15.

Poehlein, A., Böer, T., Steensen, K., & Daniel, R. (2018). Draft genome sequence of the hydrogenogenic carboxydotroph *Moorella stamsii* DSM 26271. *Genome Announcements*, *6*(18), e00345-18. https://doi.org/10.1128/genomeA.00345-18.

Price, M. N., Dehal, P. S., & Arkin, A. P. (2010). FastTree 2—Approximately maximum-likelihood trees for large alignments. *PLoS One*, *5*(3), e9490. https://doi.org/10.1371/journal.pone.0009490.

Rajeev, L., Hillesland, K. L., Zane, G. M., Zhou, A., Joachimiak, M. P., He, Z., et al. (2012). Deletion of the *Desulfovibrio vulgaris* carbon monoxide sensor invokes global changes in transcription. *Journal of Bacteriology*, *194*(21), 5783–5793. https://doi.org/10.1128/JB.00749-12.

Reeve, B., Martinez-Klimova, E., de Jonghe, J., Leak, D. J., & Ellis, T. (2016). The *Geobacillus* plasmid set: A modular toolkit for thermophile engineering. *ACS Synthetic Biology*, *5*(12), 1342–1347. https://doi.org/10.1021/acssynbio.5b00298.

Roberts, G. P., Youn, H., & Kerby, R. L. (2004). CO-sensing mechanisms. *Microbiology and Molecular Biology Reviews*, *68*(3), 453–473. https://doi.org/10.1128/MMBR.68.3.453-473.2004.

Rose, J. J., Wang, L., Xu, Q., McTiernan, C. F., Shiva, S., Tejero, J., et al. (2017). Carbon monoxide poisoning: Pathogenesis, management, and future directions of therapy. *American Journal of Respiratory and Critical Care Medicine*, *195*(5), 596–606. https://doi.org/10.1164/rccm.201606-1275CI.

Rother, M., & Metcalf, W. W. (2004). Anaerobic growth of *Methanosarcina acetivorans* C2A on carbon monoxide: An unusual way of life for a methanogenic archaeon. *Proceedings of the National Academy of Sciences of the United States of America*, *101*(48), 16929–16934. https://doi.org/10.1073/pnas.0407486101.

Saeki, K., & Kumagai, H. (1998). The rnf gene products in *Rhodobacter capsulatus* play an essential role in nitrogen fixation during anaerobic DMSO-dependent growth in the dark. *Archives of Microbiology*, *169*(5), 464–467. https://doi.org/10.1007/s002030050598.

Sant'Anna, F. H., Lebedinsky, A. V., Sokolova, T. G., Robb, F. T., & Gonzalez, J. M. (2015). Analysis of three genomes within the thermophilic bacterial species *Caldanaerobacter subterraneus* with a focus on carbon monoxide dehydrogenase evolution and hydrolase diversity. *BMC Genomics*, *16*(1), 757. https://doi.org/10.1186/s12864-015-1955-9.

Schade, G. W., & Crutzen, P. J. (1999). CO emissions from degrading plant matter (II). *Tellus Series B: Chemical and Physical Meteorology*, *51*(5), 909–918. https://doi.org/10.3402/tellusb.v51i5.16503.

Schoelmerich, M. C., & Müller, V. (2019). Energy conservation by a hydrogenase-dependent chemiosmotic mechanism in an ancient metabolic pathway. *Proceedings of the National Academy of Sciences of the United States of America*, *116*(13), 6329–6334. https://doi.org/10.1073/pnas.1818580116.

Schuchmann, K., & Müller, V. (2014). Autotrophy at the thermodynamic limit of life: a model for energy conservation in acetogenic bacteria. *Nature Reviews. Microbiology, 12*(12), 809–821. https://doi.org/10.1038/nrmicro3365.

Schut, G. J., Lipscomb, G. L., Nguyen, D. M. N., Kelly, R. M., & Adams, M. W. W. (2016). Heterologous production of an energy-conserving carbon monoxide dehydrogenase complex in the hyperthermophile *Pyrococcus furiosus*. *Frontiers in Microbiology*, 7, 29. https://doi.org/10.3389/fmicb.2016.00029.

Shelobolina, E. S., Nevin, K. P., Blakeney-Hayward, J. D., Johnsen, C. V., Plaia, T. W., Krader, P., et al. (2007). *Geobacter pickeringii* sp. nov., *Geobacter argillaceus* sp. nov. and *Pelosinus fermentans* gen. nov., sp. nov., isolated from subsurface kaolin lenses. *International Journal of Systematic and Evolutionary Microbiology, 57*(1), 126–135. https://doi.org/10.1099/ijs.0.64221-0.

Shelver, D., Kerby, R. L., He, Y., & Roberts, G. P. (1995). Carbon monoxide-induced activation of gene expression in *Rhodospirillum rubrum* requires the product of *cooA*, a member of the cyclic AMP receptor protein family of transcriptional regulators. *Journal of Bacteriology, 177*(8), 2157–2163. https://doi.org/10.1128/jb.177.8.2157-2163.1995.

Shelver, D., Kerby, R. L., He, Y., & Roberts, G. P. (1997). CooA, a CO-sensing transcription factor from *Rhodospirillum rubrum*, is a CO-binding heme protein. *Proceedings of the National Academy of Sciences of the United States of America, 94*(21), 11216–11220. https://doi.org/10.1073/pnas.94.21.11216.

Singer, S. W., Hirst, M. B., & Ludden, P. W. (2006). CO-dependent H_2 evolution by *Rhodospirillum rubrum*: Role of CODH:CooF complex. *Biochimica et Biophysica Acta—Bioenergetics, 1757*(12), 1582–1591. https://doi.org/10.1016/j.bbabio.2006.10.003.

Sinha, P., Roy, S., & Das, D. (2015). Role of formate hydrogen lyase complex in hydrogen production in facultative anaerobes. *International Journal of Hydrogen Energy, 40*(29), 8806–8815. https://doi.org/10.1016/j.ijhydene.2015.05.076.

Slepova, T. V., Sokolova, T. G., Kolganova, T. V., Tourova, T. P., & Bonch-Osmolovskaya, E. A. (2009). *Carboxydothermus siderophilus* sp. nov., a thermophilic, hydrogenogenic, carboxydotrophic, dissimilatory Fe(III)-reducing bacterium from a Kamchatka hot spring. *International Journal of Systematic and Evolutionary Microbiology, 59*(2), 213–217. https://doi.org/10.1099/ijs.0.000620-0.

Slepova, T. V., Sokolova, T. G., Lysenko, A. M., Tourova, T. P., Kolganova, T. V., Kamzolkina, O. V., et al. (2006). *Carboxydocella sporoproducens* sp. nov., a novel anaerobic CO-utilizing/H_2-producing thermophilic bacterium from a Kamchatka hot spring. *International Journal of Systematic and Evolutionary Microbiology, 56*(4), 797–800. https://doi.org/10.1099/ijs.0.63961-0.

Soboh, B., Linder, D., & Hedderich, R. (2002). Purification and catalytic properties of a CO-oxidizing:H_2-evolving enzyme complex from *Carboxydothermus hydrogenoformans*. *European Journal of Biochemistry, 269*(22), 5712–5721. https://doi.org/10.1046/j.1432-1033.2002.03282.x.

Sokolova, T. G., González, J. M., Kostrikina, N. A., Chernyh, N. A., Slepova, T. V., Bonch-Osmolovskaya, E. A., et al. (2004). *Thermosinus carboxydivorans* gen. nov., sp. nov., a new anaerobic, thermophilic, carbon-monoxide-oxidizing, hydrogenogenic bacterium from a hot pool of Yellowstone National Park. *International Journal of Systematic and Evolutionary Microbiology, 54*(6), 2353–2359. https://doi.org/10.1099/ijs.0.63186-0.

Sokolova, T. G., González, J. M., Kostrikina, N. A., Chernyh, N. A., Tourova, T. P., Kato, C., et al. (2001). *Carboxydobrachium pacificum* gen. nov., sp. nov., a new anaerobic, thermophilic, CO-utilizing marine bacterium from Okinawa Trough. *International Journal of Systematic and Evolutionary Microbiology, 51*(1), 141–149. https://doi.org/10.1099/00207713-51-1-141.

Sokolova, T. G., Hanel, J., Onyenwoke, R. U., Reysenbach, A.-L., Banta, A., Geyer, R., et al. (2007). Novel chemolithotrophic, thermophilic, anaerobic bacteria *Thermolithobacter ferrireducens* gen. nov., sp. nov. and *Thermolithobacter carboxydivorans* sp. nov. *Extremophiles, 11*(1), 145–157. https://doi.org/10.1007/s00792-006-0022-5.

Sokolova, T. G., Henstra, A. M., Sipma, J., Parshina, S. N., Stams, A. J. M., & Lebedinsky, A. V. (2009). Diversity and ecophysiological features of thermophilic carboxydotrophic anaerobes. *FEMS Microbiology Ecology*, *68*(2), 131–141. https://doi.org/10.1111/j.1574-6941.2009.00663.x.

Sokolova, T. G., Jeanthon, C., Kostrikina, N. A., Chernyh, N. A., Lebedinsky, A. V., Stackebrandt, E., et al. (2004). The first evidence of anaerobic CO oxidation coupled with H_2 production by a hyperthermophilic archaeon isolated from a deep-sea hydrothermal vent. *Extremophiles*, *8*(4), 317–323. https://doi.org/10.1007/s00792-004-0389-0.

Sokolova, T. G., Kostrikina, N. A., Chernyh, N. A., Kolganova, T. V., Tourova, T. P., & Bonch-Osmolovskaya, E. A. (2005). *Thermincola carboxydiphila* gen. nov., sp. nov., a novel anaerobic, carboxydotrophic, hydrogenogenic bacterium from a hot spring of the Lake Baikal area. *International Journal of Systematic and Evolutionary Microbiology*, *55*(5), 2069–2073. https://doi.org/10.1099/ijs.0.63299-0.

Sokolova, T. G., Kostrikina, N. A., Chernyh, N. A., Tourova, T. P., Kolganova, T. V., & Bonch-Osmolovskaya, E. A. (2002). *Carboxydocella thermautotrophica* gen. nov., sp. nov., a novel anaerobic, CO-utilizing thermophile from a Kamchatkan hot spring. *International Journal of Systematic and Evolutionary Microbiology*, *52*(6), 1961–1967. https://doi.org/10.1099/00207713-52-6-1961.

Sokolova, T., & Lebedinsky, A. (2013). CO-oxidizing anaerobic thermophilic prokaryotes. In T. Satyanarayana, J. Littlechild, & Y. Kawarabayasi (Eds.), *Thermophilic microbes in environmental and industrial biotechnology* (pp. 203–231): Springer.

Srinivas, T. N. R., Vijaya Bhaskar, Y., Bhumika, V., & Anil Kumar, P. (2013). *Photobacterium marinum* sp. nov., a marine bacterium isolated from a sediment sample from Palk Bay, India. *Systematic and Applied Microbiology*, *36*(3), 160–165. https://doi.org/10.1016/j.syapm.2012.12.002.

Suzuki, Y., Kishigami, T., Inoue, K., Mizoguchi, Y., Eto, N., Takagi, M., et al. (1983). *Bacillus thermoglucosidasius* sp. nov., a new species of obligately thermophilic bacilli. *Systematic and Applied Microbiology*, *4*(4), 487–495. https://doi.org/10.1016/S0723-2020(83)80006-X.

Svetlichny, V. A., Sokolova, T. G., Gerhardt, M., Ringpfeil, M., Kostrikina, N. A., & Zavarzin, G. A. (1991). *Carboxydothermus hydrogenoformans* gen. nov., sp. nov., a CO-utilizing thermophilic anaerobic bacterium from hydrothermal environments of Kunashir Island. *Systematic and Applied Microbiology*, *14*(3), 254–260. https://doi.org/10.1016/S0723-2020(11)80377-2.

Tan, X., Loke, H.-K., Fitch, S., & Lindahl, P. A. (2005). The tunnel of acetyl-coenzyme A synthase/carbon monoxide dehydrogenase regulates delivery of CO to the active site. *Journal of the American Chemical Society*, *127*(16), 5833–5839. https://doi.org/10.1021/ja043701v.

Tavares, A. F. N., Teixeira, M., Romão, C. C., Seixas, J. D., Nobre, L. S., & Saraiva, L. M. (2011). Reactive oxygen species mediate bactericidal killing elicited by carbon monoxide-releasing molecules. *Journal of Biological Chemistry*, *286*(30), 26708–26717. https://doi.org/10.1074/jbc.M111.255752.

Taylor, M. P., Esteban, C. D., & Leak, D. J. (2008). Development of a versatile shuttle vector for gene expression in *Geobacillus* spp. *Plasmid*, *60*(1), 45–52. https://doi.org/10.1016/j.plasmid.2008.04.001.

Techtmann, S. M., Colman, A. S., Murphy, M. B., Schackwitz, W. S., Goodwin, L. A., & Robb, F. T. (2011). Regulation of multiple carbon monoxide consumption pathways in anaerobic bacteria. *Frontiers in Microbiology*, *2*, 147. https://doi.org/10.3389/fmicb.2011.00147.

Techtmann, S. M., Colman, A. S., & Robb, F. T. (2009). 'That which does not kill us only makes us stronger': The role of carbon monoxide in thermophilic microbial consortia. *Environmental Microbiology*, *11*(5), 1027–1037. https://doi.org/10.1111/j.1462-2920.2009.01865.x.

Techtmann, S. M., Lebedinsky, A. V., Colman, A. S., Sokolova, T. G., Woyke, T., Goodwin, L., et al. (2012). Evidence for horizontal gene transfer of anaerobic carbon monoxide dehydrogenases. *Frontiers in Microbiology*, *3*, 132. https://doi.org/10.3389/fmicb.2012.00132.

Thauer, R. K., Jungermann, K., & Decker, K. (1977). Energy conservation in chemotrophic anaerobic bacteria. *Microbiology and Molecular Biology Reviews*, *41*(1), 100–180.

Toshchakov, S. V., Lebedinsky, A. V., Sokolova, T. G., Zavarzina, D. G., Korzhenkov, A. A., Teplyuk, A. V., et al. (2018). Genomic insights into energy metabolism of *Carboxydocella thermautotrophica* coupling hydrogenogenic CO oxidation with the reduction of Fe(III) minerals. *Frontiers in Microbiology*, *9*, 1759. https://doi.org/10.3389/fmicb.2018.01759.

Uffen, R. L. (1976). Anaerobic growth of a *Rhodopseudomonas* species in the dark with carbon monoxide as sole carbon and energy substrate. *Proceedings of the National Academy of Sciences of the United States of America*, *73*(9), 3298–3302. https://doi.org/10.1073/pnas.73.9.3298.

Vignais, P. M., & Billoud, B. (2007). Occurrence, classification, and biological function of hydrogenases: An overview. *Chemical Reviews*, *107*(10), 4206–4272. https://doi.org/10.1021/cr050196r.

Wareham, L. K., Southam, H. M., & Poole, R. K. (2018). Do nitric oxide, carbon monoxide and hydrogen sulfide really qualify as 'gasotransmitters' in bacteria? *Biochemical Society Transactions*, *46*(5), 1107–1118. https://doi.org/10.1042/BST20170311.

Watanabe, S., Arai, T., Matsumi, R., Atomi, H., Imanaka, T., & Miki, K. (2009). Crystal structure of HypA, a nickel-binding metallochaperone for [NiFe] hydrogenase maturation. *Journal of Molecular Biology*, *394*(3), 448–459. https://doi.org/10.1016/j.jmb.2009.09.030.

Wawrousek, K., Noble, S., Korlach, J., Chen, J., Eckert, C., Yu, J., et al. (2014). Genome annotation provides insight into carbon monoxide and hydrogen metabolism in *Rubrivivax gelatinosus*. *PLoS One*, *9*(12). e114551https://doi.org/10.1371/journal.pone.0114551.

Weghoff, M. C., & Müller, V. (2016). CO metabolism in the thermophilic acetogen *Thermoanaerobacter kivui*. *Applied and Environmental Microbiology*, *82*(8), 2312–2319. https://doi.org/10.1128/AEM.00122-16.

Wilcoxen, J., Zhang, B., & Hille, R. (2011). Reaction of the molybdenum- and copper-containing carbon monoxide dehydrogenase from *Oligotropha carboxydovorans* with quinones. *Biochemistry*, *50*(11), 1910–1916. https://doi.org/10.1021/bi1017182.

Wittenborn, E. C., Merrouch, M., Ueda, C., Fradale, L., Léger, C., Fourmond, V., et al. (2018). Redox-dependent rearrangements of the NiFeS cluster of carbon monoxide dehydrogenase. *eLife*, *7*, e39451. https://doi.org/10.7554/eLife.39451.

Wrighton, K. C., Agbo, P., Warnecke, F., Weber, K. A., Brodie, E. L., DeSantis, T. Z., et al. (2008). A novel ecological role of the Firmicutes identified in thermophilic microbial fuel cells. *The ISME Journal*, *2*(11), 1146–1156. https://doi.org/10.1038/ismej.2008.48.

Wu, M., Ren, Q., Durkin, S., Daugherty, S. C., Brinkac, L. M., Dodson, R. J., et al. (2005). Life in hot carbon monoxide: The complete genome sequence of *Carboxydothermus hydrogenoformans* Z-2901. *PLoS Genetics*, *1*(5), 563–574. https://doi.org/10.1371/journal.pgen.0010065.

Wu, L., & Wang, R. (2005). Carbon monoxide: Endogenous production, physiological functions, and Pharmacological applications. *Pharmacological Reviews*, *57*(4), 585–630. https://doi.org/10.1124/pr.57.4.3.

Xavier, J. C., Preiner, M., & Martin, W. F. (2018). Something special about CO-dependent CO_2 fixation. *The FEBS Journal*, *285*(22), 4181–4195. https://doi.org/10.1111/febs.14664.

Yoneda, Y., Kano, S. I., Yoshida, T., Ikeda, E., Fukuyama, Y., Omae, K., et al. (2015). Detection of anaerobic carbon monoxide-oxidizing thermophiles in hydrothermal environments. *FEMS Microbiology Ecology*, *91*(9), fiv093. https://doi.org/10.1093/femsec/fiv093.

Yoneda, Y., Yoshida, T., Daifuku, T., Kitamura, T., Inoue, T., Kano, S., et al. (2013). Quantitative detection of carboxydotrophic bacteria *Carboxydothermus* in a hot aquatic environment. *Fundamental and Applied Limnology*, *182*(2), 161–170. https://doi.org/10.1127/1863-9135/2013/0374.

Yoneda, Y., Yoshida, T., Kawaichi, S., Daifuku, T., Takabe, K., & Sako, Y. (2012). *Carboxydothermus pertinax* sp. nov., a thermophilic, hydrogenogenic, Fe(III)-reducing, sulfur-reducing carboxydotrophic bacterium from an acidic hot spring. *International Journal of Systematic and Evolutionary Microbiology*, *62*(7), 1692–1697. https://doi.org/10.1099/ijs.0.031583-0.

Yoneda, Y., Yoshida, T., Yasuda, H., Imada, C., & Sako, Y. (2013). A thermophilic, hydrogenogenic and carboxydotrophic bacterium, *Calderihabitans maritimus* gen. nov., sp. nov., from a marine sediment core of an undersea caldera. *International Journal of Systematic and Evolutionary Microbiology*, *63*(10), 3602–3608. https://doi.org/10.1099/ijs.0.050468-0.

Youn, H., Kerby, R. L., Conrad, M., & Roberts, G. P. (2004). Functionally critical elements of CooA-related CO sensors. *Journal of Bacteriology*, *186*(5), 1320–1329. https://doi.org/10.1128/JB.186.5.1320-1329.2004.

Younesi, H., Najafpour, G., & Mohamed, A. R. (2005). Ethanol and acetate production from synthesis gas via fermentation processes using anaerobic bacterium, *Clostridium ljungdahlii*. *Biochemical Engineering Journal*, *27*(2), 110–119. https://doi.org/10.1016/j.bej.2005.08.015.

Yu, H., Wu, C.-H., Schut, G. J., Haja, D. K., Zhao, G., Peters, J. W., et al. (2018). Structure of an ancient respiratory system. *Cell*, *173*(7), 1636–1649. https://doi.org/10.1016/j.cell.2018.03.071.

Zavarzina, D. G., Sokolova, T. G., Tourova, T. P., Chernyh, N. A., Kostrikina, N. A., & Bonch-Osmolovskaya, E. A. (2007). *Thermincola ferriacetica* sp. nov., a new anaerobic, thermophilic, facultatively chemolithoautotrophic bacterium capable of dissimilatory Fe(III) reduction. *Extremophiles*, *11*(1), 1–7. https://doi.org/10.1007/s00792-006-0004-7.

Zhao, Y., Caspers, M. P., Abee, T., Siezen, R. J., & Kort, R. (2012). Complete genome sequence of *Geobacillus thermoglucosidans* TNO-09.020, a thermophilic sporeformer associated with a dairy-processing environment. *Journal of Bacteriology*, *194*(15), 4118. https://doi.org/10.1128/JB.00318-12.

Zhao, Y., Haddad, M., Cimpoia, R., Liu, Z., & Guiot, S. R. (2013). Performance of a *Carboxydothermus hydrogenoformans*-immobilizing membrane reactor for syngas upgrading into hydrogen. *International Journal of Hydrogen Energy*, *38*(5), 2167–2175. https://doi.org/10.1016/j.ijhydene.2012.11.038.

CHAPTER FOUR

The versatility of *Pseudomonas putida* in the rhizosphere environment

Lázaro Molina, Ana Segura, Estrella Duque, Juan-Luis Ramos*
CSIC- Estación Experimental del Zaidín, Granada, Spain
*Corresponding author: e-mail address: juanluis.ramos@eez.csic.es

Contents

Abstract

This article addresses the lifestyle of *Pseudomonas* and focuses on how *Pseudomonas putida* can be used as a model system for biotechnological processes in agriculture, and in the removal of pollutants from soils. In this chapter we aim to show how a deep analysis using genetic information and experimental tests has helped to reveal insights into the lifestyle of Pseudomonads. *Pseudomonas putida* is a Plant Growth Promoting Rhizobacteria (PGPR) that establishes commensal relationships with plants. The interaction involves a series of functions encoded by core genes which favor nutrient mobilization, prevention of pathogen development and efficient niche colonization. Certain *Pseudomonas putida* strains harbor accessory genes that confer specific biodegradative properties and because these microorganisms can thrive on the roots of plants they can be exploited to remove pollutants via rhizoremediation, making the consortium plant/*Pseudomonas* a useful tool to combat pollution.

Advances in Applied Microbiology, Volume 110
ISSN 0065-2164
https://doi.org/10.1016/bs.aambs.2019.12.002

1. Introduction

Life on our planet is based on the continuous cycling of elements: carbon, nitrogen, sulfur, iron, etc. We are all aware that plants fix CO_2 to produce organic compounds. This photosynthetic carbon then enters the food chain through herbivorous animals, which eat plants which initiates the movement of organic matter, and in the respiratory process animals release CO_2 to the atmosphere. Less well known is that about 20% of the carbon fixed by plants is secreted in root exudates, where it is kept in part as a carbon reservoir; in addition, this organic matter in the soil serves as a source of carbon and energy for microbes to grow and again return CO_2 to the atmosphere. Clearly, microbes are essential players in the operation of the carbon cycle.

There are many ways to classify microorganisms, but in the context of this chapter we would like to simply distinguish two types of microbes: cosmopolitan and specialist. Cosmopolitan microbes can be isolated from a wide variety of niches—soil, fresh and marine waters, temperate environments, etc., while the term specialists, on the other hand, is used to refer to microbes that are isolated from a limited number of niches (salt lakes, acidic environments, hot springs, etc.).

Pseudomonas putida is one of the species in the genus *Pseudomonas* and microbes of this species have been isolated from many countries and from different niches, therefore they can be considered cosmopolitan (Timmis, 2002). They are Gram-negative microbes and can be found as biofilms on biotic and abiotic surfaces or as free-living microbes in aquatic environments (Moore et al., 2006). They are aerobic microorganisms, although some can be found in low oxygen tension environments. *Pseudomonas putida* gained quite a lot of attention in the 1970s when a number of strains of this species were shown to degrade aromatic compounds (Worsey & Williams, 1975). This attention continued in the 80s when they were shown to also degrade a number of xenobiotic compounds and were amenable to genetic manipulation in the laboratory, which allowed vertical and horizontal expansion of existing catabolic pathways (Ramos & Timmis, 1987; Reineke & Knackmuss, 1988).

One of the most studied strains of the species *P. putida* is the so-called KT2440 (Franklin, Bagdasarian, Bagdasarian, & Timmis, 1981). This strain is derived from the mt-2 strain and lacks the TOL plasmid that conferred the ability of the original strain to use toluene/xylenes and benzoate/toluates.

The genome of this strain was sequenced by a consortium of American and German groups and it was the first genome of the species (Nelson et al., 2002). The size of the genome is ~6.2Mb with a G+C content of ~61%; the genome is a mosaic of genes in the sense that there are regions where there is a clear bias of A+T richness, suggesting that these regions may have been acquired by horizontal gene transfer; in fact, in some cases they are clearly genomic islands (Nelson et al., 2002).

KT2440 has seven identical rRNA operons, with 16S rRNA being a component of each operon. From a taxonomic point of view Norberto Palleroni was pivotal in organizing the genus *Pseudomonas* by applying the concept of 16S rRNA to the genus which was further developed by Moore and Palleroni in the series *Pseudomonas* (Moore & Palleroni, 2004). However, we are aware that a single gene or even multilocus sequencing analysis based on house-keeping genes is not sufficient to distinguish strains of a given species or even strains of different species within a genus. We then raised a question, where to place the limits which decide if a strain belongs to the species *P. putida* or not? To solve this, we decided to determine the pan-genome of the species, for this we used the nine complete genome sequences available at the date of the analysis; all these microbes have on average genome of ~6Mb, encode ~6000 genes, had been classified as *P. putida* and were isolated from different countries or different niches. Udaondo, Molina, Segura, Duque, and Ramos (2016) found that all *P. putida* strains shared about 3000 genes. This set of genes was called the core genome. The core genes basically define the physiology of the species and a number of general functions. We should note that among core genes are a number which encode proteins involved in adhesion to surfaces to produce biofilms, chemotactic operons, information for central metabolism, catabolic pathways for certain aromatic compounds, stress resistance determinants, RNA polymerase and sigma factors, among others. Some of these genes and their functions are briefly analyzed below. There is a group of another ~3000 genes that are shared by at least two strains but not by all of the strains of the species. This set of genes is called the accessory genes and they confer niche specificity and certain catabolic properties. For instance, some *P. putida* strains use toluene as a carbon source, but there are different pathways described in the species for the metabolism of this compound. Some of the pathways are borne by plasmids, such as those on the TOL plasmid (Franklin et al., 1981; Worsey & Williams, 1975) while others are chromosomally encoded (Mosqueda & Ramos, 2000; Wackett & Gibson, 1988). Some strains are extremely tolerant to organic solvents and this set of genes are borne on self-transmissible plasmids (Ramos et al., 2002, 2015). Furthermore, each strain has between 50 and 300 unique genes which are only present in one strain. Most of these unique genes encode hypothetical proteins, however, when the function is known, genes that encode enzymes for the metabolism of organic sulfur and transporters of ions and acids are prevalent; this set of genes expands the metabolic versatility of certain strains.

The analysis of the proteins encoded by the KT2440 genome revealed that there were 900 families of paralogs genes (Nelson et al., 2002). Nearly 50% of these paralogs are present in other bacterial species, which have genomes larger than 5 Mb, indicating that the metabolic versatility of *Pseudomonas* is shared, in part, with other environmental microorganisms (Belda et al., 2016; Nogales et al., 2019; Udaondo et al., 2016). We classified enzymes and their functions based on both the experimental knowledge derived from biochemical assays and bioinformatic approaches to identify catalytic and structural motifs (Udaondo et al., 2016). The biochemical versatility of *Pseudomonas* speaks to the plasticity of these microorganisms and explains their ability to adapt to different niches.

2. *Pseudomonas putida*, a good rhizosphere colonizer

We and others have found that *P. putida* proliferates and is an abundant microbe in the soil close to the roots (rhizosphere) of plants; in fact, this is one of the primary niches for this species (Molina et al., 2000; Molina, Ramos, Ronchel, Molin, & Ramos, 2000; Roca et al., 2013; Sutra, Risède, & Gardan, 2000). Fig. 1 summarizes the basic commensal relationship between *Pseudomonas putida* and plants. Plants fix CO_2 from the air and use the root system to assimilate a number of nitrogen sources, phosphate, sulfur and micronutrients. As mentioned earlier, approximately 20% of the total CO_2 fixed by plants is released to the soil as root exudates. These exudates contain sugars (mainly glucose), organic acids (citric, lactic, others), amino acids (most of the 20 proteinogenic amino acids are present in plant exudates), pectin, proteins and other chemicals. Plants also excrete a number of secondary metabolites and high-molecular weight molecules, such as polysaccharides and proteins (reviewed by Bais, Weir, Perry, Gilroy, & Vivanco, 2006). This milieu makes the rhizosphere a nutrient-rich environment in which *P. putida* can thrive.

Pseudomonas putida strains have been identified in soils (Mulet, Bennasar, Lalucat, & García-Valdés, 2009), by isolation on plates, by direct sequencing

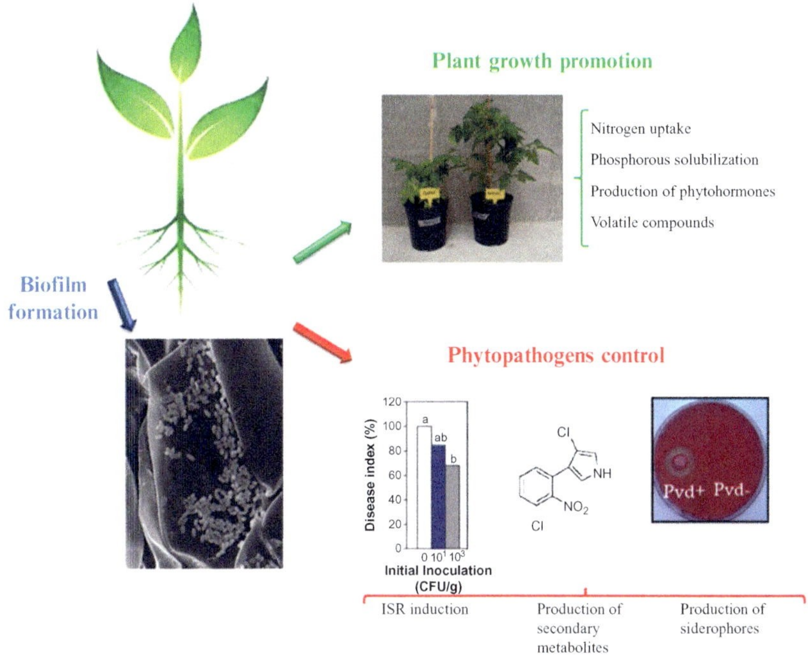

Fig. 1 Establishment of *Pseudomonas putida* in the rhizosphere and its role as a plant-growth promoting rhizobacteria (PGPR).

of DNA from rhizospheric environments (Mulet et al., 2009; Nagayama, Sugawara, Endo, et al., 2015) or by analyzing the production of specific metabolites from these environments (Oni et al., 2019). This bacterial group is able to colonize and persist in root environments of different plants in higher number (two orders of magnitude) than in soil not influenced by plants, termed bulk soil (Molina, Ramos, Duque, et al., 2000). This enhancement in cell numbers observed in the rhizospheric soil—soil under the influence of root exudates (Hiltner, 1904)—compared with that found in bulk soil is called the rhizospheric effect (Morgan & Whipps, 2001). This effect, besides being determined by external factors (presence of nutrients, inhibitors, etc.), also depends on the intrinsic characteristics of *Pseudomonas putida* strains, such as the capacity to efficiently use the molecules existing in the exudates as carbon, nitrogen, iron, phosphorous or sulfur sources, the ability to migrate and establish in the rhizosphere, and the avoidance of competition by other microbes.

The rhizosphere environment is very dynamic in the sense that the composition of root exudates changes at the different parts of the plant root, varies during plant growth, and is different between plant genotypes; this influences the establishment of microbial communities (Micallef, Channer, Shiaris, & Colón-Carmona, 2009). The existence of a high microbial diversity in the rhizosphere is another difficulty that microorganisms face when trying to survive in this habitat; microbes not only compete for the available nutrients and microsites (Duffy, 2001), but some produce antibiotics (Haas & Défago, 2005), use Type 6 secretion systems to eliminate competitors (Bernal, Llamas, & Filloux, 2018), interfere with bacterial behavior by producing quorum quenching molecules (Molina et al., 2003) or volatile organic compounds (Fincheira & Quiroz, 2018). Therefore, life in the rhizosphere involves a state of continuous alert.

3. Chemotaxis of *Pseudomonas putida* toward rhizosphere

The dynamic environment in the rhizosphere is driven by the fluxes of substances in this niche; from root to soil (out-flow—active and passive release of various groups of rhizodeposits), from soil to the root (in-flow—uptake of water and nutrients by roots), from soil/exudates to microbes (uptake of the organic compounds and nutrients by microorganisms) and release of CO_2 by both roots and microorganisms. Physical and chemical processes of sorption, desorption, precipitation and dissolution complicate this scenario (Nguyen, 2003). These fluxes decrease as the distance from the root increases, establishing concentration gradients of compounds and other environmental stimuli from the roots to the bulk soil (Kuzyakov & Razavi, 2019).

Motile bacteria such as *Pseudomonas putida* are able to sense these changes in the environmental composition and move to follow these gradients (attracted or repelled for higher concentrations) using a phenomenon termed—chemotaxis (Adler, 1966). *P. putida* move in response to these gradients using swimming motility (mediated by flagella) and twitching motility (mediated by type IV pili) (Sampedro, Parales, Krell, & Hill, 2015). Flagella can rotate following a counterclockwise–clockwise–pause pattern. The positive response, migration toward a higher concentration of nutrients, is due to a predominant counterclockwise rotation of flagella. A decrease of nutrient concentration activates the pause and clockwise rotation pattern which means the cells stop and change the direction of movement (Taylor & Koshland, 1974). Chemotaxis mediated by twitching motility is based on extension–tethering–retraction–extension of type IV pili. In this case, the mechanisms behind this positive response toward a given compound still requires further investigation; it has been hypothesized that pseudomonads move as a consequence of pilus retraction and the switching of the sites of pilus extension from one pole of the cell to the other (Shi & Sun, 2002). Both processes, swimming and twitching motility, are regulated by homologous signal transduction mechanisms that start with the perception of the signal/stimulus or ligand by the periplasmic portion of a membrane protein called a chemoreceptor (methyl-accepting chemotaxis proteins or MCPs). In the presence of lower levels of ligand, the cytoplasmic part of the chemoreceptor promotes autophosphorylation of the histidine kinase CheA (Fig. 2), eliciting the downstream phosphorylation of the response regulator CheY. CheA is bound to the receptor and the adaptor protein CheW. The activated CheY binds to the flagellar motor protein FliM, inducing the clockwise rotation of the flagellum, and the cell stops and changes its orientation. At higher concentrations of ligand, the phosphorylation of CheY is not produced, the flagella rotate in a counterclockwise pattern and the cell moves to the higher concentrations of nutrients/stimuli (Darnton, Turner, Rojevsky, & Berg, 2007; Turner, Ryu, & Berg, 2000). The activity of this transduction system can also be modulated by methylation of the chemoreceptor via the action of the methyltransferase CheR which produces an increase in the phosphorylation levels of

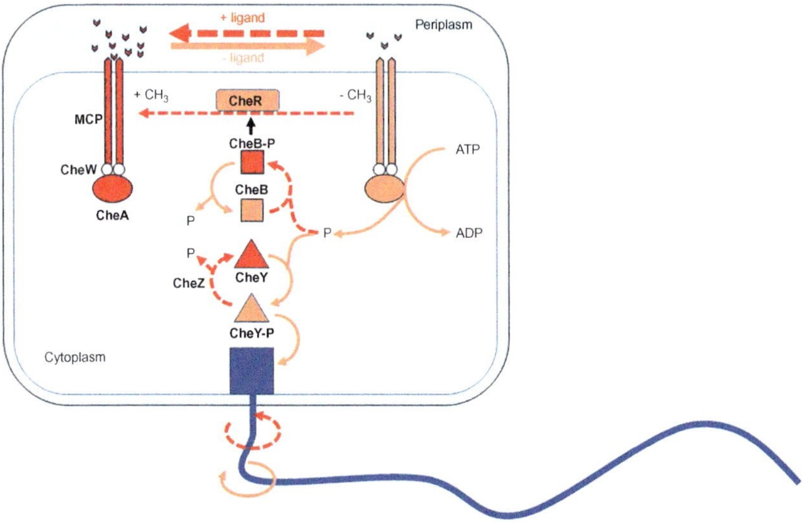

Fig. 2 Chemotaxis in *Pseudomonas putida*. Continuous lines, reactions performed at low concentration of ligand; discontinuous lines, reactions performed at high concentration of ligand. P, phosphate. CH3 methylation processes.

CheA. This phenotype can be reverted by demethylation using the methylesterase CheB. These proteins are crucial because they allow transient states in the chemotactic process to be correlated with changes in the ligand concentration in the environment (Barnakov, Barnakova, & Hazelbauer, 1999) (Fig. 2).

The specificity of the chemotactic response is determined by the periplasmic ligand binding region of the chemoreceptors (Ferrandez, Hawkins, Summerfield, & Harwood, 2002). Genomic analyses revealed that the number of these receptors is highly variable in the studied microorganisms and it was also found that these numbers are related to the bacterial lifestyle but there is an average of 14 chemoreceptors per microorganism (Alexandre, Greer-Phillips, & Zhulin, 2004). The numbers are higher in bacteria able to colonize more diverse environments than specialist microorganisms (Lacal et al., 2010; Lacal, Garcia-Fontana, Munoz-Martinez, Ramos, & Krell, 2010). *P. putida* KT2440 has 27 different chemoreceptors (Garcia-Fontana et al., 2013), some of them have been characterized: McpS (Lacal, Alfonso, et al., 2010), McpQ (Martin-Mora et al., 2016) and McpR (Parales et al., 2013) mediate chemotaxis toward different tricarboxylic acid (TCA) cycle intermediates, McpP responds to C2- and C3-organic acids (Garcia et al., 2015), McpA responds to L-amino acids and McpU senses polyamines (Corral-Lugo et al., 2016). On the other hand, McpG is specific for gamma-aminobutyrate (Reyes-Darias et al., 2015) while McpH responds to metabolizable purines (Fernández, Morel, Corral-Lugo, & Krell, 2016) and PcaY_PP to salycilate (Fernández, Matilla, Ortega, & Krell, 2017). In addition, there are three paralogous receptors for energy taxis (Sarand et al., 2008) as well as chemosensory pathways which regulate type IV pili-based motility (Corral-Lugo et al., 2016). The remaining 13 chemoreceptors are still of unknown function.

A necessary requisite for efficient rhizosphere colonization is bacterial chemotaxis toward plant root exudates (Allard-Massicotte et al., 2016; de Weert et al., 2002; Scharf, Hynes, & Alexandre, 2016). *P. putida* KT2440 shows a chemotactic response to exudates obtained from *Zea mays* plants and the activation of expression of 10 chemoreceptors, for example, McpP, McpQ and PcaY_PP C6 which respond to the metabolites listed above (Fernández et al., 2017, López-Farfán, Reyes-Darias, Matilla, & Krell, 2019). Other chemoreceptors have been determined as necessary for optimal root colonization; mutants in the GABA specific McpG (Reyes-Darias et al., 2015) and the polyamine specific McpU (Corral-Lugo et al., 2016) were affected in rhizosphere colonization capacity.

Organic acids are preferred energy-source compounds of *Pseudomonas putida* when it is confronted with mixtures of carbon sources (Molina, Rosa, Nogales, & Rojo, 2019). These chemicals are produced by plants in higher quantities when they are in the young-growth stage and at flowering time (Bowsher, Ali, Harding, Tsai, & Donovan, 2016). GABA, polyamides and salicylate are molecules produced by plants when they are confronted by biotic or abiotic stresses (Bouche & Fromm, 2004; Kuiper, Bloemberg, Noreen, Thomas-Oates, & Lugtenberg, 2001; Verma, Ravindran, & Kumar, 2016). Chemotaxis toward these compounds and others which act as defensive molecules against pests (e.g., benzoxazinoids), suggests that plants recruit specific rhizobacteria during vulnerable plant growth stages (Neal, Ahmad, Gordon-Weeks, & Ton, 2012).

4. Biofilm formation

In addition to the chemotactic responses, *Pseudomonas putida* has developed other strategies to survive and persist in the rhizosphere. Members of this species are able to form communities which adhere to biotic and abiotic surfaces, in which, cells are encased in a shared self-produced polymeric extracellular matrix formed by exopolysaccharides, nucleic acids, and proteins (reviewed by Costerton, Lewandowski, Caldwell, Korber, & Lappin-Scott, 1995). These structures are termed biofilms and provide these communities with improved environmental fitness, including increased resistance to water limitation, and reduction in the presence of toxicants or predators (López-Sánchez et al., 2016). Biofilm development is a process that occurs as a series of highly regulated steps: surface attachment, micro-colony formation, maturation and individual cell dispersal (Fig. 3). After contact with the root surface, *P. putida* is able to attach apically to this surface in a reversible fashion, which progresses into irreversible lateral interactions. Cells then proliferate and form clonal micro-colonies producing the biofilm

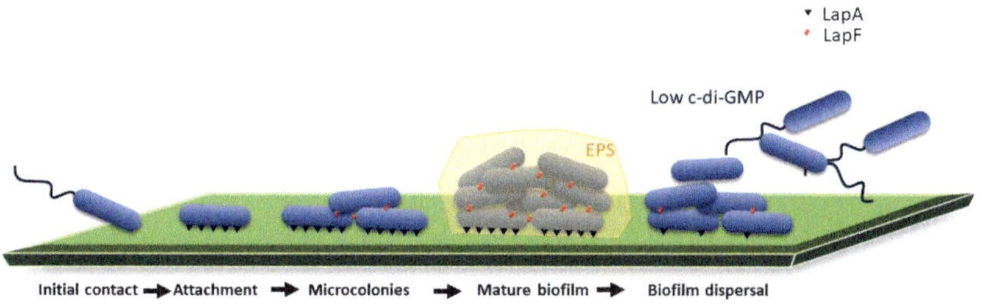

Fig. 3 Biofilm evolution in *Pseudomonas putida*.

matrix. Mature biofilms are driven by flagella dependent migration processes, where micro-colonies evolve into macro-colonies interconnected by channels that allow the transport of nutrients, water and waste (Monds & O'Toole, 2009).

Mutational insertion analyses indicated that flagella are involved in the initial attachment of *P. putida* cells. Specific binding sites for bacteria appear to exist on the root surface, and motile cells are able to reach these sites (Turnbull, Morgan, Whipps, & Saunders, 2001). Several studies demonstrated the role of the high molecular weight adhesion protein, LapA, in the irreversible attachment to both biotic and abiotic materials (Espinosa-Urgel, Salido, & Ramos, 2000). This protein, the largest one found in *P. putida* (8682 amino acids), contains four domains: the first one presents identity to the N-terminal part of the RTX toxin of the fish pathogen *Aeromonas salmonicida*; there are also two domains composed of long multiple and quasi-perfect repeats, which constitute more than three-quarters of the total length of the protein; and the last one contains several Ca^{2+}-binding motifs, similar to those identified in haemolysins and other secreted proteins known to participate in bacteria–eukaryote interactions. LapA, which is found to be loosely associated with the cell surface and present in the culture supernatant, is secreted by the LapEBC, a type I secretion system (Hinsa, Espinosa-Urgel, Ramos, & O'Toole, 2003). A second adhesion protein, LapF, the second largest protein in *P. putida* (6310 amino acids), mediates cell-cell interactions, providing support for micro-colony formation, helping bacteria to anchor to each other, and maturation of the biofilm. LapF presents three domains; the first one comprising 152 amino acids, followed by a repetitive region covering over 85% of the protein, and the last one contains a predicted calcium binding region. On the genes adjacent to LapF, as is the case of LapA, there is a type I secretion system, LapHIJ (Martínez-Gil, Yousef-Coronado, & Espinosa-Urgel, 2010). A series competition experiments between the wild-type and *lapA* and *lapF* mutants in the root of maize plants inoculated at the same cell density were reported and it was found that the wild-type cells dominated over the adhesion-deficient mutants after 1 week of inoculation, indicating that wild-type cells are more competitive than the mutants.

Pseudomonas putida KT2440 has four gene clusters involved in the production of exopolysaccharides (EPS), all of which are components of the biofilm matrix: *alg* (alginate), *bcs* (cellulose-like), the *P. putida* specific *pea* (putida exopolysaccharide A) and *peb* (putida exopolysaccharide B) (Nilsson et al., 2011). The products of the *pea* and *peb* clusters have a role in the stabilization of the biofilm structure, whereas those of *alg* and *bcs* are not essential for biofilm formation and stabilization under standard conditions, although alginate participates in biofilm architecture under water-limiting conditions (Chang et al., 2007) and *bcs* in rhizosphere colonization (Nielsen, Li, & Halverson, 2011).

The survival of the biofilm structure depends of the nutritional level of the cell. Under starvation conditions, the induction of the SpoT protein, a guanosine 3′-diphosphate 5′-diphosphate (ppGpp) synthetase, increases the levels of the phosphodiesterase BifA. This protein acts by reducing the levels of the second messenger signal cyclic diguanylate (*c*-di-GMP) (Díaz-Salazar et al., 2017), a ubiquitous bacterial second messenger that mediates the transition between the planktonic and sessile lifestyles, and is also involved in directing other diverse cellular processes, including motility, protein secretion, cell cycle progression (Boyd & O'Toole, 2012). The reduction of *c*-di-GMP levels induces the activity of the LapA-specific periplasmic protease LapG, which is inhibited by the transmembrane protein LapD in the presence of high *c*-di-GMP levels, and the subsequent proteolytic cleavage of LapA causes biofilm dispersal (Gjermansen, Ragas, Sternberg, Molin, & Tolker-Nielsen, 2005). Another effect of the depletion of *c*-di-GMP levels is the cessation of LapA and cellulose synthesis accompanied by the stimulation of flagellar synthesis via the general regulator FleQ (Xiao et al., 2016) coupling biofilm dispersal with the de novo synthesis of the flagellar apparatus in preparation for the resumption of a planktonic lifestyle (Díaz-Salazar et al., 2017). Biofilm formation under favorable nutritional conditions and dispersal from these structures under starvation or stress conditions are part of the *P. putida* strategy to survive under varying environmental conditions.

5. Metabolism of *Pseudomonas putida* in the rhizosphere

The genome of *Pseudomonas* encodes 350 membrane transporters which encompass almost 5% of all genes (Nelson et al., 2002). These transporters are critical for uptake of nutrients provided by the plant. There are 90 transporter systems related to uptake of amino acids, organic acids, glucose, gluconate and opins, which provides the ability to use the wide variety of compounds produced by plants (Nelson et al., 2002).

Plant exudates change the composition of the soil, producing an enrichment of nutrients; substances found within this microzone are primary metabolites such as organic acids, sugars, amino acids, and secondary metabolites (some toxic chemicals) such as coumarins, flavonoids, and aromatic compounds (Hutsch, Augustin, & Merbach, 2000); in addition, proteins, polysaccharide and lipids are also found in the rhizosphere. *P. putida* strains, use primary metabolites efficiently and are tolerant to toxic secondary compounds. The adaptation of *P. putida* goes even further, because this microbe is able to use some of these biocidal molecules, such as benzoxazinoids, which are chemotactic molecules (Neal et al., 2012). Plants respond to the presence of *P. putida* with the elicitation of the induced systemic resistance response, augmenting the production of these benzoxazinoids (Matilla et al., 2010; Neal & Ton, 2013). This is a clear indication of specificity in the *P. putida*-plant interaction.

Pseudomonas putida efficiently uses organic acid exudates from plants. In the exudates of tomato plants nearly 50 μg of organic acids per mg of dry weight were measured, with citric acid representing almost 50% of the total organic acids. Other organic acids found in root exudates are; malic, malonic, acetic, fumaric, succinic, lactic, tartaric and oxalic (Kravchenko, Azarova, Shaposhnikov, Makarova, & Tikhonovich, 2003; Neumann & Römheld, 1999; Pearse, Veneklaas, Cawthray, Bolland, & Lambers, 2006), Their concentrations vary during plant growth and with stress conditions (Bowsher et al., 2016).

In complex media with different carbon sources *P. putida* strains show a preferential use of organic acids and certain amino acids over glucose, which in turn is generally preferred over hydrocarbons and aromatic compounds. This hierarchy in the use of the different carbon sources present in a complex media is mediated by catabolic repression (reviewed by Rojo, 2010). The preference for organic acids, corroborated by transcriptomic and metabolomic studies, provides evidence that *P. putida* has a metabolism specialized toward the use of these molecules in the plant rhizosphere (Martins dos Santos, Heim, Moore, Strätz, & Timmis, 2004; Molina et al., 2019). This specialization permits *P. putida* strains to colonize the rhizosphere at high levels, co-habiting with other root microorganisms such as *Enterobacteriaceae* which show preference in the use of carbohydrates as carbon source (Rojo, 2010). In addition to this metabolic specialization in the metabolism of organic acids, the fact that these molecules are also good chemo-attractants for *P. putida* toward root exudates, particularly when citrate is complexed with iron, magnesium and calcium—the most frequent form of this abundant compound in the rhizosphere (Martin-Mora et al., 2016).

Free amino acids represent between 10 and 40% of the soluble organic nitrogen in the rhizosphere and their net concentration is in the micromolar level (Moe, 2013). Meta-analyses of the relative content of proteinogenic amino acids revealed than alanine and glutamic acid were the most abundant in the exudates of 22 different plants (more than 10% of the total amino acidic composition), followed by aspartate, glycine, histidine, leucine, serine, threonine, valine, arginine and asparagine (over 5%), methionine and cysteine were the least abundant amino acids in root exudates. Non proteinogenic amino acids have also been found in variable concentrations in the rhizosphere, namely, α- and ϒ-aminobutyric acid (AABA and GABA), ornithine, β-alanine, citrulline, ethanolamine, methionine sulfoxide and methylamine and polyamides (reviewed by Vranova, Rejsek, Skene, & Formanek, 2011). *Pseudomonas. putida* possesses transporters for most of these proteinogenic and non proteinogenic amino acids (Nelson et al., 2002). These transporters exhibit a high affinity (Km at the range of nanomolar to low micromolar), concentrations at which these molecules are found in the rhizosphere (Fan, Miller, & Rodwell, 1972). These findings suggest that *P. putida* is efficient at taking up amino acids from the environment, providing a competitive advantage to this microorganism in the rhizosphere.

Transcriptomic assays and in vivo induced gene expression (IVET) experiments on *P. putida* surviving in the rhizosphere showed that the genes related to amino acid transport (polyamides, branched-chain amino acids, basic amino acids, proline, GABA and general amino acid transporters), aspartate/glutamate, alanine, GABA, lysine, leucine and proline catabolism (Fernández, Conde, Duque, & Ramos, 2013; Matilla, Espinosa-Urgel, Rodríguez-Herva, Ramos, & Ramos-González, 2007) are active in the root of plants. These experiments corroborated the use of the most abundant amino acids found in the rhizosphere and the use of the non proteic amino acid GABA by *P. putida*. Furthermore, as is the case of organic acids, *P. putida* survives in the rhizosphere using metabolites that plants produce when exposed to abiotic stress, i.e., proline, leucine, polyamides (Bouche & Fromm, 2004; Gargallo-Garriga et al., 2018).

Insertional mutagenesis experiments revealed that the degradative genes of lysine were expressed at high levels in the rhizosphere of *Zea mays* plants (Espinosa-Urgel & Ramos, 2001). *P. putida* normally uses L-enantiomers of amino acids as carbon and/or nitrogen sources, but its growth rate is slower for the case of L-lysine than for D-lysine (Daniels et al., 2010; Radkov & Moe, 2013). *P. putida*, in contrast to other related microbes, such as *P. aeruginosa*, *P. fluorescens* or *P. syringae*, has optimized the capacity to degrade both enantiomers by means of three interconnected catabolic reactions: from D-lysine to 2-aminoadipate (the AMA pathway), those from L-lysine to glutaric acid (the AMV/*dav* pathway), and from L-lysine to δ-aminovalerate via cadaverine/1-piperidine (the cadaverine pathway). These pathways are interconnected at the first step by a lysine racemase able to interconvert D- and L-lysine and also at the lower end of the pathways in which aminoadipate is channeled to yield glutarate (Revelles, Espinosa-Urgel, Fuhrer, Sauer, & Ramos, 2005). The relative abundance of the D-enantiomer is only 2–15% of the L-form, and can be use as a tracer of soil aging (Amelung, Zhang, & Flach, 2006), and in general is more recalcitrant than L-lysine. The capacity of *P. putida* to degrade D-lysine is another indicator of the adaptation of this microorganism to survive in soil environments using carbon sources that are not consumed by other microorganisms.

In general, *Pseudomonas* are characterized by being microorganisms efficient in the degradation of arginine, for which they use several metabolic pathways. One of them is the route of deamination of the arginine that yields citrulline which is further converted to ornithine, a step that allows the synthesis of ATP by phosphorylation at the substrate level. This reaction allows cells to be motile in the absence of oxygen. Arginine degradation pathways are a key system for central cell metabolism and are regulated in detail by the ArgR protein (Yang & Lu, 2007).

Carbohydrates constitute 5–25% of the organic matter present in most of soils; the proportion of organic matter is higher in vegetated soils than non-vegetated soils (Cheshire, 1979). Plant and natural microbiota structural components, in the form of polysaccharides, are considered the origin of most of the carbohydrates found in soil (Lowe, 1978). These complex polysaccharides are able to bind to soil particles and generate stable aggregates that form complexes with metal ions for humus synthesis (John, Yamashita, Ludwig, & Flessa, 2005). The quantity of carbohydrate is increased in the rhizosphere by the production of root mucilage (high-molecular-weight polysaccharides whose main neutral sugar components are galactose, fucose and glucose), liberated by detached and dead plant and microbial cells (Chaboud, 1983). *P. putida* strains do not use these polymers, but they can degrade them at very low rates in the absence of other energy sources (Molina et al., 2019). The most abundant soluble component provided by root exudates is glucose (approx. 20 μM), followed by other hexoses such as fructose and maltose and the pentose xylose, but their concentrations are too low to support rapid bacterial growth (Lugtenberg, Kravchenko, & Simons, 1999). In addition, *P. putida* has a limited carbohydrate use profile, these strains are able to grow in glucose and fructose, but they do not use pentoses as carbon sources (Daniels et al., 2010). Transcriptomics assays demonstrated the activation of genes related to polysaccharide transport (PP_3132) and degradation (the periplasmic β-glucosidase BglX) in *P. putida* KT2440 during rhizosphere colonization (Matilla et al., 2007). This indicates that KT2440 is also using glucose polymers produced by plants or other microorganisms for its growth in the rhizosphere.

The metabolism of glucose by *P. putida* strains has been recently reviewed (Udaondo, Ramos, Segura, Krell, & Daddaoua, 2018). This sugar enters the periplasmic space mainly via the OprB porin and could be directly transported to the cytoplasm and/or sequentially oxidized to gluconate and 2-ketogluconate in the periplasm to obtain reducing power without the use of the intracellular metabolic machinery. The strength of this periplasmic flux depends mostly on the external concentration of glucose (Molina et al., 2019). Gluconate and 2-ketogluconate, which at high glucose concentrations, are secreted into the media for further recycling (Molina et al., 2019), or are incorporated into the cytoplasm and, together with glucose, generate 6-phosphogluconate (6PG) via a three-pronged metabolic system (Del Castillo et al., 2007). Thereafter, 6PG constitutes the starting point of the Entner–Doudoroff (ED) pathway, used predominantly by *P. putida* for hexose degradation (Del Castillo et al., 2007; Nikel, Max Chavarría, Fuhrer, Sauer, & de Lorenzo, 2015). Although, these strains possess most of the genes that encode the three pathways for the use of glucose; the pentose phosphate (PP), an incomplete Embden-Meyerhof-Parnas (EMP) and ED pathways (Del Castillo et al., 2007; Sudarsan, Dethlefsen, Blank, Siemann-Herzberg, & Schmid, 2014), the ED pathway is comparatively less costly for the production of central C3 metabolites (pyruvate and glyceraldehyde-3-phosphate) (Noor, Eden, Milo, & Alon, 2010). The predominance of the ED pathway could be explained by the lack of the glycolytic EMP enzyme 6-phosphofructo-1-kinase, which catalyzes the ATP-dependent conversion of fructose-6-phosphate into fructose-1,6-biphosphate (Chavarría, Nikel, Pérez-Pantoja, & de Lorenzo, 2013). The existence of an incomplete EMP pathway and an ineffective PP pathway in this microorganism can be explained because the triose-phosphates generated by the ED pathway can be used for the recycling of hexoses-6-phosphate

(Nikel et al., 2015). As a consequence, an "excess" of reducing power in the form of NADPH is produced, protecting *P. putida* strains against endogenously and exogenously generated oxidative stress (Imlay, 2003). In addition, the recycled fructose-6-phosphate is the precursor of exopolysaccharides, molecules that protect these microorganisms from desiccation (Chavarria et al., 2013).

6. Plant growth promoting properties

Pseudomonas putida establishes a commensal relationship with plants. As described above, plants feed *P. putida* in the rhizosphere through root exudates and in turn the microbe helps the plants to grow by producing plant hormone precursors, favoring the mobilization of nutrients and producing antibiotics that prevent the growth of pathogens. Fig. 1 shows some of the positive effects of *P. putida* on stimulation of plant growth. For instance, BIRD1 stimulates the formation of secondary roots which help to increase the plant's root surface and concomitantly nutrient assimilation. The increase in nutrient uptake results in a significantly larger sized plant in early stages of growth and also because *Pseudomonas* prevents the growth of phytopathogens the general state of the plants is better and the quality and quantity of the fruit is also improved. One of the key molecules behind these positive effects is the overproduction of indole acetic acid (IAA), a well-known phytohormone that *P. putida* BIRD1 produces through two parallel pathways (Roca et al., 2013). *P. putida* also produces organic acids and phosphatases that help to mobilize insoluble inorganic phosphate and release phosphate from organophosphorous compounds, respectively, making phosphate available for plant uptake. In fact, having *P. putida* associated with some crops can save as much as 50% on phosphorous rich fertilizers which in turn has an enormous environmental benefit (Roca et al., 2013). Furthermore, Molina, Ramos, and Espinosa-Urgel (2006) demonstrated the importance of capturing iron from soil for the colonization of roots by *P. putida*. In fact, a feature that facilitates proliferation of *P. putida* in the root is that they produce specific siderophores to solubilize iron, and that in addition to their specific siderophore transport systems, they are endowed with external membrane receptors able to capture iron chelated by foreign siderophores (xenosiderophores) that are produce by other bacterial species (Roca et al., 2013; Molina et al., 2006). This set of iron capturing systems makes *P. putida* an authentic "thief" of iron in soil. This "skill" not only helps them to acquire iron to satisfy their own needs, but prevents the use of iron by other microbes which in turn inhibits the growth of phytopathogens, a property that makes *P. putida* an indirect biocontrol agent.

Pseudomonas putida can also play a role as a direct biocontrol agent; the presence of this microorganism in the rhizosphere induces the plant systemic response, protecting the plant host against pathogen infection and proliferation (Matilla et al., 2010), a process that is characterized by the production of high levels of reactive oxygen species (ROS) (Baxter, Mittler, & Suzuki, 2014). For this reason, proliferation in this environment requires a high capacity of ROS removal. *P. putida* is endowed with a large number of oxidative stress response genes encoding superoxide dismutases, catalases, betaine (aldehyde) dehydrogenase and other enzymes, which play a role in the detoxification of abundant free radicals in the rhizospheric environment (Matilla et al., 2010).

7. Phytorremediation

Because of their ability to flourish in the rhizosphere, *P. putida* strains have been used in several rhizoremediation projects for the elimination of contaminants in soils. Removal of soil pollutants is a very complex and expensive procedure; while physical and/or chemical treatments are very efficient, they are, most of the time not economically viable. Therefore, in recent years, biological treatments have been explored as an economically and environmentally friendly alternative (Segura & Ramos, 2013). These biological treatments can be divided into: (i) *on-site* techniques (such as land-farming or composting) and (ii) in situ techniques such as bioaugmentation or phytoremediation (Colleran, 1997).

The term phytoremediation includes different strategies in which plants alone or a combination of microbes and plants are used for the elimination of contaminants. Phytovolatilization (uptake of soil contaminants by plants, transformation into volatile compounds and evaporation through leaves), phytoextraction (adsorption of contaminants to roots with or without translocation to aerial parts; plants are then harvested and disposed of) or phytostabilization (immobilization of contaminants into soil particles) have been described as processes in which plants are involved (Segura & Ramos, 2013). In general, all of the approaches above do not result in the elimination of the pollutants, they are either entrapped in the plant or are dispersed in the atmosphere. In a few cases, plants are also able to metabolize the contaminants in a similar way as the mammalian liver, i.e., activation by cytochrome P450 monooxygenases and conjugation with glutathione, glucosyl or malonyl moieties, conjugates that are later sequestered in vacuoles or plant cell walls (Burken & Schnoor, 2004). However, there is no doubt that microbes have a tremendous potential to be used in the elimination of contaminants because of the versatility of their metabolism. Mineralization of contaminants with diverse chemical structures, such as mono- and aromatic hydrocarbons, polycyclic aromatic hydrocarbons (PAHs), lindane (hexachlorohexane), triazines, or polychlorinated biphenyls (PCBs) by bacteria have been described since the 1980s (Kanaly & Harayama, 2010; Rylott, Lorenz, & Bruce, 2011), although limitations exist in the establishment of microbes in bulk soil.

The inability of the plants to completely mineralize the toxic chemicals and the difficulties of the introduction of degradative bacteria in soils, can be resolved by the utilization of a combination of both organisms (plants and microbes) in the elimination of soil contaminants in a strategy that is generally named rhizoremediation. (Segura & Ramos, 2013).

7.1 Rhizoremediation

Microbes in the rhizosphere are more abundant and are more metabolically active than those in bulk soil (Bulgarelli et al., 2012; Kuiper, Lagendijk, Bloemberg, & Lugtenberg, 2004). In the context of the elimination of contaminants, Siciliano et al. (2000, 2003) demonstrated that plants growing on contaminated soils could select bacterial degradative genotypes into their rhizosphere but that this selection depended on the plant used and the type of contaminant. Other authors have demonstrated that plants can secrete compounds that induce bacterial degradative pathways for contaminants (Gilbert & Crowley, 1997; Segura, Hernández-Sánchez, Marqués, & Molina, 2017) and promote bacterial degradation (Martin, George, Price, Ryan, & Tibbett, 2014). In addition, plant roots provide a large surface on which microbes can proliferate, transporting microbes through the soil, and by facilitating oxygen exchange, they provide oxygen for aerobic bacteria at depths where they normally cannot persist. Oxygen is also important for the initial attack of mono- and dioxygenases, enzymes that mediate many reactions in the degradation of toxic aromatic compounds. These findings provide the bases for the utilization of plant-bacteria combinations in the elimination of soil contaminants (Kotoky, Rajkumari, & Pandey, 2018).

In the last decade, a number of research articles have reported the successful utilization of plant and bacteria combinations for the elimination of contaminants (Correa-García, Pande, Séguin, St-Arnaud, & Yergeau, 2018;

Rodriguez-Conde, Molina, González, García-Puente, & Segura, 2016; Vergani et al., 2017, among many others) most of them in controlled environments; not many studies have been reported under natural conditions. In this chapter we will comment on two examples of rhizoremediation using *Pseudomonas* strains in natural set-ups. In both examples, the combination of bacteria, exogenously added, indigenous or plant growth promoting rhizobacteria (PGPR), with plants help the final outcome of the soil remediation and both examples provide solutions for some of the difficulties found in rhizoremediation strategies under natural conditions.

7.1.1 TNT rhizoremediation

The most widely used nitro aromatic compound is 2,4,6-trinitrotoluene (TNT). The manufacture, processing, and packaging of TNT at munitions plants has resulted in high concentrations of contaminants in soil and ground waters. TNT is a xenobiotic compound, meaning that it exhibits structural elements or substituents that are rarely found in natural products. Probably because only a few nitro aromatic compounds have a natural origin (chloramphenicol, nitropyoluteorin, oxypyrrolnitrin, and phidolopin), bacteria have not evolved mechanisms for their efficient degradation which has resulted in TNT being a highly recalcitrant compound (Esteve-Núñez et al., 2001).

The symmetric location of the nitro groups on the aromatic ring limits the attack by the dioxygenase enzymes which are typically involved in the microbial metabolism of aromatic compounds. However, some bacterial strains, among them Pseudomonads, are able to use TNT as a nitrogen source via removal of the nitro groups. The removal of nitro groups occurs via the formation of the Meisenheimer complex under aerobic conditions, with the concomitant reduction of the released nitrite to ammonium (Caballero & Ramos, 2006; Caballero, Lázaro, Ramos, & Esteve-Nuñez, 2005; Wittich, Ramos, & van Dillewijn, 2009). In *P. pseudoalcaligenes* it has been described that the hydroxylaminodinitrotoluenes are intermediates of the degradation. Strict anaerobic bacteria (*Clostridium* and *Desulfovibrio*) and several fungi can also degrade TNT. TNT degradation has been thoroughly investigated and reviewed (Spain, 1995; Esteve-Núñez et al., 2001; Stenuit & Agathos, 2010), although many questions regarding its mineralization or the identification of recalcitrant intermediates are still open (Fig. 4).

Because the TNT contaminated areas are in factories and abandoned munition stores, in situ bioremediation of TNT is the most cost-effective method for cleaning up. Van Dillewijn et al. (2008) carried out the

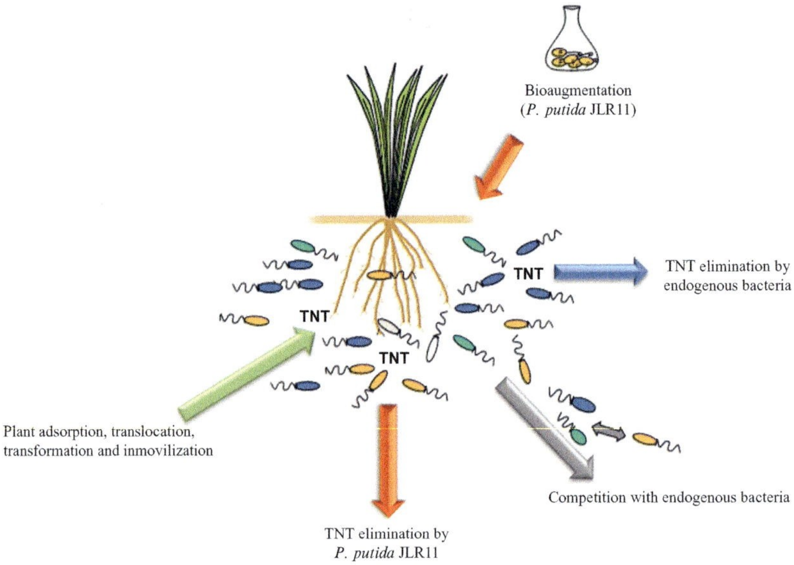

Fig. 4 Strategies to enhance pollutant degradation in the environment.

comparison among different biological treatments in a 300 m^2 plot within the grounds of a demunition plant. Although the plot was tilled mechanically to disperse the contaminant as uniformly as possible, some areas showed TNT concentrations up to 400 $mg Kg^{-1}$ of soil, while in other areas the presence of TNT was undetectable. Several experimental set-ups were organized to evaluate the different bioremediation techniques. To analyze natural attenuation, pots were filled with unamended soil and buried in the experimental plot; for bioaugmentation, the pots were filled with soil inoculated with 10^6 colony forming units (CFUs) per gram of soil of *P. putida* JLR11. This strain is able to grow using TNT as the sole nitrogen source and it is closely related to *P. putida* KT2440, a good root colonizer. The *P. putida* JLR11 genome contains the same determinants as *P. putida* KT2440 to carry out root colonization (Pascual et al., 2015). To analyze the effectivity of phytoremediation, sterile germinated maize seedlings were planted into pots with sterilized potting material, while for the evaluation of rhizoremediation, the sterile germinated maize seedlings were inoculated with 10^6 CFUs of *P. putida* JLR11 per seedling before plantation. After 1 week these plantlets were transplanted into the contaminated soil.

Natural attenuation, elimination of TNT by indigenous bacteria, was very poor during the 120 days of the experiment; the degradation product, 4-amino-2,6-dinitrotoluene (4ADNT), was detected at low concentrations

(1.5 mg kg^{-1} of soil) and remained constant throughout the experiment, suggesting limited action of the indigenous microbiota on TNT under these conditions. When *P. putida* JLR11 was added to the soil, the TNT concentration was decreased more than in non-inoculated soils and the concentration of 4-ADNT was also slightly higher (2.8 mg kg^{-1} of soil). The result of this bioaugmentation was not spectacular, probably as a consequence of the limited survival of *P. putida* JLR11 in bulk soil (viable cells decreased up to four orders of magnitude below the initial number). In the presence of plants, the TNT concentrations were significantly lower than in natural attenuation and bioaugmentation experiments, suggesting that the plants had a major influence on the elimination of TNT and its degradation product (4-ADNT) from soils. Almost 96% of the TNT disappeared after 2 months of planting and the 4-ADNT concentration dropped to approximately 0.3 mg Kg^{-1} of soil. Although *P. putida* JLR11 persisted in planted soils at medium cell densities (10^3–10^5 CFU g^{-1} soil) their presence in soil did not significantly influence the final outcome of the remediation. These findings suggest that phytoremediation is a major factor in bioremediation of TNT, however it should be considered that in hydroponic assays it has been demonstrated that TNT was sequestered by plants, but not degraded. Furthermore, it has been shown that bacteria with TNT degrading activity, mostly belonging to the *Pseudomonadaceae* and *Xanthomonadaceae* families, can be isolated from the same plot (George, Eyers, Stenuit, & Agathos, 2008). These indigenous bacteria could also contribute to degradation of TNT in the rhizosphere of maize.

Other biotechnological approaches have been assayed using the molecular knowledge of *P. putida* JLR11. One of these biotechnological applications was the utilization of the *P. putida* JLR11 gene *pnrA*, (which encodes a nitroreductase that reduces TNT to 4-hydroxylamino-2,6-dinitrotoluene at a very high rate; Caballero et al., 2005) for the construction of aspen transgenic plants (Van Dillewijn et al., 2008). The expression of this gene in plants allowed them to tolerate higher TNT concentrations (up to 57 mg TNT L^{-1} in hydroponic media and more than 1000 mg TNT kg^{-1} soil) than non-transgenic plants, (11 mg TNT L^{-1} and 500 mg TNT kg^{-1}). Transgenic plants also showed improved uptake of TNT over wild-type plants. Together, these results suggest that the expression of the *P. putida* nitroreductase *pnrA*, improved the capacity of aspen to tolerate, grow, and more importantly, eliminate TNT from contaminated soils.

One of the problems of carrying out bioremediation in heavily contaminated environments is the fact that many organisms cannot thrive under these conditions; improving natural resistance toward contaminants is very important. Finding and using naturally tolerant bacterial strains and favoring plant growth under harsh conditions are key factors for successful rhizoremediation. Furthermore, as economic issues play a decisive role in the decision to remediate polluted sites, especially for industries, the possibility of growing plants that could have added value, for example, be used as raw material in the production of biofuels, gives an economic incentive to undertake the bioremediation.

7.1.2 Rhizoremediation of burnt soils

Fires alter the vegetation cover, changing nutrient availability and organic matter content, decreasing microbial cell density and altering soil microbiota composition. The combustion of wood produces toxic hydrocarbons such as BTEX (benzene, toluene, ethylbenzene and xylene) and polycyclic aromatic hydrocarbons (PAHs) that are then deposited into the soil. Furthermore, after the fire, there is a reduction of water infiltration and rainfall retention which increases the erosion and degradation of soil. After the fires, different meteorological phenomena (such as rain and wind) can contribute to the dispersal of the ashes that reach the aquifers and surface waters, spreading through the environment and posing a health risk for flora, fauna as well as for humans. Clearly, restoration of soils after fires is required, especially in areas where they are frequent (e.g., the Mediterranean basin).

Degradation of BTEX and PAHs by bacteria has been extensively reviewed in the literature (Ghosal, Ghosh, Dutta, & Ahn, 2016; Jindrová, Chocová, Demnerová, & Brenner, 2002; Lin, Van Verseveld, & Röling, 2002; Varjani, Gnansounou, & Pandey, 2017). Dioxygenases and monoxygenases are the main enzymes responsible for the activation and cleavage of the aromatic ring that initiates the mineralization of BTEX and PAHs in aerobic bacteria. Among the many bacteria identified as degraders of these types of compounds, several *Pseudomonas* strains have been characterized.

Although many of the bacteria with the ability to degrade aromatic hydrocarbons are also able to tolerate them at relatively high concentrations (Huertas, Duque, Marqués, & Ramos, 1998; Inoue, Yamamoto, & Horikoshi, 1991), they do not survive well when inoculated in bulk soils and, therefore rhizoremediation has been considered to be a good strategy for bioremediation of soils contaminated with hydrocarbons. However, as mentioned above, for a successful rhizoremediation, plants have to also be able to thrive in these contaminated environments, as such, systems have to be designed to improve their growth. PGPRs are rhizobacteria that promote seed germination, root growth and, in general, plant health through diverse mechanisms, such as, production of plant hormones, phosphate solubilization, and siderophore production (Pérez-Montaño et al., 2014).

Because of the difficulties in growing plants in contaminated environments, Pizarro-Tobías et al. (2015) used a combination of PGPR and degradative bacteria for the restoration of a burnt soil in the Natural Park—Los Montes de Málaga (Malaga, Spain). In these field assays, *P. putida* BIRD-1, *P. putida* KT2440 containing the pWW0 plasmid and indigenous bacteria (with the capacity to degrade PAHs) were used in combination with rapidly growing pasture seeds. *P. putida* BIRD-1 is a rhizobacteria that adheres to plant roots and colonizes the rhizosphere to high cell densities. This bacterium is capable of solubilizing insoluble inorganic phosphate through acid production, is able to solubilize iron by producing pyoverdine as siderophore, and it is able to produce the well-known phytohormone indole-3-acetic acid. Furthermore, the genome of *P. putida* BIRD-1 encodes a wide range of proteins that help it to deal with reactive oxygen stress generated in the plant rhizosphere (Roca et al., 2013). *P. putida* KT2440 is an excellent root colonizer (Molina, Ramos, Duque, et al., 2000) and the pWW0 plasmid encodes genes for the degradation of several BTEX (Greated, Lambertsen, Williams, & Thomas, 2002; Worsey & Williams, 1975). For the burnt soil restoration analysis, four different soil treatments were analyzed; (1) no treatment; (2) only plants; (3) only microbial consortium (bioremediation); and (4) plants with microbial consortium (rhizoremediation). As expected, the survival of the two *P. putida* strains was higher when associated with plants than in bulk soil, demonstrating that the presence of plants increases the success of the reintroduction of laboratory strains into soils. Despite the average concentration of BTEX being $150 \pm 20\,\mu g\,kg^{-1}$ at the beginning of the assay, because of the volatility of the BTEX compounds, 2 months after treatment BTEX concentrations were undetectable under all conditions. However, the PAH concentration was reduced by 40% in treatment with plants and bioremediation experiments and by 60% in rhizoremediation plots (initial concentration $400 \pm 30\,\mu g\,kg^{-1}$ soils of PAHs). Furthermore, vegetation development and soil coverage was higher in rhizoremediation plots than in plots in which only plants were introduced. These results emphasize the importance of the elimination of contaminants by microbes for plant growth; this together with the PGPR properties of the consortium allows better growth of plants which, in turn, reduced the erosion of soil and improved soil health.

These results support the notion that soil sustainability can be promoted through the combined action of microbes and plants to remove pollutants and to create a reservoir of carbon.

Acknowledgments

Work in Granada was supported by FEDER funds and projects from the AEI, namely, RTI2018-094370-B-I00. We thank ben pakuts for critical reading of the manuscript.

References

Adler, J. (1966). Chemotaxis in bacteria. *Science*, *153*, 708–716.

Alexandre, G., Greer-Phillips, S., & Zhulin, I. B. (2004). Ecological role of energy taxis in microorganisms. *FEMS Microbiology Reviews*, *28*, 113–126.

Allard-Massicotte, R., Tessier, L., Lecuyer, F., Lakshmanan, V., Lucier, J. F., Garneau, D., et al. (2016). *Bacillus subtilis* early colonization of arabidopsis thaliana roots involves multiple chemotaxis receptors. *MBio*, *7*, e01664-16.

Amelung, W., Zhang, X., & Flach, K. W. (2006). Amino acids in grassland soils: Climatic effects on concentrations and chirality. *Geoderma*, *130*, 207–217.

Bais, H. P., Weir, T. L., Perry, L. G., Gilroy, S., & Vivanco, J. M. (2006). The role of root exudates in rhizosphere interactions with plants and other organisms. *Annual Review of Plant Biology*, *57*, 233–266.

Barnakov, A. N., Barnakova, L. A., & Hazelbauer, G. L. (1999). Efficient adaptational demethylation of chemoreceptors requires the same enzyme-docking site as efficient methylation. *Proceedings of the National Academy of Sciences of the United States of America*, *96*(19), 10667–10672.

Baxter, A., Mittler, R., & Suzuki, N. (2014). ROS as key players in plant stress signalling. *Journal of Experimental Botany*, *65*, 1229–1240.

Belda, E., van Hek, R. G. A., Lopez-Sanchez, M. J., Cruvellier, S., Barbé, V., Fraser, C., et al. (2016). The revisited genome pof *Pseudomonas putida* KT2440 enlightens its value as a robust metabolic chassis. *Environmental Microbiology*, *18*, 3403–3424.

Bernal, P., Llamas, M., & Filloux, A. (2018). Type VI secretion systems in plant-associated bacteria. *Environmental Microbiology*, *20*, 1–15.

Bouche, N., & Fromm, H. (2004). GABA in plants: Just a metabolite? *Trends in Plant Science*, *9*, 110–115.

Bowsher, A. W., Ali, R., Harding, S. A., Tsai, C.-J., & Donovan, L. A. (2016). Evolutionary divergences in root exudate composition among ecologically-contrasting helianthus species. *PLoS One*, *11*, e0148280.

Boyd, C. D., & O'Toole, G. A. (2012). Second messenger regulation of biofilm formation: breakthroughs in understanding c-di-GMP effector systems. *Annual Review of Cell and Developmental Biology*, *28*, 439–462.

Bulgarelli, D., Rott, M., Schlaeppi, K., Ver Loren van Themaat, E., Ahmadinejad, N., Assenza, F., et al. (2012). Revealing structure and assembly cues for *Arabidopsis* root-inhabiting bacterial microbiota. *Nature*, *488*, 91–95.

Burken, J. G., & Schnoor, J. L. (2004). Phytoremediation: Plant uptake of atrazine and role of root exidates. *Journal of Environmental Engineering*, *122*, 958.

Caballero, A. A., & Ramos, J. L. (2006). A double mutant of *Pseudomonas putida* deficient in the nitroreductase PnrA and nitrite rreductase (NasB) is imparired for growth on 2,4,6-trinitrotoluene. *Environmental Microbiology*, 1306–1310.

Caballero, A., Lázaro, J. J., Ramos, J. L., & Esteve-Nuñez, A. (2005). PnrA, a new nitroreductase-family enzyme in the TNT-degrading strain *Pseudomonas putida* JLR11. *Environmental Microbiology*, *7*, 1211–1219.

Chaboud, A. (1983). Isolation, purification and chemical composition of maize root cap slime. *Plant and Soil*, *73*, 395–402.

Chang, W. S., van de Mortel, M., Nielsen, L., Nino de Guzman, G., Li, X., & Halverson, L. J. (2007). Alginate production by *Pseudomonas putida* creates a hydrated microenvironment and contributes to biofilm architecture and stress tolerance under water-limiting conditions. *Journal of Bacteriology*, *189*, 8290–8299.

Chavarría, M., Nikel, P. I., Pérez-Pantoja, D., & de Lorenzo, V. (2013). The Entner-Doudoroff pathway empowers *Pseudomonas putida* KT2440 with a high tolerance to oxidative stress. *Environmental Microbiology*, *15*, 1772–1785.

Cheshire, M. V. (1979). *Nature and Origin of Carbohydrates in Soils*. New York: Academic Press.

Colleran, E. (1997). Uses of bacteria in bioremedation. *Bioremediation Protocols*, *2*, 3–22.

Corral-Lugo, A., De La Torre, J., Matilla, M. A., Fernandez, M., Morel, B., Espinosa-Urgel, M., et al. (2016). Assessment of the contribution of chemoreceptor-based signaling to biofilm formation. *Environmental Microbiology*, *18*, 3355–3372.

Correa-García, S., Pande, P., Séguin, A., St-Arnaud, M., & Yergeau, E. (2018). Rhizoremediation of petroleum hydrocarbons: A model system for plant microbiome manipulation. *Microbial Biotechnology*, *11*, 819–832.

Costerton, J. W., Lewandowski, Z., Caldwell, D. E., Korber, D. R., & Lappin-Scott, H. M. (1995). Microbial biofilms. *Annual Review of Microbiology*, *49*, 711–745.

Daniels, C., Godoy, P., Duque, E., Molina-Henares, M. A., de la Torre, J., del Arco, J. M., et al. (2010). Global regulation of food supply by *Pseudomonas putida* DOT-T1E. *Journal of Bacteriology*, *192*, 2169–2181.

Darnton, N. C., Turner, L., Rojevsky, S., & Berg, H. C. (1756–1764). On torque and tumbling in swimming *Escherichia coli*. *Journal of Bacteriology*, *189*, 2007. https://doi.org/10.1128/JB.01501-06.

de Weert, S., Vermeiren, H., Mulders, I. H., Kuiper, I., Hendrickx, N., Bloemberg, G. V., et al. (2002). Flagella-driven chemotaxis towards exudate components is an important trait for tomato root colonization by *Pseudomonas fluorescens*. *Molecular Plant-Microbe Interactions*, *15*, 1173–1180.

Del Castillo, T., Ramos, J. L., Rodríguez-Herva, J. J., Fuhrer, T., Sauer, U., & Estrella Duque, E. (2007). Convergent peripheral pathways catalyze initial glucose catabolism in *Pseudomonas putida*: Genomic and flux analysis. *Journal of Bacteriology*, *189*, 5142–5152.

Díaz-Salazar, C., Calero, P., Espinosa-Portero, R., Jiménez-Fernández, A., Wirebrand, L., Velasco-Domínguez, M. G., et al. (2017). The stringent response promotes biofilm dispersal in *Pseudomonas putida*. *Scientific Reports*, *7*, 18055.

Duffy, B. K. (2001). Competition. In O. C. Maloy & T. D. Murray (Eds.), *Encyclopedia of Plant Pathology* (pp. 243–244). New York, NY: John Wiley & Sons, Inc.

Espinosa-Urgel, M., & Ramos, J. L. (2001). Expression of a *Pseudomonas putida* aminotransferase involved in lysine catabolism is induced in the rhizosphere. *Applied and Environmental Microbiology*, *67*, 5219–5224.

Espinosa-Urgel, M., Salido, A., & Ramos, J. L. (2000). Genetic analysis of functions involved in adhesion of *Pseudomonas putida* to seeds. *Journal of Bacteriology*, *182*, 2363–2389.

Esteve-Núñez, A., Caballero, A., & Ramos, J. L. (2001). Biological degradation of 2,4,6-trinitrotoluene. *Microbiology and Molecular Biology Reviews*, *65*, 335–352.

Fan, C. L., Miller, D. L., & Rodwell, V. W. (1972). Metabolism of basic amino acids in Pseudomonas putida: Transport of lysine, ornithine, and arginine. *Journal of Biological Chemistry*, *247*, 2283–2288.

Fernández, M., Conde, S., Duque, E., & Ramos, J. (2013). Rhizosphere target genes. *Microbial Biotechnology*, *6*, 307–313.

Fernández, M., Matilla, M., Ortega, A., & Krell, T. (2017). Metabolic value chemoattractants are preferentially recognise at broad ligand range chemoreceptors of *Pseudomonas putida* KT2440. *Frontiers in Microbiology*, *8*, 990.

Fernández, M., Morel, B., Corral-Lugo, A., & Krell, T. (2016). Identification of a chemoreceptor that specifically mediates chemotaxis toward metabolizable purine derivatives. *Molecular Microbiology*, *99*, 34–42.

Ferrandez, A., Hawkins, A. C., Summerfield, D. T., & Harwood, C. S. (2002). Cluster II *che* genes from *Pseudomonas aeruginosa* are required for an optimal chemotactic response. *Journal of Bacteriology*, *184*, 4374–4383.

Fincheira, P., & Quiroz, A. (2018). Microbial volatiles as plant growth inducers. *Microbiological Research*, *208*, 63–75.

Franklin, F. C., Bagdasarian, M., Bagdasarian, M. M., & Timmis, K. N. (1981). Molecular and functional analysis of the TOL plasmid of *Pseudomonas putida* and cloning of genes of the entire regulated aromatic ring *meta* cleavage pathway. *Proceedings of the National Academy of Sciences of the United States of America*, *78*, 7458–7462.

Garcia, V., Reyes-Darias, J. A., Martin-Mora, D., Morel, B., Matilla, M. A., & Krell, T. (2015). Identification of a chemoreceptor for C2 and C3 carboxylic acids. *Applied and Environmental Microbiology*, *81*, 5449–5457.

Garcia-Fontana, C., Reyes-Darias, J. A., Munoz-Martinez, F., Alfonso, C., Morel, B., Ramos, J. L., et al. (2013). High specificity in CheR methyltransferase function: CheR2 of *Pseudomonas putida* is essential for chemotaxis, whereas CheR1 is involved in biofilm formation. *The Journal of Biological Chemistry*, *288*, 18987–18999.

Gargallo-Garriga, A., Preece, C., Sardans, J., Oravec, M., Urban, O., & Peñuelas, J. (2018). Root exudate metabolomes change under drought and show limited capacity for recovery. *Scientific Reports*, *8*(1), 12696.

George, I., Eyers, L., Stenuit, B., & Agathos, S. N. (2008). Effect of 2,4,6-trinitrotoluene on soil bacterial communities. *Journal of Industrial Microbiology & Biotechnology*, *35*, 225–236.

Ghosal, D., Ghosh, S., Dutta, T. K., & Ahn, Y. (2016). Current state of knowledge in microbial degradation of polycyclic aromatic hydrocarbons (PAHs): A review. *Frontiers in Microbiology*, 7, 1369.

Gilbert, E. S., & Crowley, D. E. (1997). Plant compounds that induce polychlorinated biphenyl biodegradation by *Arthrobacter* sp. strain B1B. *Applied and Environmental Microbiology*, *63*, 1933–1938.

Gjermansen, M., Ragas, P., Sternberg, C., Molin, S., & Tolker-Nielsen, T. (2005). Characterization of starvation-induced dispersion in *Pseudomonas putida* biofilms. *Environmental Microbiology*, 7, 894–904.

Greated, A., Lambertsen, L., Williams, P. A., & Thomas, C. M. (2002). Complete sequence of the IncP-9 TOL plasmid pWW0 from *Pseudomonas putida*. *Environmental Microbiology*, *4*, 856–871.

Haas, D., & Défago, G. (2005). Biological control of soil-borne pathogens by fluorescent pseudomonads. *Nature Reviews Microbiology*, *3*, 307–319.

Hiltner, L. (1904). Uber neuere erfahrunger und probleme auf dem gebiete der bodenbakteriologie unter besonderer berucksichtigung der grundungung und brache. *Arbeiten der Deutschen Landwirtschafts-Gesellschaft*, *98*, 59–78.

Hinsa, S. M., Espinosa-Urgel, M., Ramos, J. L., & O'Toole, G. A. (2003). Transition from reversible to irreversible attachment during biofilm formation by *Pseudomonas fluorescens* WCS365 requires an ABC transporter and a large secreted protein. *Molecular Microbiology*, *49*, 905–918.

Huertas, M. J., Duque, E., Marqués, S., & Ramos, J. L. (1998). Survival in soil of different toluene-degrading *Pseudomonas* strains after solvent shock. *Applied and Environmental Microbiology*, *64*, 38–42.

Hutsch, B. W., Augustin, J., & Merbach, W. (2000). Plant rhizodeposition an important source for carbon turnover in soils. *Journal of Plant Nutrition and Soil Science*, *165*, 397–407.

Imlay, J. A. (2003). Pathways of oxidative damage. *Annual Review of Microbiology*, *57*, 395–418.

Inoue, A., Yamamoto, M., & Horikoshi, K. (1991). *Pseudomonas putida* which can grow in the presence of toluene. *Applied and Environmental Microbiology*, *57*, 1560–1562.

Jindrová, E., Chocová, M., Demnerová, K., & Brenner, V. (2002). Bacterial aerobic degradation of benzene, toluene, ethylbenzene and xylene. *Folia Microbiologia (Praha)*, *47*, 83–93.

John, B., Yamashita, T., Ludwig, B., & Flessa, H. (2005). Mechanisms and regulation of organic matter storageof organic carbon in aggregate and density fractions of silty soils under different types of land use. *Geoderma*, *128*, 63–79.

Kanaly, R. A., & Harayama, S. (2010). Advances in the field of high molecular-weight polycyclic aromatic hydrocarbon biodegradation by bacteria. *Microbial Biotechnology*, *3*, 136–164.

Kotoky, R., Rajkumari, J., & Pandey, P. (2018). The rhizosphere microbiome: Significance in rhizoremediation of polyaromatic hydrocarbon contaminated soil. *Journal of Environmental Management*, *217*, 858–870.

Kravchenko, L. V., Azarova, T. S., Shaposhnikov, A. I., Makarova, N. M., & Tikhonovich, I. A. (2003). Root exudates of tomato plants and their effect on the growth and antifungal activity of *Pseudomonas* strains. *Microbiology*, *72*, 37–41.

Kuiper, I., Bloemberg, G. V., Noreen, S., Thomas-Oates, J. E., & Lugtenberg, B. J. J. (2001). Increased uptake of putrescine in the rhizosphere inhibits competitive root colonization by pseudomonas fluorescens strain WCS365. *Molecular Plant-Microbe Interactions*, *14*(9), 1096–1104. 2001.

Kuiper, I., Lagendijk, E. L., Bloemberg, G. V., & Lugtenberg, B. J. (2004). Rhizoremediation: A beneficial plant–microbe interaction. *Molecular Plant-Microbe Interactions*, *17*, 6–15.

Kuzyakov, Y., & Razavi, B. S. (2019). Rhizosphere size and shape: Temporal dynamics and spatial stationarity. *Soil Biology and Biochemistry*, *135*, 343–360.

Lacal, J., Alfonso, C., Liu, X., Parales, R. E., Morel, B., Conejero-Lara, F., et al. (2010). Identification of a chemoreceptor for tricarboxylic acid cycle intermediates: Differential chemotactic response towards receptor ligands. *The Journal of Biological Chemistry*, *285*, 23126–23136.

Lacal, J., Garcia-Fontana, C., Munoz-Martinez, F., Ramos, J. L., & Krell, T. (2010). Sensing of environmental signals: Classification of chemoreceptors according to the size of their ligand binding regions. *Environmental Microbiology*, *12*, 2873–2884.

Lin, B., Van Verseveld, H. W., & Röling, W. F. (2002). Microbial aspects of anaerobic BTEX degradation. *Biomedical and Environmental Sciences*, *15*, 130–144.

López-Farfán, D., Reyes-Darias, J. A., Matilla, M. A., & Krell, T. (2019). Concentration dependent effect of plant root exudates on the chemosensory systems of *Pseudomonas putida* KT2440. *Frontiers in Microbiology*, *10*, 78.

López-Sánchez, A., Leal-Morales, A., Jiménez-Díaz, L., Platero, A. I., Bardallo-Pérez, J., Díaz-Romero, A., et al. (2016). Biofilm formation-defective mutants in *Pseudomonas putida*. *FEMS Microbiology Letters*, *363*. fnw127.

Lowe, L. E. (1978). Carbohydrates in soil. In M. Schnitzer & S. U. Khan (Eds.), *Developments in Soil Science* (pp. 65–93). 8. (pp. 65–93). Elsevier.

Lugtenberg, B. J., Kravchenko, L. V., & Simons, M. (1999). Tomato seed and root exudate sugars: composition, utilization by Pseudomonas biocontrol strains and role in rhizosphere colonization. *Environmental Microbiology*, *1*, 439–446.

Martin, B. C., George, S. J., Price, C. A., Ryan, M. H., & Tibbett, M. (2014). The role of root exuded low molecular weight organic anions in facilitating petroleum hydrocarbon degradation: Current knowledge and future directions. *Science of the Total Environment*, *472*, 642–653.

Martínez-Gil, M., Yousef-Coronado, F., & Espinosa-Urgel, M. (2010). LapF, the second largest *Pseudomonas putida* protein, contributes to plant root colonization and determines biofilm architecture. *Molecular Microbiology*, *77*, 549–561.

Martin-Mora, D., Reyes-Darias, J. A., Ortega, A., Corral-Lugo, A., Matilla, M. A., & Krell, T. (2016). McpQ is a specific citrate chemoreceptor that responds preferentially to citrate/metal ion complexes. *Environmental Microbiology*, *18*, 3284–3295.

Martins dos Santos, V. A. P., Heim, S., Moore, E. R., Strätz, M., & Timmis, K. N. (2004). Insights into the genomic basis of niche specificity of Pseudomonas putida KT2440 Environ. *Microbiologica*, *6*, 1264–1286.

Matilla, M. A., Espinosa-Urgel, M., Rodríguez-Herva, J. J., Ramos, J. L., & Ramos-González, M. I. (2007). Genomic analysis reveals the major driving forces of bacterial life in the rhizosphere. *Genome Biology*, *8*, R179.

Matilla, M. A., Ramos, J. L., Bakker, P. A., Doornbos, R., Badri, D. V., Vivanco, J. M., et al. (2010). *Pseudomonas putida* KT2440 causes induced systemic resistance and changes in Arabidopsis root exudation. *Environmental Microbiology Reports*, *2*, 381–388.

Micallef, S. A., Channer, S., Shiaris, M. P., & Colón-Carmona, A. (2009). Plant age and genotype impact the progression of bacterial community succession in the Arabidopsis rhizosphere. *Plant Signaling & Behavior*, *4*(8), 777–780.

Moe, L. A. (2013). Amino acids in the rhizosphere: From plants to microbes. *American Journal of Botany*, *100*, 1692–1705.

Molina, L., Constantinescu, F., Michel, L., Reimmann, C., Duffy, B., & Défago, G. (2003). Degradation of pathogen quorum-sensing molecules by soil bacteria: A preventive and curative biological control mechanism. *FEMS Microbiology Ecology*, *45*, 71–81.

Molina, L., Ramos, C., Duque, E., Ronchel, M. C., García, J. M., Wyke, L., et al. (2000). Survival of *Pseudomonas putida* KT2440 in soil and in the rhizosphere of plants under greenhouse and environmental conditions. *Soil Biology and Biochemistry*, *32*, 315–321.

Molina, M. A., Ramos, J. L., & Espinosa-Urgel, M. (2006). A two-partner secretion sytem is involved in seed colonization and iron uptake by *Pseudomonas putida* KT2440. *Environmental Microbiology*, *8*, 639–644.

Molina, L., Ramos, C., Ronchel, M. C., Molin, S., & Ramos, J. L. (2000). Construction of an efficient biologically contained pseudomonas putida strain and its survival in outdoor assays. *Applied and Environmental Microbiology*, *64*, 2072–2078.

Molina, L., Rosa, R. L., Nogales, J., & Rojo, F. (2019). Pseudomonas putida KT2440 metabolism undergoes sequential modifications during exponential growth in a complete medium as compounds are gradually consumed. *Environmental Microbiology*, *21*, 2375–2390.

Monds, R. D., & O'Toole, G. A. (2009). The developmental model of microbial biofilms: Ten years of a paradigm up for review. *Trends in Microbiology*, *17*, 73–87.

Moore, E. R. B., & Palleroni, N. J. (2004). Taxonomy of pseudomonads: Experimental approaches. In J. L. Ramos (Ed.), *Pseudomonas, Vol. 4*. (pp. 3–44). Amsterdam: Springer.

Moore, E. R. B., Tyndall, B. J., Martins dos Santos, V., Piepper, D., Ramos, J. L., & Palleroni, N. J. (2006). Non-medical *Pseudomonas*. *Prokaryotes*, *6*, 646–673.

Morgan, J. A. W., & Whipps, J. M. (2001). Methodological approaches to the study of rhizosphere carbon flow and microbial population dynamics. In A. Pinton, Z. Varanini, &

P. Nannipieri (Eds.), *The Rhizosphere. Biochemistry and Organic Substances at the Soil-Plant Interface* (pp. 373–409). New York, USA: Marcel Dekker.

Mosqueda, G., & Ramos, J. L. (2000). A set of genes encoding a second efflux pump in *Pseudomonas putida* DOT-T1E is linked to *tod* genes for toluene metabolism. *Journal of Bacteriology, 182*, 937–943.

Mulet, M., Bennasar, A., Lalucat, J., & García-Valdés, E. (2009). An rpoD-based PCR procedure for the identification of Pseudomonas species and for their detection in environmental samples. *Molecular and Cellular Probes, 23*, 140–147.

Nagayama, H., Sugawara, T., Endo, R., et al. (2015). Isolation of oxygenase genes for indigo-forming activity from an artificially polluted soil metagenome by functional screening using *Pseudomonas putida* strains as hosts. *Applied Microbiology and Biotechnology, 99*, 4453.

Neal, A. L., Ahmad, S., Gordon-Weeks, R., & Ton, J. (2012). Benzoxazinoids in root exudates of maize attract *Pseudomonas putida* to the rhizosphere. *PLoS One*, 7(4), e35498.

Neal, A. L., & Ton, J. (2013). Systemic defense priming by *Pseudomonas putida* KT2440 in maize depends on benzoxazinoid exudation from the roots. *Plant Signaling & Behavior, 8*, e22655.

Nelson, K. E., Weinel, C., Paulsen, I. T., Dodson, R. J., Hilbert, H., Martins dos Santos, V. A., et al. (2002). Complete genome sequence and comparative analysis of the metabolically versatile *Pseudomonas putida* KT2440. *Environmental Microbiology, 4*, 799–808.

Neumann, G., & Römheld, V. (1999). Root excretion of carboxylic acids and protons in phosphorous-deficient plants. *Plant and Soil, 211*, 121–130.

Nguyen, C. (2003). Rhizodeposition of organic C by plants: Mechanisms and controls. *Agronomie, 23*, 375–396.

Nielsen, L., Li, X., & Halverson, L. J. (2011). Cell–cell and cell-surface interactions mediated by cellulose and a novel exopolysaccharide contribute to *Pseudomonas putida* biofilm formation and fitness under water-limiting conditions. *Environmental Microbiology, 13*, 1342–1356.

Nikel, P. I., Max Chavarría, M., Fuhrer, T., Sauer, U., & de Lorenzo, V. (2015). *Pseudomonas putida* KT2440 strain metabolizes glucose through a cycle formed by enzymes of the Entner-Doudoroff, Embden-Meyerhof-Parnas, and pentose phosphate pathways. *The Journal of Biological Chemistry, 290*, 25920–25927.

Nilsson, M., Chiang, W., Fazli, M., Gjermansen, M., Givskov, M., & Tolker-Nielsen, T. (2011). Influence of putative exopolysaccharide genes on Pseudomonas putida KT2440 biofilm stability. *Environmental Microbiology, 13*, 1357–1369.

Nogales, J. N., Mueller, J., Gudmundsson, S., Canalejo, F., Duque, E., Monk, J., et al. (2019). High-quality genome-scale metabolic modelling of Pseudomonas putida highlights its broad metabolic capabilities. *Environmental Microbiology* in presshttps://doi.org/10.1111/1462-2920.14843.

Noor, E., Eden, E., Milo, R., & Alon, U. (2010). Central carbon metabolism as a minimal biochemical walk between precursors for biomass and energy. *Molecular Cell, 39*, 809–820.

Oni, F. E., Geudens, N., Omoboye, O. O., Bertier, L., Hua, H. G., Adiobo, A., et al. (2019). Fluorescent *Pseudomonas* and cyclic lipopeptide diversity in the rhizosphere of cocoyam (*Xanthosoma sagittifolium*). *Environmental Microbiology, 21*, 1019–1034.

Parales, R. E., Luu, R. A., Chen, G. Y., Liu, X., Wu, V., Lin, P., et al. (2013). *Pseudomonas putida* F1 has multiple chemoreceptors with overlapping specificity for organic acids. *Microbiology, 159*, 1086–1096.

Pascual, J., Udaondo, Z., Molina, L., Segura, A., Esteve-Núñez, A., Caballero, A., et al. (2015). Draft genome sequence of *Pseudomonas putida* JLR11, a facultative anaerobic 2,4,6-Trinitrotoluene biotransforming bacterium. *Genome Announcements, 3*(5), e00904-15.

Pearse, S. J., Veneklaas, E. J., Cawthray, G., Bolland, M. D. A., & Lambers, H. (2006). Triticum aestivum shows a greater biomass response to a supply of aluminium phosphate than Lupinus albus, despite releasing fewer carboxylates into the rhizosphere. *New Phytologist*, *169*, 515–524.

Pérez-Montaño, F., Alías-Villegas, C., Bellogín, R. A., del Cerro, P., Espuny, M. R., Jiménez-Guerrero, I., et al. (2014). Plant growth promotion in cereal and leguminous agricultural important plants: From microorganism capacities to crop production. *Microbiological Research*, *169*, 325–336.

Pizarro-Tobias, P., Fernandez, M., Niqui, J. L., Solano, J., Duque, E., Ramos, J. L., et al. (2015). Restoration of a Mediterranean forest after a fire: Bioremediation and rhizoremediation field-scale trial. *Microbiology and Biotechnology*, *8*, 77–92. https://doi.org/10.1111/1751-7915.12138.

Radkov, A. D., & Moe, L. A. (2013). Amino acid racemization in *Pseudomonas putida* KT2440. *Journal of Bacteriology*, *195*(22), 5016–5024.

Ramos, J. L., Cuenca, M. S., Molina-Santiago, C., Segura, A., Duque, E., Gómez-García, R., et al. (2015). Mechanisms of solvent tolerance mediated by interplay of cellular factors in *Pseudomonas putida*. *FEMS Microbiology Reviews*, *39*, 555–566.

Ramos, J. L., Duque, E., Gallegos, M. T., Godoy, P., Ramos-González, M. I., Rojas, A., et al. (2002). Mechanisms of solvent tolerance in Gram-negative bacteria. *Annual Review of Microbiology*, *56*, 743–768.

Ramos, J. L., & Timmis, K. N. (1987). Experimental evolution of catbolic pathways. *Microbiological Sciences*, *4*, 228–237.

Reineke, W., & Knackmuss, H. J. (1988). Microbial degradation of haloaromatics. *Annual Review of Microbiology*, *42*, 263–287.

Revelles, O., Espinosa-Urgel, M., Fuhrer, T., Sauer, U., & Ramos, J. L. (2005). Multiple and interconnected pathways for *L*-lysine catabolism in Pseudomonas putida KT2440. *Journal of Bacteriology*, *187*, 7500–7510.

Reyes-Darias, J. A., Garcia, V., Rico-Jimenez, M., Corral-Lugo, A., Lesouhaitier, O., Juarez-Hernandez, D., et al. (2015). Specific gamma-aminobutyrate chemotaxis in pseudomonads with different lifestyle. *Molecular Microbiology*, *97*, 488–501.

Roca, A., Pizarro-Tobías, P., Udaondo, Z., Fernández, M., Matilla, M. A., Molina-Henares, M. A., et al. (2013). Analysis of the plant growth-promoting properties encoded by the genome of the rhizobacterium *Pseudomonas putida* BIRD-1. *Environmental Microbiology*, *15*, 780–794.

Rodriguez-Conde, S., Molina, L., González, P., García-Puente, A., & Segura, A. (2016). Degradation of phenanthrene by *Novosphingobium* sp. HS2a improved plant growth in PAHs-contaminated environments. *Applied Microbiology and Biotechnology*, *100*, 10627–10636.

Rojo, F. (2010). Carbon catabolite repression in *Pseudomonas*: Optimizing metabolic versatility and interactions with the environment. *FEMS Microbiology Reviews*, *34*, 658–684.

Rylott, E. L., Lorenz, A., & Bruce, N. C. (2011). Biodegradation and biotransformation of explosives. *Current Opinion in Biotechnology*, *22*, 434–440.

Sampedro, I., Parales, R. E., Krell, T., & Hill, J. E. (2015). *Pseudomonas* chemotaxis. *FEMS Microbiology Reviews*, *39*, 17–46.

Sarand, I., Osterberg, S., Holmqvist, S., Holmfeldt, P., Skarfstad, E., Parales, R. E., et al. (2008). Metabolism-dependent taxis towards (methyl)phenols is coupled through the most abundant of three polar localized Aer-like proteins of *Pseudomonas putida*. *Environmental Microbiology*, *10*, 1320–1334.

Scharf, B. E., Hynes, M. F., & Alexandre, G. M. (2016). Chemotaxis signaling systems in model beneficial plant-bacteria associations. *Plant Molecular Biology*, *90*, 549.

Segura, A., Hernández-Sánchez, V., Marqués, S., & Molina, L. (2017). Insights in the regulation of the degradation of PAHs in *Novosphingobium* sp. HR1a and utilization of this regulatory system as a tool for the detection of PAHs. *Science of the Total Environment, 590–591*, 381–393.

Segura, A., & Ramos, J. L. (2013). Plant–bacteria interactions in the removal of pollutants. *Current Opinion in Biotechnology, 24*, 467–473.

Siciliano, S. D., Germida, J. J., Banks, K., & Greer, C. W. (2003). Changes in microbial community composition and function during a polyaromatic hydrocarbon phytoremediation field trial. *Applied and Environmental Microbiology, 69*, 483–489.

Siciliano, S. D., Roy, R., & Greer, C. W. (2000). Reduction in denitrification activity in field soils exposed to long term contamination by 2,4,6-trinitrotoluene (TNT). *FEMS Microbiology Ecology, 32*, 61–68.

Stenuit, B. A., & Agathos, S. N. (2010). Microbial 2,4,6-trinitrotoluene degradation: could we learn from (bio)chemistry for bioremediation and vice versa? *Applied Microbiology and Biotechnology, 88*, 1043–1064.

Sudarsan, S., Dethlefsen, S., Blank, L. M., Siemann-Herzberg, M., & Schmid, A. (2014). The functional structure of central carbon metabolism in *Pseudomonas putida* KT2440. *Applied and Environmental Microbiology, 80*, 5292–5303.

Sutra, L., Risède, J., & Gardan, L. (2000). Isolation of fluorescent pseudomonads from the rhizosphere of banana plants antagonistic towards root necrosing fungi. *Letters in Applied Microbiology, 31*, 289–293.

Taylor, B. L., & Koshland, D. E. (1974). Reversal of flagellar rotation in monotrichous and peritrichous bacteria-generation of changes in direction. *Journal of Bacteriology, 119*, 640–642.

Timmis, K. N. (2002). *Pseudomonas putida:* A cosmopolitan opportunist per excellence. *Environmental Microbiology, 4*, 779–781.

Turnbull, G. A., Morgan, J. A., Whipps, J. M., & Saunders, J. R. (2001). The role of motility in the in vitro attachment of *Pseudomonas putida* PaW8 to wheat roots. *FEMS Microbiology Ecology, 35*, 57–65.

Turner, L., Ryu, W. S., & Berg, H. C. (2000). Real-time imagingof fluorescent flagellar filaments. *Journal of Bacteriology, 182*, 2793–2801.

Udaondo, Z., Molina, L., Segura, A., Duque, E., & Ramos, J. L. (2016). Analysis of the core genome and pangenome of *Pseudomonas putida. Environmental Microbiology, 18*, 3268–3283.

Udaondo, Z., Ramos, J. L., Segura, S., Krell, T., & Daddaoua, A. (2018). Regulation of carbohydrate degradation pathways in Pseudomonas involves a versatile set of transcriptional regulators. *Microbial Biotechnology, 11*, 442–454.

Van Dillewijn, P., Couselo, J. L., Corredoira, E., Delgado, A., Wittich, R. M., Ballester, A., et al. (2008). Bioremediation of 2,4,6-trinitrotoluene by bacterial nitroreductase expressing transgenic aspen. *Environmental Science & Technology, 42*, 7405–7410.

Varjani, S. J., Gnansounou, E., & Pandey, A. (2017). Comprehensive review on toxicity of persistent organic pollutants from petroleum refinery waste and their degradation by microorganisms. *Chemosphere, 188*, 280–291.

Verma, V., Ravindran, P., & Kumar, P. (2016). Plant hormone-mediated regulation of stress responses. *BMC Plant Biology, 16*, 86.

Vranova, V., Rejsek, K., Skene, K., & Formanek, P. (2011). Non-protein amino acids: Plant, soil and ecosystem interactions. *Plant and Soil, 342*, 31–48.

Wackett, L. P., & Gibson, D. T. (1988). Degradation of trichloroethylene by toluene dioxygenase in whole-cell studies with *Pseudomonas putida* F1. *Applied and Environmental Microbiology, 54*, 1703–1708.

Wittich, R. M., Ramos, J. L., & van Dillewijn, P. (2009). Microorganism and explosives: Mechanisms of nitrogen release from TNT for use as an N-source for growth. *Environmental Science & Technology*, *43*, 2773–2776.
Xiao, Y., Nie, H., Liu, H., Luo, X., Chen, W., & Huang, Q. (2016). C-di-GMP regulates the expression of *lapA* and bcs operons via FleQ in *Pseudomonas putida* KT2440. *Environmental Microbiology Reports*, *8*, 659–666.
Yang, Z., & Lu, C.-D. (2007). Functional genomics enables identification of genes of the arginine transaminase pathway in *Pseudomonas aeruginosa*. *Journal of Bacteriology*, *189*, 3945–3953.

Further reading

Jiménez-Fernández, A., López-Sánchez, A., Calero, P., & Govantes, F. (2015). The c-di-GMP phosphodiesterase BifA regulates biofilm development in *Pseudomonas putida*. *Environmental Microbiology Reports*, 7, 78–84.
Jones, D. L., & Darrah, P. R. (1994). Role of root derived organic acids in the mobilization of nutrients from the rhizosphere. *Plant and Soil*, *166*, 247.
La Rosa, R., Behrends, V., Williams, H. D., Bundy, J. G., & Rojo, F. (2016). Influence of the Crc regulator on the hierarchical use of carbon sources from a complete medium. *Environmental Microbiology*, *18*, 807–818.
Ramos, J. L., Molina, L., & Segura, A. (2009). Removal of organic toxic chemicals in the rhizosphere and phyllosphere of plants. *Microbial Biotechnology*, *2*, 144–146.
Shi, W. Y., & Sun, H. (2002). Type IV pilus-dependent motility and its possible role in bacterial pathogenesis. *Infection and Immunity*, *70*, 1–4.
Spain, J. C. (1995). Biodegradation of nitroaromatic compounds. *Annual Review of Microbiology*, *49*, 523–555.
Terzaghi, E., Vergani, L., Mapelli, F., Borin, S., Raspa, G., Zanardini, E., et al. (2019). Rhizoremediation of weathered PCBs in a heavily contaminated agricultural soil: Results of a biostimulation trial in semi field conditions. *Science of the Total Environment*, *686*, 484–496.
Vergani, L., Mapelli, F., Zanardini, E., Terzaghi, E., Di Guardo, A., Morosini, C., et al. (2017). Phyto-rhizoremediation of polychlorinated biphenyl contaminated soils: An outlook on plant-microbe beneficial interactions. *Science of the Total Environment*, *575*, 1395–1406.
Weyens, N., van der Lelie, D., Artois, T., Smeets, K., Taghavi, S., Newman, L., et al. (2009). Bioaugmentation with engineered endophytic bacteria improves contaminant fate in phytoremediation. *Environmental Science & Technology*, *43*, 9413–9418.
Worsey, M. J., & Williams, P. A. (1975). Metabolism of toluene and xylenes by *Pseudomonas* ([sic]*putida* (*arvilla*)) mt-2: Evidence for a new function of the TOL plasmid. *Journal of Bacteriology*, *124*, 7–13.

CHAPTER FIVE

Glutathione: A powerful but rare cofactor among Actinobacteria

Anna C. Lienkamp[a], Thomas Heine[b], Dirk Tischler[a,*]
[a]Microbial Biotechnology, Faculty of Biology and Biotechnology, Ruhr University Bochum, Bochum, Germany
[b]Environmental Microbiology, Faculty of Chemistry and Physics, TU Bergakademie Freiberg, Freiberg, Germany
[*]Corresponding author: e-mail address: dirk.tischler@rub.de

Contents

Abstract

Glutathione (γ-L-glutamyl-L-cysteinylglycine, GSH) is a powerful cellular redox agent. In nature only the L,L-form is common among the tree of life. It serves as antioxidant or redox buffer system, protein regeneration and activation by interaction with thiol groups, unspecific reagent for conjugation during detoxification, marker for amino acid or peptide transport even through membranes, activation or solubilization of compounds during degradative pathways or just as redox shuttle. However, the role of

Advances in Applied Microbiology, Volume 110
ISSN 0065-2164
https://doi.org/10.1016/bs.aambs.2019.12.003

GSH production and utilization in bacteria is more complex and especially little is known for the Actinobacteria. Some recent reports on GSH use in degradative pathways came across and this is described herein. GSH is used by transferases to activate and solubilize epoxides. It allows funneling epoxides as isoprene oxide or styrene oxide into central metabolism. Thus, the distribution of GSH synthesis, recycling and application among bacteria and especially Actinobacteria are highlighted including the pathways and contributing enzymes.

Abbreviations

ADP adenosine diphosphate
ATP adenosine triphosphate
BSH bacillithiol
CoA coenzyme A
DNA deoxyribonucleic acid
FAD flavin adenine dinucleotide
Fig figure
GCS glutamate cysteine ligase
GFA glutathione-dependent formaldehyde-activating enzyme
GK γ-glutamyl kinase
Glo glyoxalase
GLR glutathione reductase
GMB 2-glutathionyl-2-methyl-3-butenal
GMBA 2-glutathionyl-2-methyl-3-butenoate
GPX glutathione peroxidase
GR glutathione reductase
GRX glutaredoxin
GS glutathione synthetase
GSH reduced glutathione
GSSG oxidized glutathione (dimer, glutathione disulfide)
GST glutathione *S*-transferase
HGMB 1-hydroxy-2-glutathionyl-2-methyl-3-butene
IsoMO isoprene monooxygenase
LD_{50} lethal dose, 50%
MSH mycothiol
NADH nicotinamide adenine dinucleotide
NADPH nicotinamide adenine dinucleotide phosphate
NAD(P)H NADH or NADPH
NS-GPX seleno-independent GPX
PHGPX phospholipid-hydroperoxide glutathione peroxidase
Prx peroxiredoxin
ROS reactive oxygen species
rRNA ribosomal ribonucleic acid
Seleno-GPX seleno-dependent GPX
SOI styrene oxide isomerase

1. Glutathione: An introduction

In 1921, glutathione (γ-L-glutamyl-L-cysteinylglycine; GSH) was first isolated from yeast and mammalian tissues and characterized by Nobel laureate and funder of the Biochemistry Department at the University of Cambridge, Sir F.G. Hopkins. He already recorded it to be reversibly oxidable by disulfide bond formation between two molecules (oxidized dimer, GSSG) (Fig. 1). The reduced form seemed to be predominant in

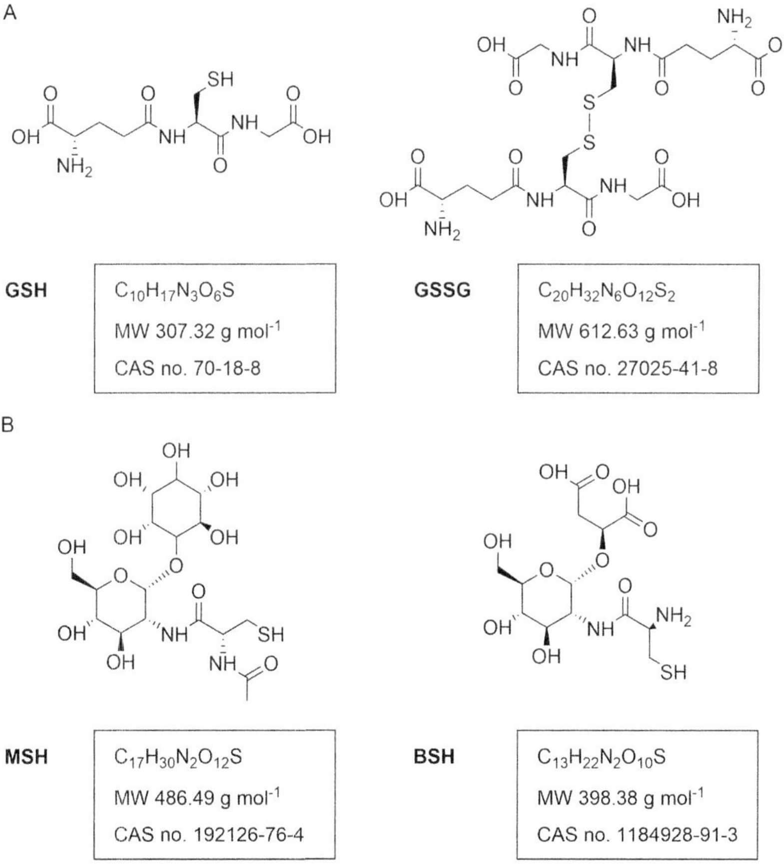

Fig. 1 Selected low molecular weight thiols present in bacteria. (A) Glutathione (γ-L-glutamyl-L-cysteinylglycine, GSH) and its disulfide (GSSG). (B) Mycothiol (AcCys-GlcN-Ins, MSH) and Bacillithiol (Cys-GlcN-mal, BSH).

living cells (Hopkins, 1921; Hopkins & Dixon, 1922; Simoni, Hill, & Vaughan, 2002) and utilized for coping with oxidative stress and maintaining a reduced environment. Nevertheless, the exact composition of GSH as a tripeptide of L-glutamate, L-cysteine and glycine remained controversial until 1929 (Figs. 1 and 6) (Hopkins, 1927; Hopkins & Harris, 1929; Hunter & Eagles, 1927). Glutathione and its metabolism most likely evolved as redox system along with oxygenic photosynthesis (Deponte, 2013; Fahey, Buschbacher, & Newton, 1987; Ondarza, Rendón, & Ondarza, 1983) and, in the course of time, it became an omnipresent low molecular weight thiol in animals, plants and bacteria (Fahey, Brown, Adams, & Worsham, 1978; Noctor et al., 2012; Smirnova & Oktyabrsky, 2005; Stenersen, Kobro, Bjerke, & Arend, 1987).

It is not as prone to autoxidation, which can lead to the generation of toxic intermediates, like cysteine and bacillithiol (BSH, Cys-GlcN-mal), but not as stable as mycothiol (MSH, AcCys-GlcN-Ins). The two latter serve as GSH analogs in some firmicutes and the vast majority of Actinobacteria, respectively (Fig. 1). Hence, GSH is also a feasible way for cells to store cysteine (Den Hengst & Buttner, 2008; Fahey, 2013; Garg, Soni, Singh Paliya, Verma, & Jadaun, 2013; Helmann, 2011; Lu, 2009; Newton, Buchmeier, & Fahey, 2008).

In nature only the L,L-form of GSH can be found. An unusual γ-peptide linkage between glutamate and cysteine protects it from degradation through peptidases. The sulfhydryl group of its cysteine allows reduced GSH to serve as nucleophile and react with a broad range of electrophilic substances. This is typical during conjugation reactions and thus occurs in a variety of detoxification routes. Reduced (GSH) and oxidized (GSSG) glutathione are commercially available as white powder and both are very soluble in water. GSH has an LD_{50} of $6000\,mg\,kg^{-1}$ (mouse, intravenous) and due to being an important cofactor in humans its relevance regarding clinical applications as well as dietary supplement is under frequent investigation (Allen & Bradley, 2011; Biswas & Rahman, 2009; Sen, Atalay, & Hänninen, 1994). Although oxidized and reduced GSH function as redox buffer in living cells the electrochemical potential of this couple, even though often mentioned, is not applicable for biological systems due to their complex nature (Flohé, 2013).

Considering its abundancy today, the versatile roles of GSH and the manifold of enzymes interacting with and utilizing it do not surprise. GSH metabolism has been extensively studied in mammalian cells but the investigations rapidly expanded including its role in plants and bacteria.

Depending on organism, organ and compartment, GSH plays a role in a multitude of processes including detoxification and excretion, oxidative stress response, transport, signal transduction, gene expression, cysteine storage, cell proliferation, apoptosis, protein interactions, formation of disulfide bonds and protein folding, to only name a few (Anderson, 1998; Deponte, 2013; Pastore, Federici, Bertini, & Piemonte, 2003; Sies, 1999; Smirnova & Oktyabrsky, 2005). Consequently, GSH has pharmaceutical relevance and is a potent and non-expensive co-substrate in a multitude of enzymatic reactions leaving many options for industrial applications. Though a lot of research already has been done, it becomes obvious that there is much more to discover about this versatile compound.

As the reduced form of GSH represents the main unit of occurrence in cells, it needs to be mentioned that it can be obtained via biosynthesis (see Section 3.1) but also due to recycling steps (see Section 3.2). Besides from the ATP-dependent GSH biosynthesis route from L-glutamate and L-cysteine, it can be obtained from the oxidized dimer (GSSG) upon NAD(P)H glutathione reductase activity. The latter reaction is also important from another perspective as GSH is known to interact together with L-ascorbic acid toward stress response caused by oxidative agents (Winkler, Orselli, & Rex, 1994). Thus, in this course the use of L-ascorbic acid leads to dehydroascorbic acid which can be recycled by GSH. This leads to the formation of GSSG which can be reduced as described above. Thus both GSH and L-ascorbic acid act together to maintain a reductive environment in cells (Winkler et al., 1994).

Due to its versatile roles and distribution, it makes sense to summarize the most important aspects. The distribution of glutathione (GSH) among organisms is given in Table 1 and a phylogenetic analysis of prokaryotic glutathione synthetases highlights that (Fig. 2). The latter is also often used in combination with (meta)genome mining approaches to predict GSH biosynthesis and use for various reasons.

Glutathione is the major thiol in all eukaryotes and, except for some parasites, most of them are able to synthesize it (Copley & Dhillon, 2002). In archaeal genomes, various glutathione-dependent proteins were identified (Allocati et al., 2012; Fahey, 2013), for instance, transferases, peroxidases, and reductases. For instance, transferases, peroxidases and reductases. However, none of so far investigated archaea contained enzymes required for the biosynthesis of glutathione (see Section 3.1). Only the precursor γ-glutamylcysteine can be synthesized and is supposed to function as the major thiol (Allocati et al., 2012; Fahey, 2013).

Table 1 Distribution of glutathione among the tree of life.

Domain	Kingdom	Phylum (class)	Glutathione present	GshB (homolog) on genome	Reference
Eukaryota	Animals	–	●	●	Wu, Fang, Yang, Lupton, and Turner (2004)
	Plants	–	●	●	Hasanuzzaman, Nahar, Anee, and Fujita (2017)
	Fungi	–	●	●	Pócsi, Prade, and Penninckx (2004)
Archaea	–	–	○	○	Allocati, Federici, Masulli, and Di Ilio (2012) and Fahey (2013)
Bacteria	–	Acidobacteria	–	●	–
		Actinobacteria	●	●	Johnson, Newton, Fahey, and Rawat (2009)
		Aquificae	–	●	–
		Armatimonadetes	–	○	–
		Bacteroidetes	–	●	–
		Caldiserica	–	○	–
		Chlamydiae	–	○	–
		Chlorobi	●	●	Fahey et al. (1987)
		Chloroflexi	–	●	–
		Chrysiogenetes	–	○	–
		Coprothermobacterota	–	○	–
		Cyanobacteria	●	●	Cameron and Pakrasi (2010)

	Deferribacteres		–	○	–
	Deinococcus-Thermus		–	○	–
	Dictyoglomi		–	○	–
	Elusimicrobia		–	●	–
	Fibrobacteres		–	●	–
	Firmicutes		●	●	Janowiak and Griffith (2005) and Potter, Trappetti, and Paton (2012)
	Fusobacteria		–	●	–
	Gemmatimonadetes		–	●	–
	Lentisphaerae		–	●	–
	Nitrospirae		–	●	–
	Planctomycetes		–	●	–
	Proteobacteria	(α)	●	●	Fahey et al. (1987)
		(β)	●	●	Fahey et al. (1987)
		(γ)	●	●	Kato, Tanaka, Nishioka, Kimura, and Oda (1988) and Vergauwen, de Vos, and van Beeumen (2006)
		(δ)	–	●	–
		(ε)	–	●	–

Continued

Table 1 Distribution of glutathione among the tree of life.—Cont'd

Domain	Kingdom	Phylum (class)	Glutathione present	GshB (homolog) on genome	Reference
		Spirochaetes	–	●	–
		Synergistetes	–	○	–
		Tenericutes	–	○	–
		Thermodesulfobacteria	–	●	–
		Thermotogae	–	○	–
		Verrucomicrobia	–	●	–

Not specified or determined (–), yes (●), no (○).

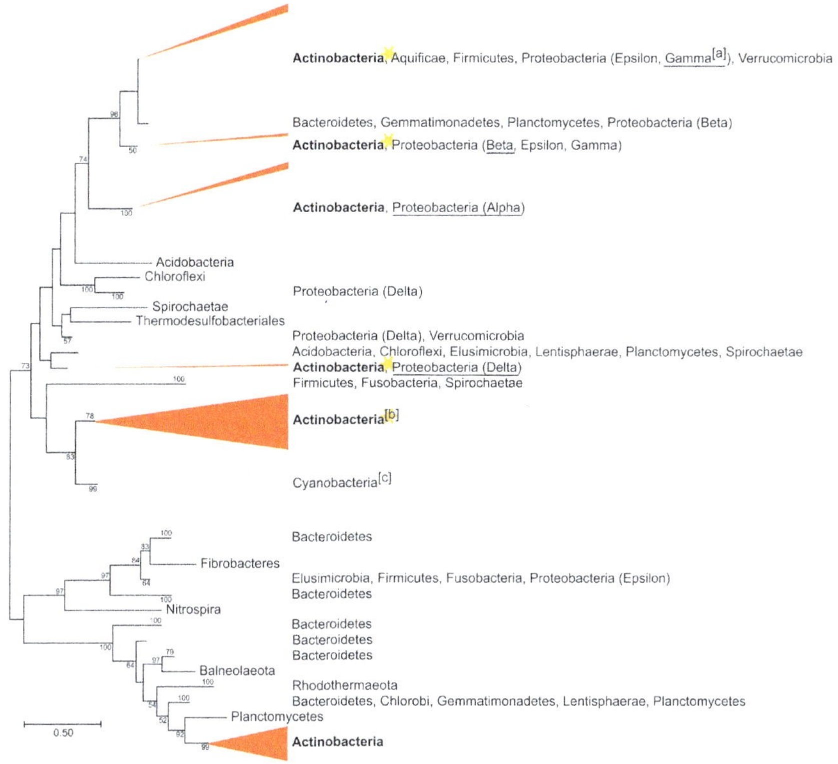

Fig. 2 Phylogenetic tree of (putative) glutathione synthetases (GshBs). The GshBs P04425[a] (Proteobacteria), KJF19172[b] (Actinobacteria) and O32463[c] (Cyanobacteria) used as query sequences for the identification of homologous enzymes in each bacterial phylum via BLAST search. The recognition limit for a positive hit was set to an e-value higher than 10^{-6} (Pearson, 2013). The condensed branches of the tree containing the query sequences are marked with the respective footnote. Underlined phyla represent the main proportion of a condensed branch. Condensed branches containing actinobacterial GshB sequences are shaded in orange, while the quantity of shading approximates the proportion of Actinobacteria of the condensed branches. A yellow star indicates branches were also GshA sequences were found on the genomes. The evolutionary history was inferred by using the Maximum Likelihood method and JTT matrix-based model (Jones, Taylor, & Thornton, 1992). The tree with the highest log likelihood (−119,301.57) is shown. The percentage of trees in which the associated taxa clustered together is shown next to the branches. Initial tree(s) for the heuristic search were obtained automatically by applying Neighbor-Join and BioNJ algorithms to a matrix of pairwise distances estimated using a JTT model, and then selecting the topology with superior log likelihood value. The tree is drawn to scale, with branch lengths measured in the number of substitutions per site. This analysis involved 360 amino acid sequences. There was a total of 427 positions in the final dataset. Evolutionary analyses were conducted in MEGA X (Kumar, Stecher, Li, Knyaz, & Tamura, 2018).

In bacteria, the thiol content depends on the phylogenetic origin. Various thiols have been identified besides glutathione, including mycothiol (MSH, Actinobacteria) and bacillithiol (BSH, Firmicutes), among others (Fig. 1). For some phyla, so far, no low molecular weight thiol has been identified, while others produce one thiol, and some contain two or more (Fahey, 2013). Overall, glutathione has been identified in representatives of seven phyla. However, a database mining approach revealed that glutathione biosynthesis genes, in particular, glutathione synthetase (GshB, see Section 3.1) homologs, are encoded on genomes of 23 phyla (Table 2 and Fig. 2). Therefore, a BLAST search with known GshBs was done as described in the legend of Fig. 2.

The phylogenetic tree separates into two major branches, while all so far characterized GshBs are located on one of these. There is apparently no differentiation of the GshBs according to phyla. This is in accordance with previous studies suggesting a horizontal gene transfer of glutathione biosynthesis genes not only between different phyla but also among different domains of life (Copley & Dhillon, 2002). This seems to have happened several times in history, as, in case of Actinobacteria, GshB-homologs can be found in seven branches. In view of the nodes of the branches, it is likely that Actinobacteria have acquired the respective genes from cyanobacteria and proteobacteria, and maybe from some other phyla. However, a glutamate cysteine ligase (GshA) can only be detected in four of the respective branches and therefore not all Actinobacteria might be able to synthesize glutathione or use a different pathway (see Section 3.1).

In conclusion it can be stated that glutathione is omnipresent and used in versatile reactions in the biological world. In most of these application cases, GSH forms autocatalytically or by means of an enzyme a covalent linkage to a target molecule. This is very well investigated in various detoxification reactions. However, the role of GSH production and utilization in bacteria is more complex and especially little is known for the old and diverse class of Actinobacteria. Here some recent reports on GSH participation in degradative pathways came across and this is described in the following. Thus, GSH is used by transferases to activate and solubilize epoxides as it is known for human phase II detoxification pathways. Actually, it might lower the potential of epoxides within respective pathways to damage DNA or proteins and to avoid the formation of reactive aldehydes from the epoxides. Therefore, it might not just play a role in the course of the degradation but also shields the host itself. Here it allows to funnel epoxides as isoprene oxide or styrene oxide into central intermediates of bacterial metabolism and thus plays a

Table 2 Selection of Actinobacteria that (putatively) utilize glutathione.

Strain	Role of glutathione		Reference
Gordonia rubripertincta CWB2	Styrene degradation	●	Heine et al. (2018)
	Glutathione transport	○	–
	Potassium efflux	○	–
Aeromicrobium sp. Root495	Styrene degradation	○	Heine, Zimmerling, et al. (2018)
	Peroxide detoxification	○	–
Nocardioides sp. Root240	Styrene degradation	○	Heine, Zimmerling, et al. (2018)
	Peroxide detoxification	○	–
Rhodococcus sp. AD45	Isoprene degradation	●	van Hylckama, Leemhuis, Spelberg, and Janssen (2000)
	Glutathione transport	○	–
	Peroxide detoxification	○	–
	Aldehyde detoxification	○	–
Gordonia sp. i37	Isoprene degradation	●	Johnston et al. (2017)
	Peroxide detoxification	○	–
	Aldehyde detoxification	○	–
Mycobacterium sp. AT1	Isoprene degradation	●	Johnston et al. (2017)
	Peroxide detoxification	○	–
	Aldehyde detoxification	○	–

Continued

Table 2 Selection of Actinobacteria that (putatively) utilize glutathione.—Cont'd

Strain	Role of glutathione		Reference
Leifsonia sp. i49	Isoprene degradation	●	Alvarez, Exton, Timmis, Suggett, and McGenity (2009)
Micrococcus sp. i61b	Isoprene degradation	●	El Khawand et al. (2016)
Rhodococcus sp. SAORIC-690	Peroxide detoxification	○	–
	Aldehyde detoxification	○	–
Rhodococcus erythropolis S43	Peroxide detoxification	○	Retamal-Morales et al. (2017)
Rhodococcus opacus 1CP	Arsenate reductase	○	–
	Aldehyde detoxification	○	–
	Peroxide detoxification	○	–

Experimental evidence: putative on genome level (○), proven (●).

central role of degradative metabolism. Hence, the distribution of GSH biosynthesis, recycling and application among bacteria and especially Actinobacteria are highlighted including the pathways and contributing enzymes. More pathways incorporating GSH are expected to be uncovered.

2. Actinobacteria and glutathione

Actinobacteria have been described to possess GSH-dependent metabolic pathways for isoprene and styrene (Heine, Zimmerling, et al., 2018; McGenity, Crombie, & Murrell, 2018). Both of these need glutathione for the processing of an epoxide formed by a monooxygenase and are to date the only experimentally proven uses of glutathione in Actinobacteria. In addition, some putative metabolic activities linked to glutathione can be deduced from genome analysis as various glutathione-dependent proteins are encoded in Actinobacteria. For example, homologous proteins can be found for detoxification and transport processes. A list of (putative) glutathione-dependent activities of respective Actinobacteria is given in

Table 2. In the following, we describe the individual pathways and their distribution among Actinobacteria. However, we also refer to counterparts in other organisms to provide a complete view.

2.1 Microbial pathways comprising glutathione in Actinobacteria

The major thiol in Actinobacteria, mycothiol, consisting of an acetylated cysteine linked to glucosamine and inositol (Fig. 1), was initially described in 1994 (Newton et al., 1996, 2008; Sakuda, Zhou, & Yamada, 1994). Already in 1990, Grund and coworkers demonstrated glutathione dependence of the gentisate degradation pathway in *Streptomyces ghanaensis* ATCC 14672 (Grund, Knorr, & Eichenlaub, 1990). However, this strain does not contain glutathione in the cell and therefore it is unlikely to be used *in vivo* (Johnson et al., 2009). The first verified detection of glutathione in an Actinobacterium was 1998 in *Rhodococcus* sp. AD45 by van Hylckama, Kingma, van den Wijngaard, and Janssen (1998). *Rhodococcus* sp. AD45 origins from an enrichment of isoprene degraders from freshwater sediment (van Hylckama et al., 1998). Isoprene is one of the most produced biogenic volatile compounds and released by various plants but also animals and prokaryotes (McGenity et al., 2018). It is very reactive and influences the earth's climate through various reactions (Carrión et al., 2018; Crombie, Mejia-Florez, McGenity, & Murrell, 2019). Due to its intrinsic reactivity, isoprene can be decomposed through various physicochemical processes (e.g., photochemical oxidation, reaction with radicals or ozone) in the atmosphere or aqueous solutions (Crombie et al., 2019; Srivastva, Singh, Bhardwaj, & Dubey, 2018). In the isoprene cycle, also the microbial community is supposed to be a significant sink (Crombie et al., 2018; Gray, Helmig, & Fierer, 2015; Srivastva et al., 2018) and Actinobacteria, mainly *Rhodococcus* strains, seem to be one of the most dominant isoprene degraders in various habitats (Carrión et al., 2018; Crombie et al., 2018; McGenity et al., 2018). To date only one biological degradation pathway for isoprene is known to aim its utilization as source of carbon and energy, which depends on glutathione (Srivastva et al., 2018).

2.2 Isoprene degradation: The first validated utilization of glutathione in Actinobacteria

This glutathione-dependent degradation route has been identified in various microorganisms since its first recognition. It is initiated by a multicomponent soluble diiron monooxygenase, namely, isoprene monooxygenases (IsoMO),

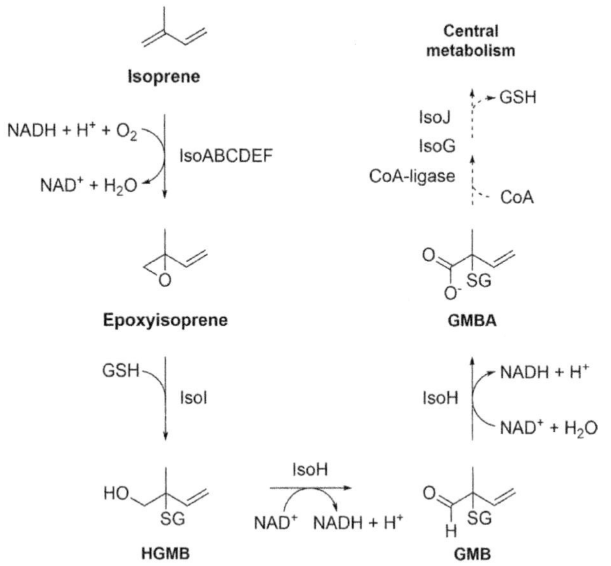

Fig. 3 Isoprene degradation in *Rhodococcus* sp. AD45. Isoprene is converted by an isoprene monooxygenase (IsoABCDEF) to epoxyisoprene. Addition of GSH by IsoI, a glutathione *S*-transferase, generates 1-hydroxy-2-glutathionyl-2-methyl-3-butene (HGMB). Subsequent conversions by a dehydrogenase (IsoH) leads to the formation of 2-glutathionyl-2-methyl-3-butenal (GMB) and 2-glutathionyl-2-methyl-3-butenoate (GMBA) which enter the central metabolism by not yet fully characterized steps (CoA-ligase, IsoG, IsoJ) (McGenity et al., 2018; van Hylckama et al., 2000).

homologous to methane, toluene and alkene monooxygenases (Leahy, Batchelor, & Morcomb, 2003; Srivastva et al., 2018). IsoMOs are composed of six proteins that enable the epoxidation of the methyl-substituted double bond of isoprene (Fig. 3) (Carrión et al., 2018; Johnston et al., 2017). In the next step, a glutathione *S*-transferase (IsoI) appears on the scene to open the epoxide ring with glutathione forming the adduct 1-hydroxy-2-glutathionyl-2-methyl-3-butene (HGMB). HGMB is further processed by a dehydrogenase (IsoH) and the metabolites are subsequently funneled into the central metabolism. However, it remains unclear at which point glutathione is withdrawn from reaction intermediates. Thus, it remains to be clarified if the enzyme does accept the glutathione adduct as substrate or not. This may answer at which stage glutathione is released.

A second glutathione *S*-transferase IsoJ is encoded next to IsoI on the isoprene degradation cluster (Fig. 4). IsoJ is supposed to be involved in the removal of glutathione from a downstream metabolite (van Hylckama et al., 2000). However, while transcriptome analysis supports the presence

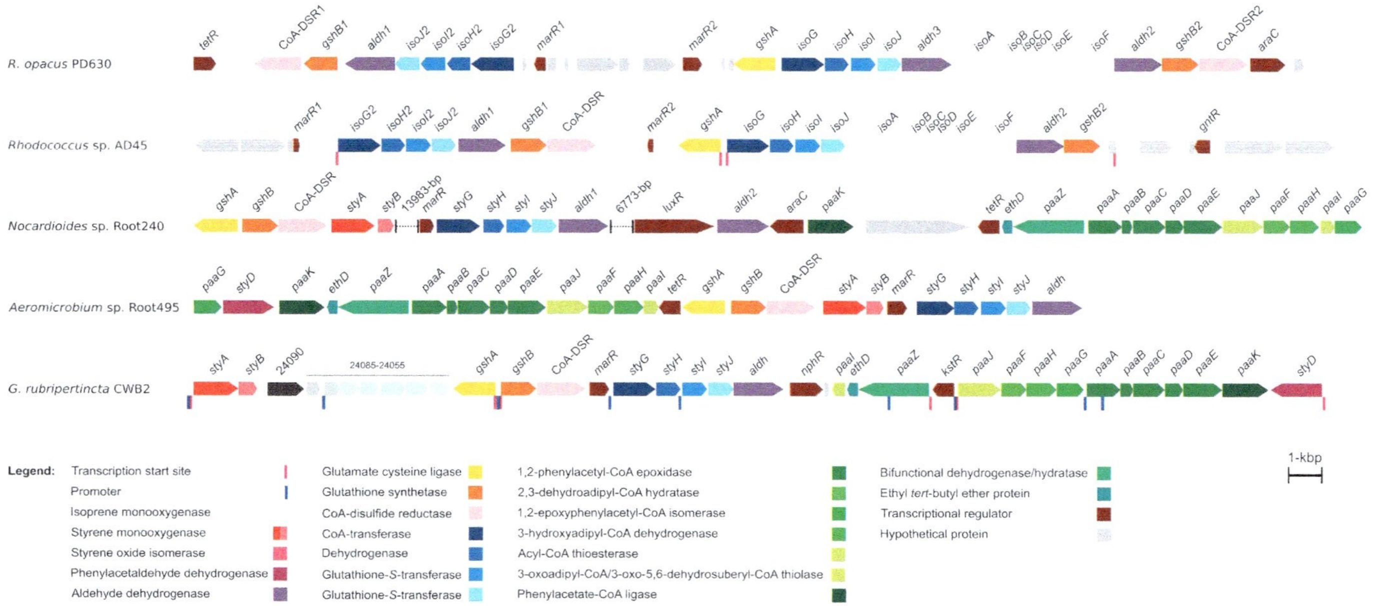

Fig. 4 Comparison of isoprene and styrene degradation gene clusters from Actinobacteria. While the genes encoding for the monooxygenases and the lower degradation pathway are different, the genes required for the glutathione-dependent handling of the epoxides show homology and similar arrangement between isoprene and styrene degraders. The transcription start sites of the isoprene degradation gene cluster in strain AD45 were identified previously (Crombie et al., 2015). In strain CWB2, nine putative promoter sequences were proposed by using CNNPromoter (Umarov & Solovyev, 2017) and putative transcription start sites were estimated from transcriptome data (Heine, Zimmerling, et al., 2018) using ReadXplorer (Hilker et al., 2016). Due to the different arrangement of the open reading frames and regulatory elements of styrene and isoprene degradation clusters, it is likely that they are transcribed differently. The color scheme is deduced from Heine, Zimmerling, et al. (2018).

of IsoJ after isoprene induction, the specific role remains to be proven (Crombie et al., 2015). The respective isoprene degradation gene cluster looks similar in different strains and has also recently been identified in a proteobacterium (Crombie et al., 2018). Further, many isoprene degraders possess additional copies of the glutathione-dependent enzymes and/or the glutathione biosynthesis proteins, which are presumably both expressed during isoprene exposure (Crombie et al., 2015).

2.3 Styrene degradation: Expanding the portfolio of glutathione-dependent reactions

The first steps of the recently identified degradation pathway of styrene resembles one detoxification route of this aromatic compound in humans. Herein, an unspecific epoxidation of styrene is catalyzed by cytochrome P450 monooxygenases. The reactive oxirane is rendered harmless by conjugation to glutathione in the liver by a mu class glutathione *S*-transferase (Pacifici, Warholm, Guthenberg, Mannervik, & Rane, 1987). In *Gordonia rubripertincta* CWB2, (*S*)-styrene oxide is formed by a styrene monooxygenase (Heine, Zimmerling, et al., 2018). This monooxygenase belongs to flavin-dependent two-component monooxygenases, which are involved in various metabolic processes (Ellis, 2010; Heine, van Berkel, Gassner, van Pée, & Tischler, 2018). Crude cell extract studies showed that styrene oxide is rapidly decomposed as soon as glutathione is present (Heine, Zimmerling, et al., 2018). Genome analysis revealed that the putative gene cluster contains homologous genes as found in strain AD45 and other isoprene degraders (Fig. 4). Especially, the toolset (glutathione *S*-transferases and aldehyde dehydrogenases) that is required for conversion of the epoxide to an intermediate suitable for the central metabolism appears to be very similar. Therefore, it is likely that the degradation pathway is partially similar, although, comparable to the isoprene pathway, details have to be experimentally verified (Fig. 5).

However, the utilization of glutathione for the degradation of styrene is remarkable as all so far known pathways follow different routes (Tischler, 2015). As stated before, this could offer additional benefit for the Actinobacterium as potentially toxic epoxides can be intercepted and processed at a later stage. This might allow handling larger amounts of this carbon and energy source.

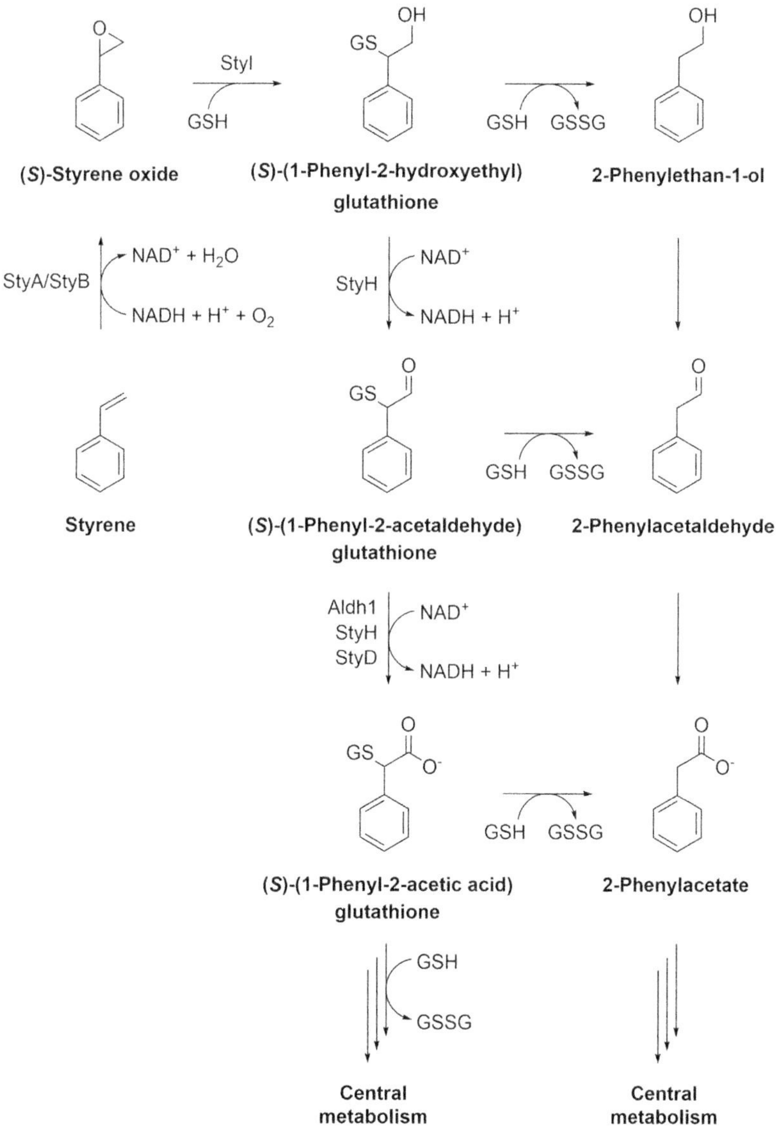

Fig. 5 Glutathione dependent styrene degradation pathway in *Gordonia rubripertincta* CWB2. Styrene is converted into (*S*)-styrene oxide by a monooxygenase (StyA/StyB). The opening of the epoxide ring is then conducted by a glutathione *S*-transferase (StyI). From here, multiple enzymatic steps (StyH, StyD, Aldh1) are proposed to lead to the formation of phenylacetic acid which enters the central metabolism (Heine, Zimmerling, et al., 2018; Tischler, 2015).

2.4 The evolutionary history of glutathione utilization in Actinobacteria

Degradation of isoprene and styrene are to date the only verified metabolic roles of glutathione in Actinobacteria. Thus, it is interesting to have a look on the genomic background of the respective gene (clusters) to find possible acquisition routes of these biocatalytic tools. In the case of the isoprene degraders, the monooxygenase sequences (*isoA*) do not group congruently according to their 16S rRNA genes in phylogenetic analysis (Johnston et al., 2017). Quite contrary to that, *isoA*s from diverse strains form groups indicating horizontal transfer of the respective glutathione-dependent metabolic gene clusters (Crombie et al., 2018). In *Rhodococcus* sp. AD45 and *Gordonia rubripertincta* CWB2 the metabolic gene clusters are located on a plasmid containing transposases. A horizontal gene transfer of the plasmid of strain CWB2 is further supported by a genomic island analysis of the whole genome. The complete plasmid was herein identified as alien (Heine, Zimmerling, et al., 2018).

These findings indicate that the ability of Actinobacteria to utilize glutathione is young from an evolutionary point of view (Johnson et al., 2009). An additional hint is the phylogenetic analysis of glutathione biosynthesis genes. Already in 2002, Copley and coworkers proposed a complex evolutionary history of the γ-glutamylcysteine ligase and glutathione synthetase. They found various examples for horizontal gene transfer, including transfer from prokaryote to eukaryote (Copley & Dhillon, 2002). At that time, the finding of glutathione utilization by Actinobacteria was relatively new and thus they were not specifically investigated in that study (Copley & Dhillon, 2002). However, a phylogenetic analysis of glutathione synthetases from Actinobacteria and homologs from various bacterial phyla supports an ongoing exchange of these genes (Allocati, Federici, Masulli, & Di Ilio, 2009; Allocati et al., 2012) (Fig. 2). In cases of Actinobacteria (and likely some other phyla), this horizontal gene transfer seems to have happened several times independently as there is apparently no definite grouping.

The proven ability to produce and utilize glutathione leads to the assumption that Actinobacteria might already be able to use this thiol for other cellular processes or that they will assimilate the required tools by gene transfer in the future. Analysis of actinobacterial genomes indicate the existence of additional glutathione-dependent functions like detoxification and transport (see Table 2). For instance, *Rhodococcus* sp. SAORIC-690 encodes for about 20 proteins that are annotated as glutathione-dependent and at least eight glutathione *S*-transferase homologs (Fig. 9). Such transferases are often needed for detoxification of reactive compounds just as glutathione peroxidases (reduction of peroxides or H_2O_2), which are encoded in various Actinobacteria (see Sections 4.1 and 4.2). Further, lactoylglutathione lyase and hydroxyacylglutathione hydrolase homologs were found, which belong to the glyoxalase system (Suttisansanee & Honek, 2011). This system catalyzes the detoxification of reactive aldehydes just as glutathione-dependent formaldehyde-activating enzymes, *S*-(hydroxymethyl)glutathione dehydrogenases and *S*-formylglutathione hydrolases, which were also found (Hopkinson et al., 2015). However, the latter three were not encoded together on one actinobacterial genome suggesting that this route cannot be used so far. However, a lyase that utilizes mycothiol instead of glutathione is discussed for Actinobacteria indicating that there might be different routes of detoxification (Hand, Auzanneau, & Honek, 2006; Suttisansanee & Honek, 2011). A glutathione-dependent arsenate reductase was found on the genome of *Rhodococcus opacus* 1CP. This enzyme is part of the arsenate resistance machinery and catalyzes the reduction to arsenite (Mukhopadhyay & Rosen, 2002).

Besides detoxification processes, *Gordonia rubripertincta* CWB2 encodes for glutathione regulated potassium efflux proteins as well as ABC-transporters (GsiABCD) for the import of glutathione.

However, an active role of these glutathione-dependent proteins remains speculative, but their existence indicates that there might be much more to learn about glutathione utilization in Actinobacteria.

3. Enzymes providing glutathione

3.1 Biosynthesis of glutathione

The most investigated pathway for the biosynthesis of glutathione is an ATP-dependent two-step process. The first step is conducted through the γ-glutamylcysteine synthetase or glutamate cysteine ligase (γ-GCS, GshA in prokaryotes, Gsh1 in eukaryotes, EC 6.3.2.2) wherein the γ-carboxyl-group of L-glutamate is linked to the amino group of L-cysteine generating γ-L-glutamyl-L-cysteine at the expense of ATP. The final generation of γ-L-glutamyl-L-cysteinylglycine through the addition of the amino group of glycine via a common peptide-bond to the carboxy-group of the cysteine moiety, again alongside with ATP consumption, is completed via the glutathione synthetase (GS, GshB in prokaryotes, Gsh2 in eukaryotes, EC 6.3.2.3) (Fig. 6) (Meister & Anderson, 1983). GshA and

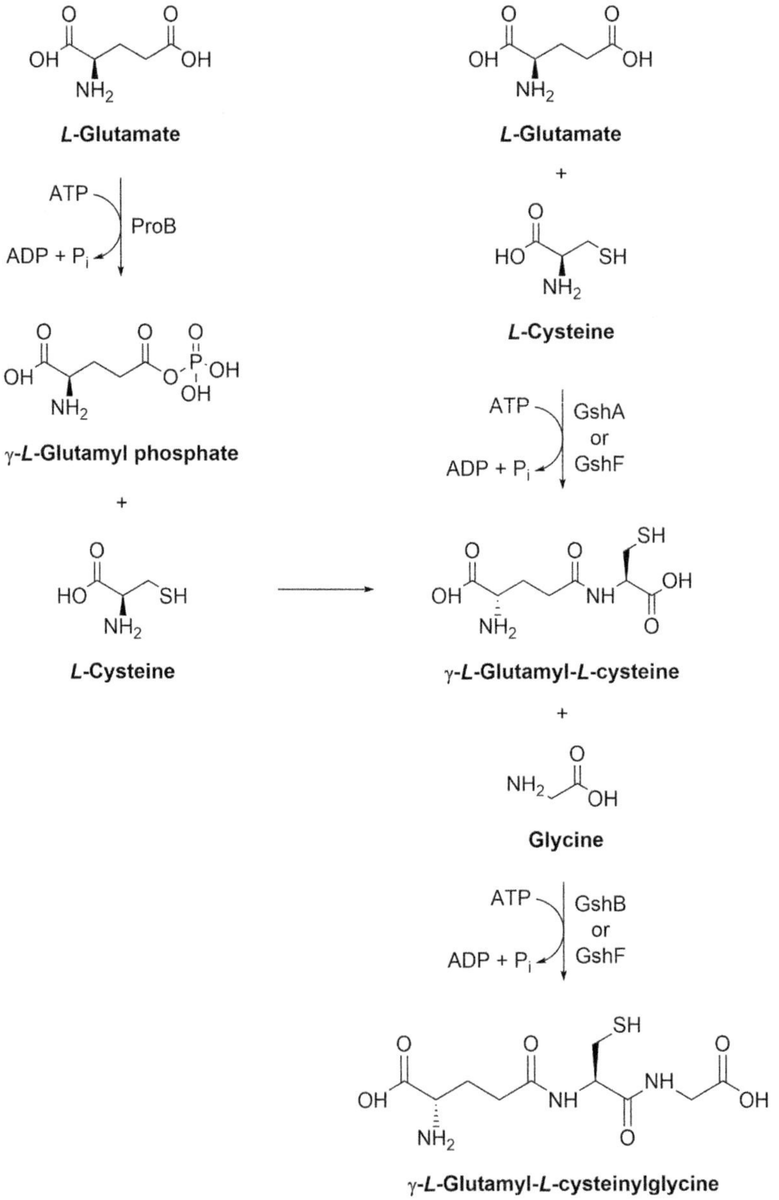

Fig. 6 Known pathways for glutathione biosynthesis. A γ-glutamylcysteine synthetase (GshA) forms a γ-linkage between L-glutamate and L-cysteine. Subsequently, glycine is added to the cysteine moiety via a glutathione synthetase (GshB) (right). These steps can also be carried out by a bifunctional enzyme (GshF) (right). Alternatively, a γ-glutamyl kinase (ProB) can convert L-glutamate to γ-L-glutamyl phosphate which reacts with L-cysteine and thus also forms the intermediate γ-L-glutamyl-L-cysteine (left) which is then further converted to GSH by GshB (enters right).

GshB seem to have evolved independently and consecutively, with the former being the older, and imply an extensive impact of horizontal gene transfer on the evolutionary progression and distribution (Copley & Dhillon, 2002). This would be consistent with the two other described ways of GSH biosynthesis. One way comprises a bifunctional protein consisting of a combination of an N-terminal GshA fused to a protein from the ATP-grasp superfamily at the C-terminus which was first reported by Copley and Dhillon (2002) through *in silico* analysis followed by isolation and characterization in different bacteria (Gopal et al., 2005; Janowiak & Griffith, 2005). This GshF (γ-GCS-GS, also EC 6.3.2.2) carries out both synthesis steps generating GSH from the respective amino acids while drawing the required energy from ATP (Fig. 6). Besides biosynthesis, some bacteria additionally or exclusively rely on GSH import from the media (Sherrill & Fahey, 1998; Thomas, 1984; Vergauwen, Pauwels, Vaneechoutte, & van Beeumen, 2003; Vergauwen et al., 2006). Another form of GSH synthesis applies for organisms lacking GshA. Here, the first enzyme of the proline biosynthesis pathway, a γ-glutamyl kinase (GK, Pro1 in eukaryotes, ProB in prokaryotes, EC 2.7.2.11) (Smith, Deutch, & Rushlow, 1984), can partly circumvent the first step by generating γ-glutamyl phosphate. This intermediate, in combination with L-cysteine, can form γ-L-glutamyl-L-cysteine which is then further converted via GshB to GSH (Fig. 6) (Spector, Labarre, & Toledano, 2001; Tang et al., 2015; Veeravalli, Boyd, Iverson, Beckwith, & Georgiou, 2011).

The presence of GSH in an Actinobacterium was first proven for the isoprene degrading *Rhodococcus* sp. AD45 in 1998 (van Hylckama et al., 1998). Genes for the potential GSH biosynthesis proteins GshA, GshB1 and GshB2 were identified on the plasmid carrying the genes for isoprene degradation (van Hylckama et al., 2000). Moreover, the transcription of these genes was shown to be upregulated upon exposure to isoprene or epoxyisoprene (Crombie et al., 2015). Recently, the presence of putative GshA and GshB was also shown on transcriptome and proteome level in another Actinobacterium, *Gordonia rubripertincta* CWB2 (Heine, Zimmerling, et al., 2018). It seems that GSH biosynthesis in Actinobacteria is conducted via the GshA and GshB dependent route although no explicit research exists on this topic so far. However, GshF homologs were only detected in two Actinobacteria (CPW45951 and RRF89859, query sequence: Q8DXM9) and therefore it seems unlikely that synthesis occurs along this pathway.

3.2 Glutathione recycling (GSSG-reductase)

Glutathione reductases (GR, GLR, GSH- or GSSG-reductases, EC 1.8.1.7) are flavin adenine dinucleotide (FAD)-dependent proteins and omnipresent

$$GSSG + NADPH + H^+ \xrightleftharpoons{GR} 2\ GSH + NADP^+$$

Fig. 7 Glutathione reductase (GR) catalyzed reduction of glutathione disulfide to glutathione.

as GSH itself. Their different isoforms are highly conserved across GSH-utilizing organisms generally forming homodimers of approximately 110 kDa of which each subunit binds one FAD. In prokaryotes, only a few have been characterized so far. GRs maintain the ratio of reduced and oxidized GSH in cells by catalyzing the NADPH-dependent reduction of GSSG to two molecules of GSH (Fig. 7) and thereby contributing to the cells redox state (Deponte, 2013; Ji, Barnwell, & Grunden, 2015; Massey & Williams, 1965; Racker, 1955; Schulz, Schirmer, Sachsenheimer, & Pai, 1978; Yamamoto, Kamio, & Higuchi, 1999).

For Actinobacteria glutathione reductase activity was measured in crude extracts of *Streptomyces coelicolor* cultures during exponential growth phase upon hydrogen peroxide stress and in untreated cultures showing activities of 923 ± 226 and $508 \pm 233\,\mathrm{U\,mg^{-1}}$, respectively (Lee, Hah, & Roe, 1993). Besides generating reduced GSH from respective amino acids the ability of its regeneration from GSSG is essential for any GSH-utilizing organism which gives GRs a crucial role in any cellular GSH system.

4. Glutathione utilizing enzymes among Actinobacteria

As the previous section focused on enzymes, that provide reduced GSH, this section will focus on known or putative enzyme candidates from Actinobacteria utilizing it. Generally, GSH can take part in a broad range of reactions catalyzed by a variety of enzymes. The glyoxalases I and II (GloI, GloII) are required for the detoxification of 2-oxoaldehydes (Deponte, 2013). Glutaredoxins (Grx) function as electron carriers using GSH as donor and thus, analogous to the thioredoxin system, play a role as antioxidants (Den Hengst & Buttner, 2008). The glutathione-dependent formaldehyde-activating enzyme (GFA) takes part in formaldehyde metabolism (Hopkinson et al., 2015). GSH oxidases generate hydrogen peroxide from GSH and molecular oxygen (Kusakabe, Kuninaka, & Yoshino, 1982) and peroxiredoxins (Prx) have similar functions to glutathione peroxidases which are described below (see Section 4.2) (Le Roes-Hill, Khan, & Burton, 2011). These glutathione-dependent enzymes, to mention a few, are generally well-studied but as GSH metabolism in Actinobacteria is still scarcely investigated, the presence of these enzymes in Actinobacteria is solely speculative. Thus, this section will focus on glutathione *S*-transferases (GSTs, Section 4.1) and glutathione peroxidases (GPXs, Section 4.2), respectively.

4.1 Glutathione *S*-transferase

Glutathione *S*-transferases (GSTs, EC 2.5.1.18) are ubiquitous enzymes that generally catalyze the nucleophilic addition of GSH to electrophilic residues of a broad range of substrates (Fig. 8) or directly bind compounds without GSH-addition mediating inactivation, degradation or excretion and thus being an important part of the cellular detoxification system concerning xenobiotics, antimicrobial and other exo- and endogenous substances. In bacteria, they also function in basal metabolism, oxidative stress response and the utilization of unusual carbon sources. GSTs are widely distributed and functionally very divers. Originating from at least three different protein folds, four different families have been categorized so far, namely, microsomal, mitochondrial, cytoplasmic and bacterial GSTs of which most likely all accounted for very different functions leading to a broad spectrum of their usage in an organism. All families have representatives in prokaryotes while the occurrence of bacterial GSTs is limited to prokaryotic organisms (Allocati et al., 2009; Hayes, Flanagan, & Jowsey, 2005; Oakley, 2005; Pearson, 2005; Sheehan, Meade, Foley, & Dowd, 2001). Cytoplasmic GSTs are further divided into classes according to their sequence homology and function. This classification of GST was already outlined by Allocati and coworkers in the past (Allocati et al., 2009). They also showed in 2012 that bacterial, Gram-negative GSTs are mainly found in class beta and that Gram-positive GSTs are mainly part of class xi (Allocati et al., 2012).

Though some putative GSTs were found in actinobacterial genomes (Allocati et al., 2012) only a few have been investigated. As already outlined above (see Section 2.2), the characterized GSTs IsoI and IsoJ are part of the isoprene metabolism of *Rhodococcus* sp. AD45 (Rui, Kwon, Reardon, & Wood, 2004; van Hylckama et al., 1998, 2000; van Hylckama, Kingma, Kruizinga, & Janssen, 1999). Likewise, *Gordonia rubripertincta* CWB2 utilizes GSTs (StyI and StyJ) for styrene degradation as shown for respective crude

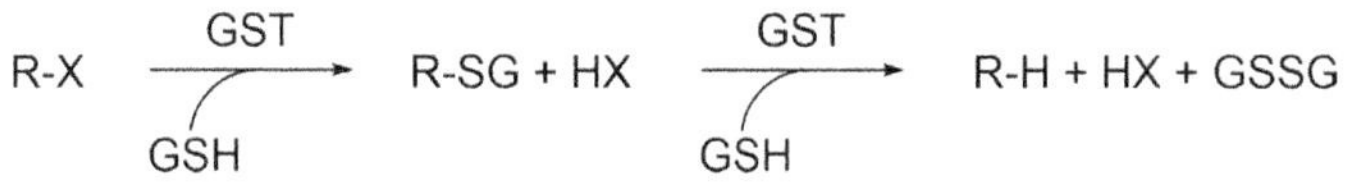

Fig. 8 Glutathione *S*-transferase (GST) catalyzed addition and removal of glutathione to a substrate with concurrent removal of a specific residue.

extracts (Heine, Zimmerling, et al., 2018). Another two GST isoforms were characterized in *Streptomyces griseus*, but since GSH was not detectable in this strain and activity was not tested for MSH their classification and role remain unclear (Dhar, Dhar, & Rosazza, 2003; Johnson et al., 2009).

An extended phylogenetic analysis with putative actinobacterial GSTs revealed that several of these could not be assigned to one of the known classes (Fig. 9). For instance, *Rhodococcus* sp. SAORIC-690 encodes for nine GST-homologs that are partially hard to classify. Further, the GSTs that are part of the isoprene and styrene degradation gene clusters also seem to form separate clusters, which was already recognized before (Allocati et al., 2009, 2012). While IsoI-like transferases are supposed to form a completely new class, IsoJ-like transferases possess some similarity to Ure2p-like proteins (Allocati et al., 2012). The phylogenetic difference of both transferases supports the finding that they are involved in different steps of the degradation pathways. Altogether, it is likely that the increasing number of actinobacterial GSTs reflects an ongoing expansion of the enzymatic toolset offering glutathione-dependent reactions.

4.2 Glutathione peroxidase

Since the emergence of oxygenic photosynthesis, the defense against oxidative stress is essential for aerobic organisms. As mentioned above (see Section 1), GSH most likely evolved as one way of repelling these stressors. Glutathione peroxidases (GPX; EC 1.11.1.9 and phospholipid-hydroperoxide glutathione peroxidases, PHGPX; EC 1.11.1.12) catalyze the GSH-dependent reduction of hydrogen peroxide to water, alkyl hydroperoxides to their corresponding alcohols or lipid hydroperoxides to the respective lipids (Fig. 10). Along with the peroxiredoxins (Prx; EC 1.11.1.15) they compose the protein family of non-heme thiol peroxidases which are generally essential for the removal of reactive oxygen species (ROS) in organisms (Brigelius-Flohé & Maiorino, 2013; Deponte, 2013; Herbette et al., 2002; Le Roes-Hill et al., 2011; Ursini, Maiorino, & Gregolin, 1985). Consequently, they can be found in all major kingdoms (Passardi et al., 2007). While most GPXs have an essential cysteine at the catalytic site (NS-GPX), especially human GPXs rely on a selenocysteine residue instead (Seleno-GPX). However, humans possess both types. For some annotated GPXs it has been postulated or even demonstrated that the reducing agent might not be GSH or that they accept other co-substrates beside it, e.g., thioredoxins or glutaredoxins, resulting in

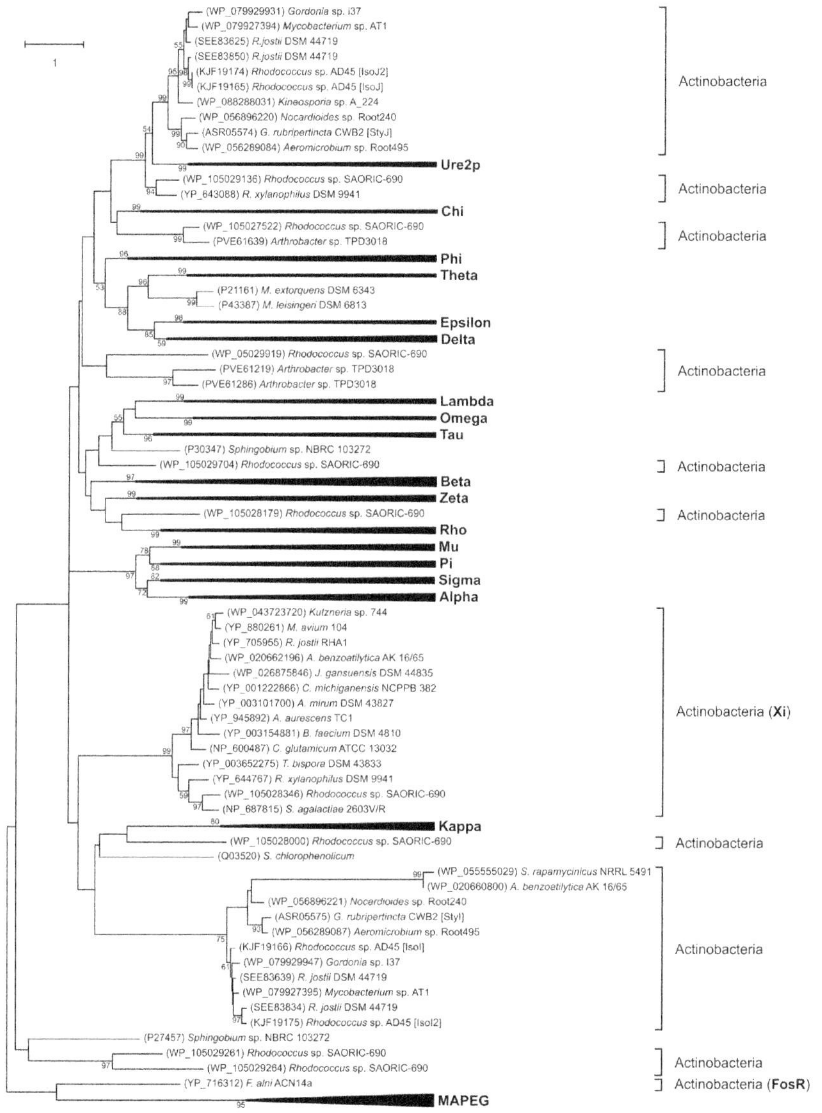

Fig. 9 Phylogenetic tree of (putative) glutathione *S*-transferases. Common GST-classes are indicated at collapsed branches and branches containing actinobacterial homologs are enclosed by brackets. Proteins, which cannot be assigned to a common GST-class, with proved function at protein level, are connected by blue branches. The evolutionary history was inferred by using the Maximum Likelihood method and JTT matrix-based model (Jones et al., 1992). The tree with the highest log likelihood (−51,981.06) is shown. The percentage of trees in which the associated taxa clustered together is shown next to the branches. Initial tree(s) for the heuristic search were obtained automatically by applying Neighbor-Join and BioNJ algorithms to a matrix of pairwise distances estimated using a JTT model, and then selecting the topology with superior log likelihood value. The tree is drawn to scale, with branch lengths measured in the number of substitutions per site. This analysis involved 122 amino acid sequences. There was a total of 490 positions in the final dataset. Evolutionary analyses were conducted in MEGA X (Kumar et al., 2018).

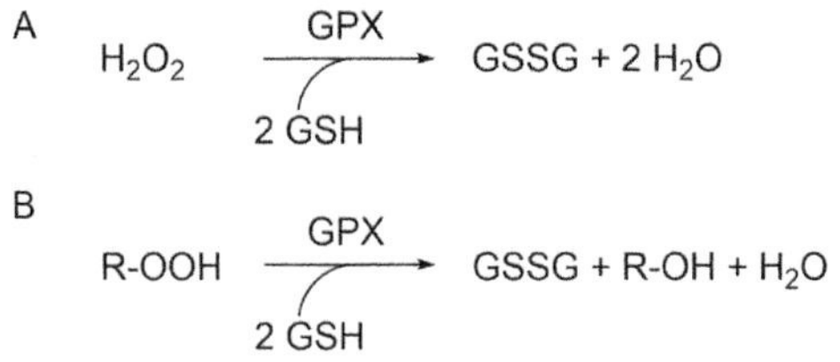

Fig. 10 Glutathione peroxidase (GPX) catalyzed reduction of hydrogen peroxide to water (A) and conversion of alkyl or lipid hydroperoxides to respective alcohols or lipids (B).

complications concerning their correct classification (Herbette, Roeckel-Drevet, & Drevet, 2007; Margis, Dunand, Teixeira, & Margis-Pinheiro, 2008; Navrot et al., 2006; Sztajer et al., 2001; Tanaka, Izawa, & Inoue, 2005).

Only very few studies have focused on prokaryotic glutathione peroxidases leave alone Actinobacteria. For example, an *in silico* analysis of the codon-usage of *Nocardia farcinica* predicts an increased expression of a putative glutathione peroxidase (Wu, Nie, & Zhang, 2006). However, so far there is no certain characterization of a GPX from an Actinobacterium.

5. Applied microbiology and biotechnology

With respect to environmental microbiology it can be stated that GSH and along the GSH-utilizing enzymes are involved in degradative pathways of naturally occurring and anthropogenically released compounds as isoprene and styrene, for example.

With respect to styrene degradation there are several pathways known (Tischler, 2015). However, the one including a glutathione *S*-transferase step is meant to be only rarely distributed among bacteria in general and thus also among Actinobacteria and thus plays not a major role for styrene metabolism—to be discussed.

In case of isoprene the role of bacteria and especially Actinobacteria was recently re-evaluated and this represents an environmental case of major importance (McGenity et al., 2018). It is the most abundantly released, biogenic, volatile organic compound. So, isoprene is produced by bacteria, fungi, animals, algae and plants. Respectively, terrestrial plants have the largest outcome and contribute most to the isoprene in the atmosphere. But it is also produced by human activity from petroleum in order to fuel various industries, thus also the atmospheric entry route has an anthropogenic origin. Isoprene is reactive and has effects on climate and health and thus its environmental role is of major relevance. It is absorbed on plant material and in soil but also to marine environments. Most directly connected to a degradative bacterial consortium to have a kind of balance between isoprene release and consumption. For the degradation, mostly Actinobacteria (*Arthrobacter*, *Nocardia*, *Rhodococcus*) have been described but also some Proteobacteria (*Alcaligenes*, *Klebsiella*, *Pseudomonas*). However, so far only some *rhodococci* are studied in detail and it is supposed that these play a key role in the isoprene cycle. Thus, more investigations are needed to understand the role of bacteria in isoprene production as well as consumption. Here, it needs to be verified if truly Actinobacteria represent the main group of degraders in terrestrial but also marine environments. So, this is of importance for microbial biotechnology and environmental microbiology with respect to soil remediation.

In case of styrene, the picture is totally different. Here the natural production is limited to some plant and fungal activities (Tischler, 2015). Most styrene is released due to anthropogenic activity and rapidly degraded by all kinds of bacteria and fungi (Oelschlägel, Zimmerling, & Tischler, 2018; Tischler, 2015). Interestingly, the styrene specific pathway including a GSH-transformation from the strain *Gordonia rubripertincta* CWB2 was found to be interesting in fine chemical synthesis (Heine, Zimmerling, et al., 2018; Oelschlägel, Kaschabek, Zimmerling, Schlömann, & Tischler, 2015; Oelschlägel et al., 2018). In this course, styrene degrading bacteria have been investigated to produce phenylacetic acid and related compounds (Table 3). Therefore, the bacteria were cultivated in the presence of styrene to induce the upper styrene metabolic pathway which converts styrene to phenylacetic acid (Tischler, 2015). The biomass obtained was then employed as biocatalyst in biotransformation studies while various styrene-like molecules were fed. In most cases, the corresponding phenylacetic acid-derivatives were not just produced but also secreted into the medium and thus accumulated. Further, this allowed an easy product determination as well as following the progress of biotransformation over time. Only slow or no further conversion of these products was determined. Especially, the halogenated phenylacetic acids were not further transformed by the used bacteria. The best producer was *Pseudomonas fluorescens* ST as it allowed to produce the corresponding acids (3-chloro-, 4-chloro-, 4-fluoro-, α-methyl- as well as 4-chloro-α-methylphenylacetic acid) at high concentrations in short times compared to the other strains. This strain could even be used for more than 300 days as an active biocatalyst. Also, this strain was highly tolerant toward the styrene

Table 3 Styrene degrading bacteria as whole-cell biocatalysts (Oelschlägel et al., 2015).

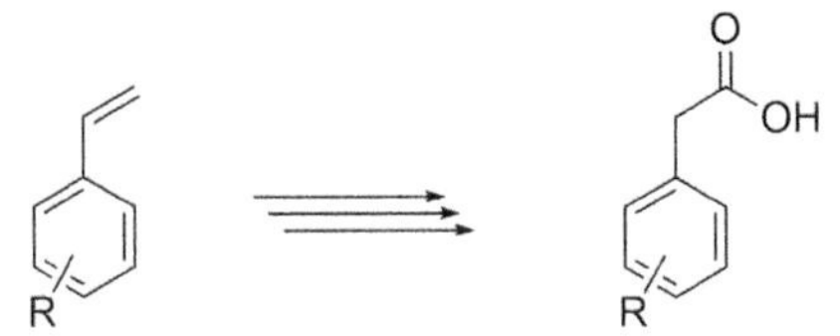

Styrene **Phenylacetic acid**

Strain	**Pathway**	**Product (yield [μmol g_{cds}^{-1}])**			
		R=3-Cl	R=4-Cl	R=4-Fl	R=α-Me
Proteobacteria					
Pseudomonas fluorescens ST	StyABCD	(235)	(520)	(780)	(480)
Sphingopyxis fribergensis Kp5.2	StyABCD	(95)	(100)	(90)	(150)
Actinobacteria					
Rhodococcus opacus 1CP	StyABCD	(15)	(30)	(5)	(10)
Gordonia rubripertincta CWB2	GSH-dependent	(15)	(70)	(35)	(30)

The StyABCD is the conventional styrene degradation pathway employing a monooxygenase, isomerase and dehydrogenase while the GSH-dependent route substitutes the isomerase by a glutathione *S*-transferase. Selected results are presented.

substrates and thus this might be a chance to develop or even to evolve a better phenylacetic acid producer. However, it needs to be mentioned that bulkier substrates were not converted or only very slowly by most of the tested strains. Here, the only exception was the strain *Gordonia rubripertincta* CWB2.

Gordonia rubripertincta CWB2 was found to be the only strain converting bulkier styrene analogs as 4-isobutyl-α-methylstyrene (Fig. 11). This substrate had been chosen on the base of a retrosynthetic approach searching for the respective substrate to allow production of 4-isobutyl-α-methylphenylacetic acid, which is also known as ibuprofen. Compared to other styrene derivatives this was also a poor substrate but it was converted (about 13 mg in 23 days) (Oelschlägel et al., 2015). And as strain CWB2 was the only one accepting and converting this substrate the metabolic reason was discussed until we solved the actual styrene degradation pathway in this Actinobacterium (Heine, Zimmerling, et al., 2018). Thus, it became obvious that the change from styrene oxide isomerase (SOI) to glutathione *S*-transferase to convert the intermediate styrene oxide was crucial.

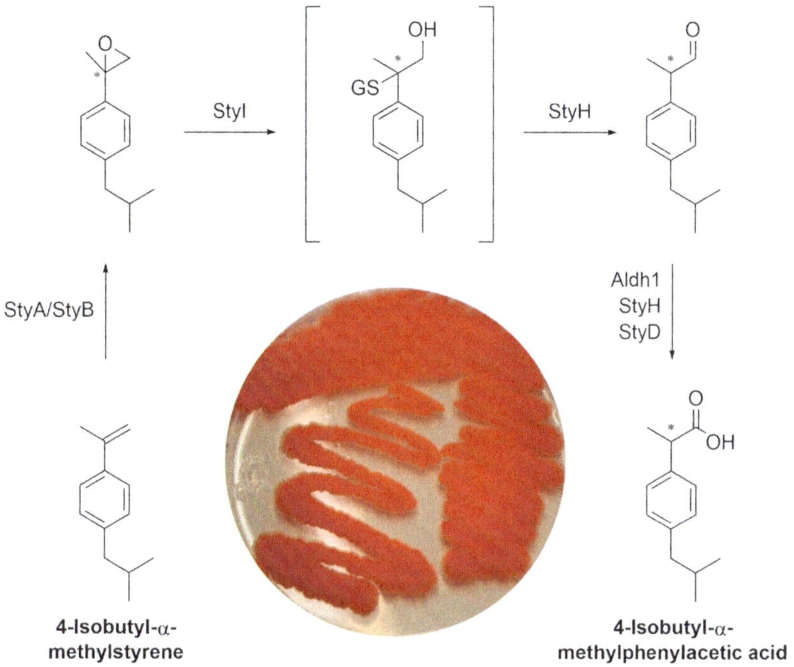

Fig. 11 An ibuprofen producing Actinobacterium. The proposed styrene degradation pathway of *Gordonia rubripertincta* CWB2 (plated on solid media in the center of the figure) enables this strain the conversion of 4-isobutyl-α-methylstyrene to 4-isobutyl-α-methylphenylacetic acid (Oelschlägel et al., 2015).

All the other strains listed in Table 3 employ an SOI and thus convert styrene oxide into phenylacetaldehyde. This enzyme seems to present the bottleneck as it has only a narrow substrate spectrum (Miyamoto, Okuro, & Ohta, 2007; Oelschlägel et al., 2015; Oelschlägel, Richter, Stuhr, Hofmann, & Schlömann, 2017) and bulkier epoxides cannot be converted. The glutathione *S*-transferase of strain CWB2 must have a relaxed substrate spectrum and this is currently under investigation. Thus, this GSH-dependent enzyme seems a promising candidate to activate various epoxides either for degradative studies or for artificial pathways toward novel products.

6. Detoxification by means of glutathione

Toxic molecules are often activated by P450 monooxygenases in organisms, prokaryotes and eukaryotes, and in order to transport or even to eliminate these reactive intermediates a conjugation reaction is needed. Here, GSH is frequently used by means of glutathione *S*-transferases and the formed adducts a further processed (Allocati et al., 2009). However, no details for detoxification reactions among Actinobacteria are reported to the best of our knowledge. But, these reactions likely occur and await elucidation.

7. Conclusion

Starting from general properties of glutathione and its distribution among organisms, it becomes evident that it can be used for several biological activities. Among the bacteria and thus Actinobacteria it may have actually three major roles: redox shuttle, detoxifying agent and participation in metabolic pathways. Actually, the latter two seem to have coevolved as the conjugation of epoxides (e.g., styrene oxide) is a typical detoxification route but also occurs here in a degradation pathway. Genome analysis and literature analyses provide hints that glutathione use is not common but somewhat frequent among Actinobacteria. We found biochemical evidence for isoprene oxide and styrene oxide conversion or genetic opportunities for various other glutathione functions mainly of detoxification and redox balancing. Thus, it can be concluded that glutathione is a powerful tool used by Actinobacteria next to mycothiol which was so far described as the glutathione alternative used by Actinobacteria.

Acknowledgments

A part of this research and especially Anna C. Lienkamp was funded by the DFG Research Training Group GRK 2341 "Microbial Substrate Conversion (MiCon)." The project was supported in the frame of the MERCUR starting grant An-2018-0044. Thomas Heine was partially funded by a grant of the European Social Fund and Saxonian Government (GETGEOWEB: 100101363). Dirk Tischler was supported by the Federal Ministry for Innovation, Science and Research of North Rhine-Westphalia (PtJ-TRI/1411ng006)-ChemBioCat.

References

Allen, J., & Bradley, R. D. (2011). Effects of oral glutathione supplementation on systemic oxidative stress biomarkers in human volunteers. *Journal of Alternative and Complementary Medicine (New York, N.Y.)*, *17*(9), 827–833. https://doi.org/10.1089/acm.2010.0716.

Allocati, N., Federici, L., Masulli, M., & Di Ilio, C. (2009). Glutathione transferases in bacteria. *The FEBS Journal*, *276*(1), 58–75. https://doi.org/10.1111/j.1742-4658.2008.06743.x.

Allocati, N., Federici, L., Masulli, M., & Di Ilio, C. (2012). Distribution of glutathione transferases in Gram-positive bacteria and Archaea. *Biochimie*, *94*(3), 588–596. https://doi.org/10.1016/j.biochi.2011.09.008.

Alvarez, L. A., Exton, D. A., Timmis, K. N., Suggett, D. J., & McGenity, T. J. (2009). Characterization of marine isoprene-degrading communities. *Environmental Microbiology, 11*(12), 3280–3291. https://doi.org/10.1111/j.1462-2920.2009.02069.x.

Anderson, M. E. (1998). Glutathione: An overview of biosynthesis and modulation. *Chemico-Biological Interactions, 111–112*, 1–14. https://doi.org/10.1016/S0009-2797(97)00146-4.

Biswas, S. K., & Rahman, I. (2009). Environmental toxicity, redox signaling and lung inflammation: The role of glutathione. *Molecular Aspects of Medicine, 30*(1–2), 60–76. https://doi.org/10.1016/j.mam.2008.07.001.

Brigelius-Flohé, R., & Maiorino, M. (2013). Glutathione peroxidases. *Biochimica et Biophysica Acta (BBA)—General Subjects, 1830*(5), 3289–3303. https://doi.org/10.1016/j.bbagen.2012.11.020.

Cameron, J. C., & Pakrasi, H. B. (2010). Essential role of glutathione in acclimation to environmental and redox perturbations in the cyanobacterium *Synechocystis* sp. Pcc 6803. *Plant Physiology, 154*(4), 1672–1685. https://doi.org/10.1104/pp.110.162990.

Carrión, O., Larke-Mejía, N. L., Gibson, L., Farhan Ul Haque, M., Ramiro-García, J., McGenity, T. J., et al. (2018). Gene probing reveals the widespread distribution, diversity and abundance of isoprene-degrading bacteria in the environment. *Microbiome, 6*(1), 219. https://doi.org/10.1186/s40168-018-0607-0.

Copley, S. D., & Dhillon, J. K. (2002). Lateral gene transfer and parallel evolution in the history of glutathione biosynthesis genes. *Genome Biology, 3*(5), research0025. https://doi.org/10.1186/gb-2002-3-5-research0025.

Crombie, A. T., Khawand, M. E., Rhodius, V. A., Fengler, K. A., Miller, M. C., Whited, G. M., et al. (2015). Regulation of plasmid-encoded isoprene metabolism in *Rhodococcus*, a representative of an important link in the global isoprene cycle. *Environmental Microbiology, 17*(9), 3314–3329. https://doi.org/10.1111/1462-2920.12793.

Crombie, A. T., Larke-Mejia, N. L., Emery, H., Dawson, R., Pratscher, J., Murphy, G. P., et al. (2018). Poplar phyllosphere harbors disparate isoprene-degrading bacteria. *Proceedings of the National Academy of Sciences of the United States of America, 115*, 13081–13086. https://doi.org/10.1073/pnas.1812668115.

Crombie, A. T., Mejia-Florez, N. L., McGenity, T. J., & Murrell, J. C. (2019). Genetics and ecology of isoprene degradation. In F. Rojo (Ed.), *Aerobic utilization of hydrocarbons, oils, and lipids* (pp. 557–571). Springer International Publishing. https://doi.org/10.1007/978-3-319-50418-6_27.

Den Hengst, C. D., & Buttner, M. J. (2008). Redox control in actinobacteria. *Biochimica et Biophysica Acta, 1780*(11), 1201–1216. https://doi.org/10.1016/j.bbagen.2008.01.008.

Deponte, M. (2013). Glutathione catalysis and the reaction mechanisms of glutathione-dependent enzymes. *Biochimica et Biophysica Acta, 1830*(5), 3217–3266. https://doi.org/10.1016/j.bbagen.2012.09.018.

Dhar, K., Dhar, A., & Rosazza, J. P. N. (2003). Glutathione S-transferase isoenzymes from *Streptomyces griseus*. *Applied and Environmental Microbiology, 69*(1), 707–710. https://doi.org/10.1128/AEM.69.1.707-710.2003.

El Khawand, M., Crombie, A. T., Johnston, A., Vavlline, D. V., McAuliffe, J. C., Latone, J. A., et al. (2016). Isolation of isoprene degrading bacteria from soils, development of *isoA* gene probes and identification of the active isoprene-degrading soil community using DNA-stable isotope probing. *Environmental Microbiology, 18*(8), 2743–2753. https://doi.org/10.1111/1462-2920.13345.

Ellis, H. R. (2010). The FMN-dependent two-component monooxygenase systems. *Archives of Biochemistry and Biophysics, 497*(1–2), 1–12. https://doi.org/10.1016/j.abb.2010.02.007.

Fahey, R. C. (2013). Glutathione analogs in prokaryotes. *Biochimica et Biophysica Acta, 1830*(5), 3182–3198. https://doi.org/10.1016/j.bbagen.2012.10.006.

Fahey, R. C., Brown, W. C., Adams, W. B., & Worsham, M. B. (1978). Occurrence of glutathione in bacteria. *Journal of Bacteriology*, *133*(3), 1126–1129.

Fahey, R. C., Buschbacher, R. M., & Newton, G. L. (1987). The evolution of glutathione metabolism in phototrophic microorganisms. *Journal of Molecular Evolution*, *25*, 81–88. https://doi.org/10.1007/BF02100044.

Flohé, L. (2013). The fairytale of the GSSG/GSH redox potential. *Biochimica et Biophysica Acta*, *1830*(5), 3139–3142. https://doi.org/10.1016/j.bbagen.2012.10.020.

Garg, A., Soni, B., Singh Paliya, B., Verma, S., & Jadaun, V. (2013). Low molecular weight thiols: Glutathione (GSH), mycothiol (MSH) potential antioxidant compound from actinobacteria. *Journal of Applied Pharmaceutical Science*, *3*, 117–120.

Gopal, S., Borovok, I., Ofer, A., Yanku, M., Cohen, G., Goebel, W., et al. (2005). A multidomain fusion protein in *Listeria monocytogenes* catalyzes the two primary activities for glutathione biosynthesis. *Journal of Bacteriology*, *187*(11), 3839–3847. https://doi.org/10.1128/JB.187.11.3839-3847.2005.

Gray, C. M., Helmig, D., & Fierer, N. (2015). Bacteria and fungi associated with isoprene consumption in soil. *Elementa: Science of the Anthropocene*, *3*(53). https://doi.org/10.12952/journal.elementa.000053.

Grund, E., Knorr, C., & Eichenlaub, R. (1990). Catabolism of benzoate and monohydroxylated benzoates by *Amycolatopsis* and *Streptomyces* spp. *Applied and Environmental Microbiology*, *56*(5), 1459–1464.

Hand, C. E., Auzanneau, F.-I., & Honek, J. F. (2006). Conformational analyses of mycothiol, a critical intracellular glycothiol in Mycobacteria. *Carbohydrate Research*, *341*(9), 1164–1173. https://doi.org/10.1016/j.carres.2006.03.020.

Hasanuzzaman, M., Nahar, K., Anee, T. I., & Fujita, M. (2017). Glutathione in plants: Biosynthesis and physiological role in environmental stress tolerance. *Physiology and Molecular Biology of Plants: An International Journal of Functional Plant Biology*, *23*(2), 249–268. https://doi.org/10.1007/s12298-017-0422-2.

Hayes, J. D., Flanagan, J. U., & Jowsey, I. R. (2005). Glutathione transferases. *Annual Review of Pharmacology and Toxicology*, *45*, 51–88. https://doi.org/10.1146/annurev.pharmtox.45.120403.095857.

Heine, T., van Berkel, W. J. H., Gassner, G., van Pée, K.-H., & Tischler, D. (2018). Two-component FAD-dependent monooxygenases: Current knowledge and biotechnological opportunities. *Biology*, 7(3), 42. https://doi.org/10.3390/biology7030042.

Heine, T., Zimmerling, J., Ballmann, A., Kleeberg, S. B., Rückert, C., Busche, T., et al. (2018). On the enigma of glutathione-dependent styrene degradation in *Gordonia rubripertincta* CWB2. *Applied and Environmental Microbiology*, *84*(9), e00154–18. https://doiorg/10.1128/AEM.00154-18.

Helmann, J. D. (2011). Bacillithiol, a new player in bacterial redox homeostasis. *Antioxidants & Redox Signaling*, *15*(1), 123–133. https://doi.org/10.1089/ars.2010.3562.

Herbette, S., Lenne, C., Leblanc, N., Julien, J.-L., Drevet, J. R., & Roeckel-Drevet, P. (2002). Two GPX-like proteins from *Lycopersicon esculentum* and *Helianthus annuus* are antioxidant enzymes with phospholipid hydroperoxide glutathione peroxidase and thioredoxin peroxidase activities. *European Journal of Biochemistry*, *269*(9), 2414–2420. https://doi.org/10.1046/j.1432-1033.2002.02905.x.

Herbette, S., Roeckel-Drevet, P., & Drevet, J. R. (2007). Seleno-independent glutathione peroxidases. More than simple antioxidant scavengers. *The FEBS Journal*, *274*(9), 2163–2180. https://doi.org/10.1111/j.1742-4658.2007.05774.x.

Hilker, R., Stadermann, K. B., Schwengers, O., Anisiforov, E., Jaenicke, S., Weisshaar, B., et al. (2016). Readxplorer 2-detailed read mapping analysis and visualization from one single source. *Bioinformatics (Oxford, England)*, *32*(24), 3702–3708. https://doi.org/10.1093/bioinformatics/btw541.

Hopkins, F. G. (1921). On an autoxidisable constituent of the cell. *Biochemical Journal, 15*(2), 286–305. https://doi.org/10.1042/bj0150286.
Hopkins, F. G. (1927). On the isolation of glutathione. *The Journal of Biological Chemistry, 72*(1), 185–187. http://www.jbc.org/content/72/1/185.short.
Hopkins, F. G., & Dixon, M. (1922). On glutathione: II. A thermostable oxidation-reduction system. *The Journal of Biological Chemistry, 54*(3), 527–563. http://www.jbc.org/content/54/3/527.short.
Hopkins, F. G., & Harris, L. J. (1929). On glutathione: A reinvestigation. *The Journal of Biological Chemistry, 84*(1), 269–320. http://www.jbc.org/content/84/1/269.short.
Hopkinson, R. J., Leung, I. K. H., Smart, T. J., Rose, N. R., Henry, L., Claridge, T. D. W., et al. (2015). Studies on the glutathione-dependent formaldehyde-activating enzyme from *Paracoccus denitrificans*. *PLoS One, 10*(12), e0145085. https://doi.org/10.1371/journal.pone.0145085.
Hunter, G., & Eagles, B. A. (1927). Glutathione: A critical study. *The Journal of Biological Chemistry, 72*(1), 147–166. http://www.jbc.org/content/72/1/147.short.
Janowiak, B. E., & Griffith, O. W. (2005). Glutathione synthesis in *Streptococcus agalactiae*. One protein accounts for gamma-glutamylcysteine synthetase and glutathione synthetase activities. *The Journal of Biological Chemistry, 280*(12), 11829–11839. https://doi.org/10.1074/jbc.M414326200.
Ji, M., Barnwell, C. V., & Grunden, A. M. (2015). Characterization of recombinant glutathione reductase from the psychrophilic antarctic bacterium *Colwellia psychrerythraea*. *Extremophiles, 19*(4), 863–874. https://doi.org/10.1007/s00792-015-0762-1.
Johnson, T., Newton, G. L., Fahey, R. C., & Rawat, M. (2009). Unusual production of glutathione in actinobacteria. *Archives of Microbiology, 191*(1), 89–93. https://doi.org/10.1007/s00203-008-0423-1.
Johnston, A., Crombie, A. T., El Khawand, M., Sims, L., Whited, G. M., McGenity, T. J., et al. (2017). Identification and characterisation of isoprene-degrading bacteria in an estuarine environment. *Environmental Microbiology, 19*, 3526–3537. https://doi.org/10.1111/1462-2920.13842.
Jones, D. T., Taylor, W. R., & Thornton, J. M. (1992). The rapid generation of mutation data matrices from protein sequences. *Computer Applications in the Biosciences: CABIOS, 8*(3), 275–282.
Kato, H., Tanaka, T., Nishioka, T., Kimura, A., & Oda, J. (1988). Role of cysteine residues in glutathione synthetase from *Escherichia coli* B. chemical modification and oligonucleotide site-directed mutagenesis. *The Journal of Biological Chemistry, 263*(24), 11646–11651.
Kumar, S., Stecher, G., Li, M., Knyaz, C., & Tamura, K. (2018). Mega X: Molecular evolutionary genetics analysis across computing platforms. *Molecular Biology and Evolution, 35*(6), 1547–1549. https://doi.org/10.1093/molbev/msy096.
Kusakabe, H., Kuninaka, A., & Yoshino, H. (1982). Purification and properties of a new enzyme, glutathione oxidase from *Penicillium* sp. K-6-5. *Agricultural and Biological Chemistry, 46*(8), 2057–2067. https://doi.org/10.1080/00021369.1982.10865382.
Le Roes-Hill, M., Khan, N., & Burton, S. G. (2011). Actinobacterial peroxidases: An unexplored resource for biocatalysis. *Applied Biochemistry and Biotechnology, 164*(5), 681–713. https://doi.org/10.1007/s12010-011-9167-5.
Leahy, J. G., Batchelor, P. J., & Morcomb, S. M. (2003). Evolution of the soluble diiron monooxygenases. *FEMS Microbiology Reviews, 27*(4), 449–479. https://doi.org/10.1016/S0168-6445(03)00023-8.
Lee, J.-S., Hah, Y.-C., & Roe, J.-H. (1993). The induction of oxidative enzymes in *Streptomyces coelicolor* upon hydrogen peroxide treatment. *Journal of General Microbiology, 139*(5), 1013–1018. https://doi.org/10.1099/00221287-139-5-1013.

Lu, S. C. (2009). Regulation of glutathione synthesis. *Molecular Aspects of Medicine*, *30*(1–2), 42–59. https://doi.org/10.1016/j.mam.2008.05.005.

Margis, R., Dunand, C., Teixeira, F. K., & Margis-Pinheiro, M. (2008). Glutathione peroxidase family—An evolutionary overview. *The FEBS Journal*, *275*(15), 3959–3970. https://doi.org/10.1111/j.1742-4658.2008.06542.x.

Massey, V., & Williams, C. H. (1965). On the reaction mechanism of yeast glutathione reductase. *The Journal of Biological Chemistry*, *240*(11), 4470–4480.

McGenity, T. J., Crombie, A. T., & Murrell, J. C. (2018). Microbial cycling of isoprene, the most abundantly produced biological volatile organic compound on Earth. *The ISME Journal*, *12*(4), 931–941. https://doi.org/10.1038/s41396-018-0072-6.

Meister, A., & Anderson, M. E. (1983). Glutathione. *Annual Review of Biochemistry*, *52*, 711–760. https://doi.org/10.1146/annurev.bi.52.070183.003431.

Miyamoto, K., Okuro, K., & Ohta, H. (2007). Substrate specificity and reaction mechanism of recombinant styrene oxide isomerase from *Pseudomonas putida* S12. *Tetrahedron Letters*, *48*(18), 3255–3257. https://doi.org/10.1016/j.tetlet.2007.03.016.

Mukhopadhyay, R., & Rosen, B. P. (2002). Arsenate reductases in prokaryotes and eukaryotes. *Environmental Health Perspectives*, *110*(Suppl. 5), 745–748. https://doi.org/10.1289/ehp.02110s5745.

Navrot, N., Collin, V., Gualberto, J., Gelhaye, E., Hirasawa, M., Rey, P., et al. (2006). Plant glutathione peroxidases are functional peroxiredoxins distributed in several subcellular compartments and regulated during biotic and abiotic stresses. *Plant Physiology*, *142*(4), 1364–1379. https://doi.org/10.1104/pp.106.089458.

Newton, G. L., Arnold, K., Price, M. S., Sherrill, C., Delcardayre, S. B., Aharonowitz, Y., et al. (1996). Distribution of thiols in microorganisms: Mycothiol is a major thiol in most actinomycetes. *Journal of Bacteriology*, *178*(7), 1990–1995. https://doi.org/10.1128/jb.178.7.1990-1995.1996.

Newton, G. L., Buchmeier, N., & Fahey, R. C. (2008). Biosynthesis and functions of mycothiol, the unique protective thiol of actinobacteria. *Microbiology and Molecular Biology Reviews*, *72*(3), 471–494. https://doi.org/10.1128/MMBR.00008-08.

Noctor, G., Mhamdi, A., Chaouch, S., Han, Y., Neukermans, J., Marquez-Garcia, B., et al. (2012). Glutathione in plants: An integrated overview. *Plant, Cell & Environment*, *35*(2), 454–484. https://doi.org/10.1111/j.1365-3040.2011.02400.x.

Oakley, A. J. (2005). Glutathione transferases: New functions. *Current Opinion in Structural Biology*, *15*(6), 716–723. https://doi.org/10.1016/j.sbi.2005.10.005.

Oelschlägel, M., Kaschabek, S. R., Zimmerling, J., Schlömann, M., & Tischler, D. (2015). Co-metabolic formation of substituted phenylacetic acids by styrene-degrading bacteria. *Biotechnology Reports (Amsterdam, Netherlands)*, *6*, 20–26. https://doi.org/10.1016/j.btre.2015.01.003.

Oelschlägel, M., Richter, L., Stuhr, A., Hofmann, S., & Schlömann, M. (2017). Heterologous production of different styrene oxide isomerases for the highly efficient synthesis of phenylacetaldehyde. *Journal of Biotechnology*, *252*, 43–49. https://doi.org/10.1016/j.jbiotec.2017.04.038.

Oelschlägel, M., Zimmerling, J., & Tischler, D. (2018). A review: The styrene metabolizing cascade of side-chain oxygenation as biotechnological basis to gain various valuable compounds. *Frontiers in Microbiology*, *9*, 490. https://doi.org/10.3389/fmicb.2018.00490.

Ondarza, R. N., Rendón, J. L., & Ondarza, M. (1983). Glutathione reductase in evolution. *Journal of Molecular Evolution*, *19*(5), 371–375. https://doi.org/10.1007/BF02101641.

Pacifici, G. M., Warholm, M., Guthenberg, C., Mannervik, B., & Rane, A. (1987). Detoxification of styrene oxide by human liver glutathione transferase. *Human Toxicology*, *6*(6), 483–489.

Passardi, F., Theiler, G., Zamocky, M., Cosio, C., Rouhier, N., Teixera, F., et al. (2007). PeroxiBase: The peroxidase database. *Phytochemistry, 68*(12), 1605–1611. https://doi.org/10.1016/j.phytochem.2007.04.005.

Pastore, A., Federici, G., Bertini, E., & Piemonte, F. (2003). Analysis of glutathione: Implication in redox and detoxification. *Clinica Chimica Acta; International Journal of Clinical Chemistry, 333*(1), 19–39. https://doi.org/10.1016/S0009-8981(03)00200-6.

Pearson, W. R. (2005). Phylogenies of glutathione transferase families. In H. Sies & L. Packer (Eds.), *Methods in enzymology. gluthione transferases and gamma-glutamyl transpeptidases Vol. 401*. (pp. 186–204). Elsevier. https://doi.org/10.1016/S0076-6879(05)01012-8.

Pearson, W. R. (2013). An introduction to sequence similarity ("Homology") searching. *Current Protocols in Bioinformatics, 42*, 3.1.1–3.1.8. https://doi.org/10.1002/0471250953.bi0301s42.

Pócsi, I., Prade, R. A., & Penninckx, M. J. (2004). Glutathione, altruistic metabolite in fungi. *Advances in Microbial Physiology, 49*, 1–76. https://doi.org/10.1016/S0065-2911(04)49001-8.

Potter, A. J., Trappetti, C., & Paton, J. C. (2012). *Streptococcus pneumoniae* uses glutathione to defend against oxidative stress and metal ion toxicity. *Journal of Bacteriology, 194*(22), 6248–6254. https://doi.org/10.1128/JB.01393-12.

Racker, E. (1955). Glutathione reductase from bakers' yeast and beef liver. *The Journal of Biological Chemistry, 217*(2), 855–865.

Retamal-Morales, G., Mehnert, M., Schwabe, R., Tischler, D., Schlömann, M., & Levicán, G. J. (2017). Genomic characterization of the arsenic-tolerant actinobacterium, *Rhodococcus erythropolis* S43. *Solid State Phenomena, 262*, 660–663. https://doi.org/10.4028/www.scientific.net/SSP.262.660.

Rui, L., Kwon, Y. M., Reardon, K. F., & Wood, T. K. (2004). Metabolic pathway engineering to enhance aerobic degradation of chlorinated ethenes and to reduce their toxicity by cloning a novel glutathione S-transferase, an evolved toluene o-monooxygenase, and gamma-glutamylcysteine synthetase. *Environmental Microbiology, 6*(5), 491–500. https://doi.org/10.1111/j.1462-2920.2004.00586.x.

Sakuda, S., Zhou, Z. Y., & Yamada, Y. (1994). Structure of a novel disulfide of 2-(N-acetylcysteinyl)amido-2-deoxy-alpha-D-glucopyranosyl-myo-inositol produced by *Streptomyces* sp. *Bioscience, Biotechnology, and Biochemistry, 58*(7), 1347–1348. https://doi.org/10.1271/bbb.58.1347.

Schulz, G. E., Schirmer, R. H., Sachsenheimer, W., & Pai, E. F. (1978). The structure of the flavoenzyme glutathione reductase. *Nature, 273*(5658), 120–124. https://doi.org/10.1038/273120a0.

Sen, C. K., Atalay, M., & Hänninen, O. (1994). Exercise-induced oxidative stress: Glutathione supplementation and deficiency. *Journal of Applied Physiology (Bethesda, Md.: 1985), 77*(5), 2177–2187. https://doi.org/10.1152/jappl.1994.77.5.2177.

Sheehan, D., Meade, G., Foley, V. M., & Dowd, C. A. (2001). Structure, function and evolution of glutathione transferases: Implications for classification of non-mammalian members of an ancient enzyme superfamily. *Biochemical Journal, 360*(Pt. 1), 1–16. https://doi.org/10.1042/0264-6021:3600001.

Sherrill, C., & Fahey, R. C. (1998). Import and metabolism of glutathione by *Streptococcus mutans*. *Journal of Bacteriology, 180*(6), 1454–1459.

Sies, H. (1999). Glutathione and its role in cellular functions. *Free Radical Biology and Medicine, 27*(9–10), 916–921. https://doi.org/10.1016/S0891-5849(99)00177-X.

Simoni, R. D., Hill, R. L., & Vaughan, M. (2002). The discovery of glutathione by F. Gowland Hopkins and the beginning of biochemistry at Cambridge University. *The Journal of Biological Chemistry, 277*(24), e13. http://www.jbc.org/content/277/24/e13.short.

Smirnova, G. V., & Oktyabrsky, O. N. (2005). Glutathione in bacteria. *Biochemistry (Moscow), 70*(11), 1199–1211. https://doi.org/10.1007/s10541-005-0248-3.

Smith, C. J., Deutch, A. H., & Rushlow, K. E. (1984). Purification and characteristics of a gamma-glutamyl kinase involved in *Escherichia coli* proline biosynthesis. *Journal of Bacteriology, 157*(2), 545–551.

Spector, D., Labarre, J., & Toledano, M. B. (2001). A genetic investigation of the essential role of glutathione: Mutations in the proline biosynthesis pathway are the only suppressors of glutathione auxotrophy in yeast. *The Journal of Biological Chemistry, 276*(10), 7011–7016. https://doi.org/10.1074/jbc.M009814200.

Srivastva, N., Singh, A., Bhardwaj, Y., & Dubey, S. K. (2018). Biotechnological potential for degradation of isoprene: A review. *Critical Reviews in Biotechnology, 38*(4), 587–599. https://doi.org/10.1080/07388551.2017.1379467.

Stenersen, J., Kobro, S., Bjerke, M., & Arend, U. (1987). Glutathione transferases in aquatic and terrestrial animals from nine phyla. *Comparative Biochemistry and Physiology Part C: Comparative Pharmacology, 86*(1), 73–82. https://doi.org/10.1016/0742-8413(87)90147-2.

Suttisansanee, U., & Honek, J. F. (2011). Bacterial glyoxalase enzymes. *Seminars in Cell & Developmental Biology, 22*(3), 285–292. https://doi.org/10.1016/j.semcdb.2011.02.004.

Sztajer, H., Gamain, B., Aumann, K. D., Slomianny, C., Becker, K., Brigelius-Flohé, R., et al. (2001). The putative glutathione peroxidase gene of *Plasmodium falciparum* codes for a thioredoxin peroxidase. *The Journal of Biological Chemistry, 276*(10), 7397–7403. https://doi.org/10.1074/jbc.M008631200.

Tanaka, T., Izawa, S., & Inoue, Y. (2005). Gpx2, encoding a phospholipid hydroperoxide glutathione peroxidase homologue, codes for an atypical 2-Cys peroxiredoxin in *Saccharomyces cerevisiae*. *The Journal of Biological Chemistry, 280*(51), 42078–42087. https://doi.org/10.1074/jbc.M508622200.

Tang, L., Wang, W., Zhou, W., Cheng, K., Yang, Y., Liu, M., et al. (2015). Three-pathway combination for glutathione biosynthesis in *Saccharomyces cerevisiae*. *Microbial Cell Factories, 14*, 139. https://doi.org/10.1186/s12934-015-0327-0.

Thomas, E. L. (1984). Disulfide reduction and sulfhydryl uptake by *Streptococcus mutans*. *Journal of Bacteriology, 157*(1), 240–246.

Tischler, D. (2015). *Microbial styrene degradation*. Springer Verlag.

Umarov, R. K., & Solovyev, V. V. (2017). Recognition of prokaryotic and eukaryotic promoters using convolutional deep learning neural networks. *PLoS One, 12*(2). e0171410https://doi.org/10.1371/journal.pone.0171410.

Ursini, F., Maiorino, M., & Gregolin, C. (1985). The selenoenzyme phospholipid hydroperoxide glutathione peroxidase. *Biochimica et Biophysica Acta (BBA)—General Subjects, 839*(1), 62–70. https://doi.org/10.1016/0304-4165(85)90182-5.

van Hylckama, V. J., Kingma, J., Kruizinga, W., & Janssen, D. B. (1999). Purification of a glutathione S-transferase and a glutathione conjugate-specific dehydrogenase involved in isoprene metabolism in *Rhodococcus* sp. strain AD45. *Journal of Bacteriology, 181*(7), 2094–2101.

van Hylckama, V. J., Kingma, J., van den Wijngaard, A. J., & Janssen, D. B. (1998). A glutathione S-transferase with activity towards cis-1, 2-dichloroepoxyethane is involved in isoprene utilization by *Rhodococcus* sp. strain AD45. *Applied and Environmental Microbiology, 64*(8), 2800–2805.

van Hylckama, V. J., Leemhuis, H., Spelberg, J. H. L., & Janssen, D. B. (2000). Characterization of the gene cluster involved in isoprene metabolism in *Rhodococcus* sp. strain AD45. *Journal of Bacteriology, 182*(7), 1956–1963.

Veeravalli, K., Boyd, D., Iverson, B. L., Beckwith, J., & Georgiou, G. (2011). Laboratory evolution of glutathione biosynthesis reveals natural compensatory pathways. *Nature Chemical Biology*, 7(2), 101–105. https://doi.org/10.1038/nchembio.499.

Vergauwen, B., de Vos, D., & van Beeumen, J. J. (2006). Characterization of the bifunctional gamma-glutamate-cysteine ligase/glutathione synthetase (GshF) of *Pasteurella multocida*. *The Journal of Biological Chemistry, 281*(7), 4380–4394. https://doi.org/10.1074/jbc.M509517200.

Vergauwen, B., Pauwels, F., Vaneechoutte, M., & van Beeumen, J. J. (2003). Exogenous glutathione completes the defense against oxidative stress in *Haemophilus influenzae*. *Journal of Bacteriology, 185*(5), 1572–1581. https://doi.org/10.1128/JB.185.5.1572-1581.2003.

Winkler, B. S., Orselli, S. M., & Rex, T. S. (1994). The redox couple between glutathione and ascorbic acid: A chemical and physiological perspective. *Free Radical Biology and Medicine, 17*(4), 333–349. https://doi.org/10.1016/0891-5849(94)90019-1.

Wu, G., Fang, Y.-Z., Yang, S., Lupton, J. R., & Turner, N. D. (2004). Glutathione metabolism and its implications for health. *The Journal of Nutrition, 134*(3), 489–492. https://doi.org/10.1093/jn/134.3.489.

Wu, G., Nie, L., & Zhang, W. (2006). Predicted highly expressed genes in *Nocardia farcinica* and the implication for its primary metabolism and nocardial virulence. *Antonie Van Leeuwenhoek, 89*(1), 135–146. https://doi.org/10.1007/s10482-005-9016-z.

Yamamoto, Y., Kamio, Y., & Higuchi, M. (1999). Cloning, nucleotide sequence, and disruption of *Streptococcus mutans* glutathione reductase gene (gor). *Bioscience, Biotechnology, and Biochemistry, 63*(6), 1056–1062. https://doi.org/10.1271/bbb.63.1056.

CPI Antony Rowe
Eastbourne, UK
May 16, 2020